Electrolyzed Water in Food: Fundamentals and Applications

Tian Ding · Deog-Hwan Oh ·
Donghong Liu
Editors

Electrolyzed Water in Food: Fundamentals and Applications

ZHEJIANG UNIVERSITY PRESS
浙江大学出版社

Springer

Editors
Tian Ding
Department of Food Science and Nutrition,
National Engineering Laboratory
of Intelligent Food Technology
and Equipment
Zhejiang University
Hangzhou, Zhejiang, China

Key Laboratory for Agro-Products
Postharvest Handling of Ministry
of Agriculture
Zhejiang Key Laboratory for Agro-Food
Processing
Hangzhou, Zhejiang, China

Fuli Institute of Food Science
Zhejiang University
Hangzhou, China

Deog-Hwan Oh
Department of Food Science
and Biotechnology
Kangwon National University
Chuncheon, Republic of Korea

Donghong Liu
Department of Food Science and Nutrition,
National Engineering Laboratory of
Intelligent Food Technology and Equipment
Zhejiang University
Hangzhou, Zhejiang, China

ISBN 978-981-13-3806-9 ISBN 978-981-13-3807-6 (eBook)
https://doi.org/10.1007/978-981-13-3807-6

Jointly published with Zhejiang University Press, Hangzhou, China
The print edition is not for sale in the Mainland of China. Customers from the Mainland of China please order the print book from: Zhejiang University Press.

Library of Congress Control Number: 2018965884

This Springer imprint is published by the registered company Springer Nature Singapore Pte Ltd.
The registered company address is: 152 Beach Road, #21-01/04 Gateway East, Singapore 189721, Singapore

Contents

Contributors

Yanhong Bai College of Food and Biological Engineering, Zhengzhou University of Light Industry, Zhengzhou, China;
Henan Key Laboratory of Cold Chain Food Quality and Safety Control, Zhengzhou, China;
Henan Collaborative Innovation Center of Food Production and Safety, Zhengzhou, China

Tian Ding Department of Food Science and Nutrition, National Engineering Laboratory of Intelligent Food Technology and Equipment, Zhejiang University, Hangzhou, Zhejiang, China;
Key Laboratory for Agro-Products Postharvest Handling of Ministry of Agriculture, Zhejiang Key Laboratory for Agro-Food Processing, Hangzhou, Zhejiang, China;
Fuli Institute of Food Science, Zhejiang University, Hangzhou, China

Fereidoun Forghani Center for Food Safety, College of Agricultural and Environmental Sciences, University of Georgia, Griffin, GA, USA

Rongwei Han College of Food Science and Engineering, Qingdao Agricultural University, Qingdao, Shandong, China

Jianxiong Hao College of Bio Science and Engineering, Hebei University of Science and Technology, Shijiazhuang, Hebei, China

Mohammed Shakhawat Hussain Department of Bioconvergence Science and Technology, College of Agriculture and Life Science, Kangwon National University, Chuncheon, Republic of Korea

Imran Khan Department of Bioconvergence Science and Technology, College of Agriculture and Life Science, Kangwon National University, Chuncheon, Republic of Korea

Xinyu Liao Department of Food Science and Nutrition, National Engineering Laboratory of Intelligent Food Technology and Equipment, Zhejiang University, Hangzhou, Zhejiang, China;
Key Laboratory for Agro-Products Postharvest Handling of Ministry of Agriculture, Zhejiang Key Laboratory for Agro-Food Processing, Hangzhou, Zhejiang, China;
Fuli Institute of Food Science, Zhejiang University, Hangzhou, China

Jiangang Ling Institution of Agricultural Products Processing, Key Laboratory of Preservation Engineering of Agricultural Products, Ningbo Academy of Agricultural Sciences, Ningbo, Zhejiang, China

Donghong Liu Department of Food Science and Nutrition, National Engineering Laboratory of Intelligent Food Technology and Equipment, Zhejiang University, Hangzhou, Zhejiang, China;
Key Laboratory for Agro-Products Postharvest Handling of Ministry of Agriculture, Zhejiang Key Laboratory for Agro-Food Processing, Hangzhou, Zhejiang, China;
Fuli Institute of Food Science, Zhejiang University, Hangzhou, China

Ruiling Lv Department of Food Science and Nutrition, National Engineering Laboratory of Intelligent Food Technology and Equipment, Zhejiang University, Hangzhou, Zhejiang, China;
Key Laboratory for Agro-Products Postharvest Handling of Ministry of Agriculture, Zhejiang Key Laboratory for Agro-Food Processing, Hangzhou, Zhejiang, China

Pradeep K. Malakar College of Food Science and Technology, Shanghai Ocean University, Shanghai, China

H. M. Murshed Department of Animal Science, Bangladesh Agricultural University, Mymensingh, Bangladesh

Liyuan Niu College of Food and Biological Engineering, Zhengzhou University of Light Industry, Zhengzhou, China;
Henan Key Laboratory of Cold Chain Food Quality and Safety Control, Zhengzhou, China;
Henan Collaborative Innovation Center of Food Production and Safety, Zhengzhou, China

Deog-Hwan Oh Department of Bioconvergence Science and Technology, College of Agriculture and Life Science, Kangwon National University, Chuncheon, Republic of Korea

S. M. E. Rahman Department of Animal Science, Bangladesh Agricultural University, Mymensingh, Bangladesh

Charles Nkufi Tango Department of Bioconvergence Science and Technology, College of Agriculture and Life Science, Kangwon National University, Chuncheon, Republic of Korea;
Department of Chemistry and Agricultural Industries, Faculty of Agronomy, University of Kinshasa, Kinshasa, Democratic Republic of the Congo

Jun Wang College of Food Science and Engineering, Qingdao Agricultural University, Qingdao, Shandong, China

Qingfa Wang College of Bio Science and Engineering, Hebei University of Science and Technology, Shijiazhuang, Hebei, China

Siqi Wang College of Food Science and Technology, Shanghai Ocean University, Shanghai, China;
Laboratory of Quality and Safety Risk Assessment for Aquatic Products on Storage and Preservation, Ministry of Agriculture, Shanghai, China;
Shanghai Engineering Research Center of Aquatic-Product Processing and Preservation, Shanghai, China

Qisen Xiang College of Food and Biological Engineering, Zhengzhou University of Light Industry, Zhengzhou, China;
Henan Key Laboratory of Cold Chain Food Quality and Safety Control, Zhengzhou, China;
Henan Collaborative Innovation Center of Food Production and Safety, Zhengzhou, China

Xiaoting Xuan Institution of Agricultural Products Processing, Key Laboratory of Preservation Engineering of Agricultural Products, Ningbo Academy of Agricultural Sciences, Ningbo, Zhejiang, China

Zhaohuan Zhang College of Food Science and Technology, Shanghai Ocean University, Shanghai, China;
Laboratory of Quality and Safety Risk Assessment for Aquatic Products on Storage and Preservation, Ministry of Agriculture, Shanghai, China;
Shanghai Engineering Research Center of Aquatic-Product Processing and Preservation, Shanghai, China

Li Zhao College of Food Science and Technology, Shanghai Ocean University, Shanghai, China;
Laboratory of Quality and Safety Risk Assessment for Aquatic Products on Storage and Preservation, Ministry of Agriculture, Shanghai, China;
Shanghai Engineering Research Center of Aquatic-Product Processing and Preservation, Shanghai, China

Yong Zhao College of Food Science and Technology, Shanghai Ocean University, Shanghai, China;
Laboratory of Quality and Safety Risk Assessment for Aquatic Products on Storage and Preservation, Ministry of Agriculture, Shanghai, China;
Shanghai Engineering Research Center of Aquatic-Product Processing and Preservation, Shanghai, China

Chapter 1
Generation of Electrolyzed Water

Xiaoting Xuan and Jiangang Ling

1.1 Introduction

Along with consumer living standards, the expense consciousness of consumers is also gradually growing, and minimally processed or fresh foods are highly popular because of their nutritional value. For the consumer, food safety is of crucial importance. However, each year, 48 million people become sick in the United States from one of 250 identified foodborne diseases, 128,000 are hospitalized, and 3000 die. The top five germs causing food poisoning in the United States are *Norovirus*, *Salmonella*, *Clostridium perfringens*, *Campylobacter*, and *Staphylococcus aureus* (Rahman et al. 2016). The best way to reduce foodborne diseases is to ensure the safety of the food supply. Hence, cleaning and sanitization to produce high-quality, microbiologically safe food before it is delivered to the market are crucially important steps of the Hazard Analysis Critical Control Point (HACCP) system.

Numerous disinfection techniques have been studied and/or used throughout the food chain. A few of these methods are the use of chemical disinfectants (hypochlorite, chlorine dioxide, hydrogen peroxide, hydrochloric acid, ozone, etc.), physical treatments (heat and irradiation, etc.), and their combinations (Koide et al. 2011; Zhang et al. 2011a, b). However, some of these techniques have bottleneck constraints when applied to minimally processed food, including remaining chemical residues, a low inactivation efficacy, adverse effects on the health of humans or the quality of food, environmental harm, and a high price (Ramos et al. 2013; Al-Haq et al. 2005). Hence, for both food providers and consumers, it increasingly demands novel sanitation technologies to ensure the safety and freshness of minimally processed foods.

X. Xuan (✉) · J. Ling
Institution of Agricultural Products Processing, Key Laboratory of Preservation Engineering of Agricultural Products, Ningbo Academy of Agricultural Sciences, Ningbo 315400, Zhejiang, China
e-mail: xuanxiaoting163@163.com

T. Ding et al. (eds.), *Electrolyzed Water in Food: Fundamentals and Applications*,
https://doi.org/10.1007/978-981-13-3807-6_1

In recent years, various nonthermal disinfection techniques have emerged and have been regarded as effective methods for decontaminating microbes on the food, including high pressure processing (Cui et al. 2016; Campus 2010), ultrasound (Li et al. 2017a, b), pulsed electric field (PEF) (Wan et al. 2009; Toepfl et al. 2006), and cold plasma (Moreau et al. 2008; Niemira 2012). In addition, the use of electrolyzed water (EW) produced by the electrolysis of dilute salt (NaCl) or hydrogen chloride (HCl) solution has been reported as an effective and broad-spectrum sterilization because of its main effective form of chlorine compounds—hypochlorous acid (HClO) (Ding et al. 2015a, b). Studies have been reported its antimicrobial activity against various microorganisms, such as *Escherichia coli*, *Listeria monocytogenes*, *Salmonella typhimurium*, *S. aureus*, and *Vibrio parahaemolyticus* (Anonymous 1997; Bari et al. 2003; Park et al. 2009; Issa-Zacharia et al. 2010; Quan et al. 2010; Zhang et al. 2011a, b; Sun et al. 2012). Anonymous (1997) has reported that the disinfection activity of HClO against *E. coli* was approximately 80-fold higher than that of an equivalent concentration of the hypochlorite ion (ClO^-) under the same conditions. EW is also generally recognized as safe (GRAS) and already regarded as a legitimate food additive in the US, Japan, and Korea (Xuan et al. 2017). It is becoming more attractive because of its easy production and low-cost materials, chlorine off-gassing, and non-corrosivity to equipment (Len et al. 2000; Jadeja and Hung 2014). Numerous studies have been conducted on the efficiency of its antimicrobial activity in fresh-cut fruits (Ding et al. 2015b; Kim and Hung 2012; Graça et al. 2011), vegetables (Koide et al. 2009; Rahman et al. 2010a, b; Issa-Zacharia et al. 2011; Xuan et al. 2016), poultry (Cao et al. 2009; Rahman et al. 2011), meat (Xuan et al. 2016; Rahman et al. 2013; Ding et al. 2010), and aquatic products (Wang et al. 2014a, b; Xuan et al. 2017). Cao et al. (2009) reported that slightly acidic electrolyzed (SAEW, ACC of 15 mg/L) could induce a bacterial reduction of 6.5 log CFU/g on *Salmonella enteritidis* on shell eggs in 3 min, and no surviving *S. enteritidis* was recovered in waste SAEW. The studies of Koide et al. (2009) were found that the disinfectant efficacy of SAEW on fresh-cut cabbage was equivalent to or even higher than that of traditional chlorine disinfectant, e.g., sodium hypochlorite solution (NaOCl).

This chapter provides an overview of the production mechanism, types of EW, and its production equipment, and then discusses the influencing factors on the properties of EW including its properties, electrolysis parameters, electrodes, water properties, and storage environments. In addition, numerous studies are summarized and discussed with their key results, followed by the advantages and disadvantages of EW during application.

1.2 Mechanism of Production

EW is produced by electrolysis of NaCl solutions using an electrolysis chamber containing a separating membrane (diaphragm or septum) between the anode and cathode electrodes (Rahman et al. 2016). The reactions on the electrodes and the generator structure for EW are shown in Figs. 1.1 and 1.2. As depicted in Fig. 1.1, by

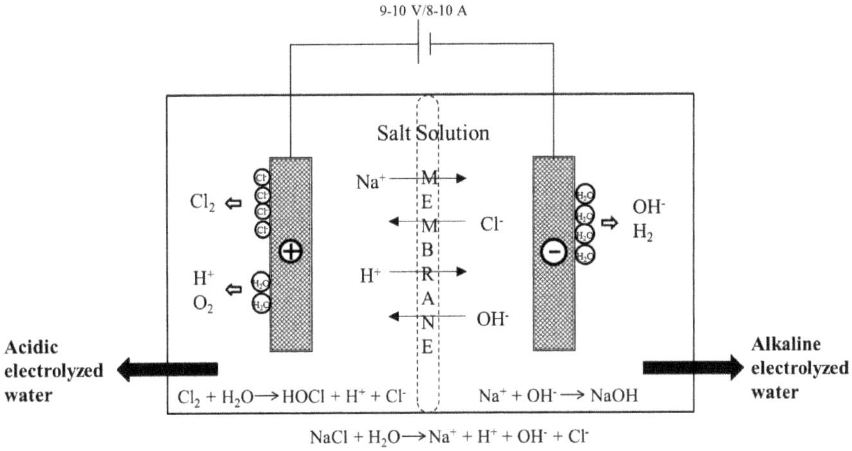

Fig. 1.1 Schematic of acidic and alkaline electrolyzed water generation

application of direct current voltages (9–10 V/8–10 A), different EW solutions are generated at the positive pole and negative pole, respectively (Al-Haq et al. 2005). Positively charged ions (hydrogen and sodium) move to the cathode where they gain electrons and then become hydrogen gas (H_2) and sodium hydroxide (NaOH). Based on the reactions on the positive pole, acidic electrolyzed water (AEW) with a low pH (2–3), high oxidation reduction potential (ORP, >1000 mV), and available chlorine concentration (ACC of 10–90 ppm) is generated. Meanwhile, the negatively charged ions (chloride and hydroxide) move to the anode where they lose electrons and then become oxygen gas (O_2), chlorine gas (Cl_2), hypochlorite ion (^-OCl), hypochlorous acid (HOCl), and hydrochloric acid (HCl). Based on the reactions on the negative pole, alkaline electrolyzed water (AlEW) with a high pH (10–13) and a low ORP (−800 to −900 mV) is generated (Hsu 2005; Hricova et al. 2008). In summary, the principle of generating AEW and AlEW is shown in Fig. 1.1 with their reactions as follows:

Positive pole:

$$2NaCl \rightarrow Cl_2 \uparrow + 2Na^+ + 2e^-$$
$$2H_2O \rightarrow O_2 \uparrow + 4H^+ + 4e^-$$
$$Cl_2 + H_2O \rightarrow HCl + HOCl$$

Negative pole:

$$2NaCl + 2OH^- \rightarrow 2NaOH + Cl^-$$
$$2H_2O + 2e^- \rightarrow 2OH^- + H_2 \uparrow$$

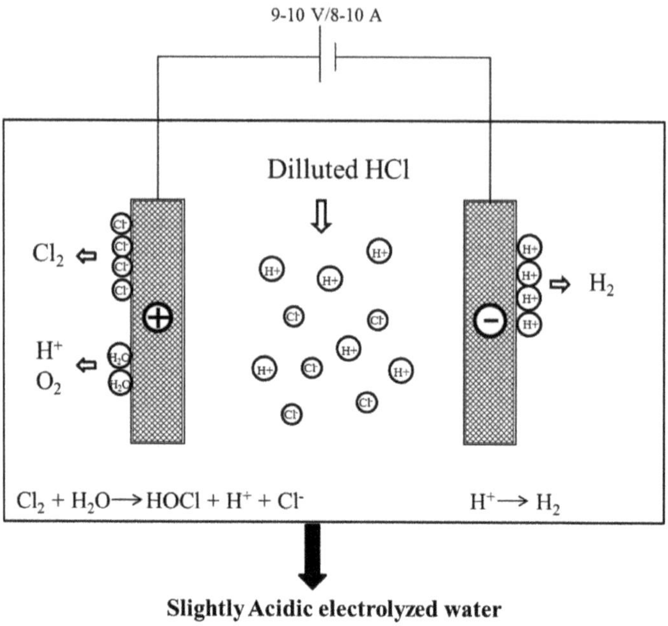

Fig. 1.2 Schematic of neutral and slightly acidic electrolyzed water generation

Moreover, studies have been published of novel EW generation equipment using an electrolysis chamber without the separating membrane between the anode and cathode electrodes. Due to the single-cell chamber, neutralization occurs when hydroxide ions (OH^-) from the negative pole contact with protons (H^+) from the positive pole and then neutral electrolyzed water (NEW) with a pH of 7–8 and an ORP of 750–900 mV is produced (Deza et al. 2007). In addition, slightly acidic electrolyzed water (SAEW) with a pH of 5.0–6.5 and an ORP of 800–900 mV is produced by electrolysis of HCl or in combination with NaCl in a EW generation equipment using an electrolysis chamber without the separating membrane (Forghani et al. 2015). The principle of generating NEW and SAEW is shown in Fig. 1.2. Ding et al. (2015a, b) initiatively added a circulating device to the generation equipment, which produces SAEW with a higher ACC (up to 200 ppm). The circulating device can transport the electrolyzed water solution back to the electrolytic cell for repetitive electrolysis as illustrated in Fig. 1.3. This type of EW can be applied in a diluted form to prolong its shelf life (Xuan et al. 2016).

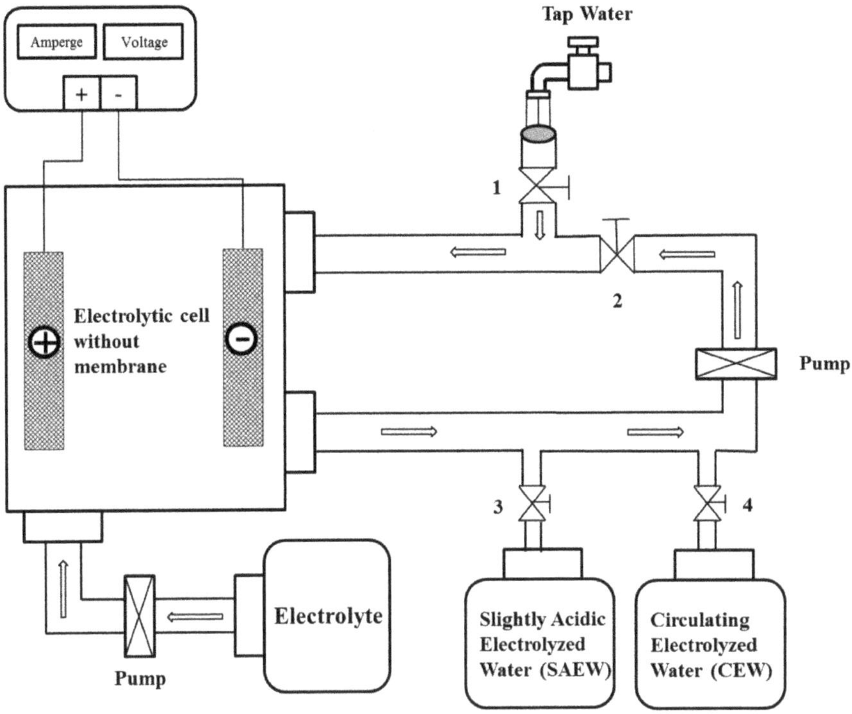

Fig. 1.3 Schematic of the generation of SAEW and CEW by using an electrolysis chamber without the separating membrane; when switches 1 and 3 are opened and 2 and 4 are closed, SAEW can be generated; otherwise, CEW can be produced

1.3 Types of EW

1.3.1 Acidic EW

The acidic EW (AEW) with a low pH (2.5–3.5), high ORP (1000–1200 mV) and free chlorine (30–90 ppm) is generated from the anode where hydrochloride, HOCl, chlorine, and oxygen gas are also formed. Fig. 1.4 illustrates the applications of EW at different pH values in various areas. As depicted in Fig. 1.4, AEW with low pH could be used for industry device. What's more, the higher amperage and voltage result in a more acidic solution with a higher ORP and a free chlorine concentration. Due to the low pH, high ORP, and the presence of HOCl, acidic EW effectively inhibits the growth of bacteria. For example, at amperage setting of 14 A, Park et al. (2004) produced acidic EW with a pH of 2.57, an ORP of 1082 mV and approximately 50 mg/L of free chlorine.

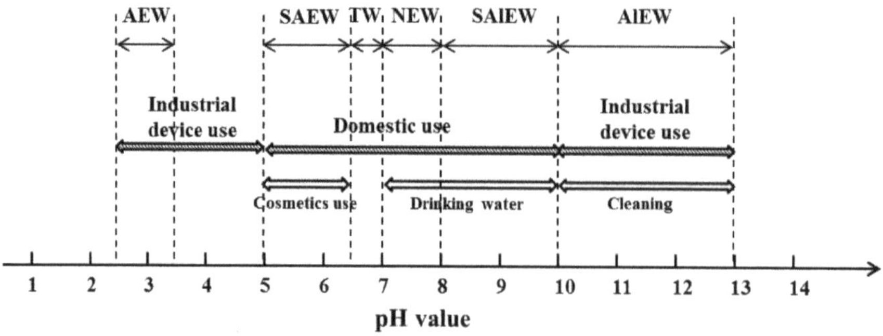

Fig. 1.4 pH ranges of various electrolyzed waters and its applications

1.3.2 Alkaline EW

The alkaline EW (AlEW) with a high pH (10–13) and low ORP (−800 to −900 mV) is generated from the cathode where sodium ions (Na^+) and hydroxyl ions (OH^-) also form sodium hydroxide (Hsu 2005; Hricova et al. 2008). The indices of voltage, amperage, and flow rate settings can greatly affect the pH and ORP of alkaline EW (Sharma and Demirci 2003). Alkaline EW possesses detergent-like functionality and an inactivation function owing to the hydroxyl ions and negative ORP (Fabrizio et al. 2002; Walker et al. 2005). Moreover, it has the potential to reduce the free radicals based on its strong reducing ORP (Al-Haq et al. 2005).

1.3.3 Neutral EW

Neutral EW (NEW) with a neutral pH (7–8) and an ORP of 750–900 mV is produced from the anode but partially mixed with hydroxide ions or generated using a single-cell chamber without the separating membrane (Al-Haq et al. 2005; Deza et al. 2007). Along with the changes in generator settings, the pH and ORP levels of neutral EW are varied, and their disinfection elements are mainly HOCl and HCl. In comparison with acidic EW, it is believed to be a less corrosive pH, higher ORP with effective disinfection and to have a longer shelf life under certain circumstances (Rahman et al. 2010a, b).

1.3.4 Slightly Acidic EW

Slightly acidic EW (SAEW) with a pH of 5.0–6.5 and an ORP of 800–900 mV is produced by electrolysis of HCl or in combination with NaCl in EW generation equipment using an electrolysis chamber without the separating membrane (Forghani et al. 2015). Its main free chlorine is HOCl, which has a disinfection efficacy for *E. coli* is 80-fold greater than that of an equivalent concentration of ClO^- under the same treatment conditions (Anonymous 1997). It is an attractive and effective method for the food industry owing to its easy production, low-cost materials, high disinfection efficacy, and broad spectrum of disinfection activity (Hao et al. 2013; Koide et al. 2011; Jadeja and Hung 2014).

1.4 Types of EW-Producing Equipment

Several EW generators produced by popular manufacturers have been accepted by the market, such as Aquaox LLC (West Palm Beach, FL, United States), Hoshizaki Electric Inc. (Aichi, Japan), Envirolyte (Northland, New Zealand), MIOX (Albuquerque, NM, United States), AMANO (Japan), and so on. Many leading manufacturers of EW machines are from Japan. Based on the structure of the EW-producing generator, most EW machines can be divided into two types. The first type contains diaphragms with two cell chambers that can generate AEW and AlEW, such as AMANO, HOSHIZAKI, Envirolyte, et al. The second type does not contain diaphragms and has single-cell chambers that can generate NEW and SAEW, such as MIOX (Al-Haq et al. 2005; Deza et al. 2007). In addition, Ding et al. (2015a, b) initiatively developed a circulating electrolyzed water generation that modified the traditional electrolyzed water generator and added a circulating device to the generation equipment. The circulating device can pump the electrolyzed water back to the electrolytic cell for repetitive electrolysis.

The properties of EW can be greatly affected by the voltage, amperage, flow rate, chlorine concentration, and other parameters (Hsu 2003). To easily control the EW generation, manufacturers allow the users to select some parameters. For Hoshizaki Electric Inc. (Aichi, Japan), users can control the amperages and/or voltages; for Nippon Intek (Japan), the chlorine concentration can be selected and automatically the flow rate and amperages/voltages are changed; for Amano (Japan), the electrolyte flow rate can be controlled and the generators adjust amperages/voltages automatically. Various EW-producing generators that have been applied in reports are summarized in Table 1.1.

1.5 Influencing Factors on the Properties of EW

EW is considered as an effective disinfectant in food decontamination and preservation. Its disinfection efficacy against different foodborne pathogens, e.g., *L. monocytogenes*, *E. coli* O157:H7, *S. aureus*, *S. typhimurium*, and *V. parahaemolyticus* have been investigated (Al-Holy and Rasco 2015; Cao et al. 2009; McCarthy and Burkhardt 2012; Wang et al. 2014a, b). As mentioned previously, EW is produced by electrolysis of NaCl solution, which is the only chemical material. It has fewer

Table 1.1 Various electrolyzed water-producing generators applied in reports

Types of EW generators	Country and model	Substrate used	Current used (A)	pH	ORP (mV)	ACC (mg/L)	Target bacteria	Subject	Reference
AEW	ROX-20TA (Hoshizaki Electric Co. Ltd., Toyoake, Aichi, Japan)	12% NaCl	14–20	2.6 ± 0.1	1140 ± 7.0	30.3 ± 3.1	*L. monocytogenes*, *E. coli* O157:H7, and *Salmonella*	Tomatoes, alfalfa seeds, and sprouts	Bari et al. (2003), Len et al. (2002), Park et al. (2004), Sharma and Demirci (2003), Keskinen et al. (2009)
NEW	Envirolyte EI-900 unit (Envirolyte Industries International Ltd., Tallin, Estonia)	25% NaCl	32 ± 2	8.27 ± 0.21	774.0 ± 0.9	60	*E. coli*, *L. monocytogenes*, *Pseudomonas aeruginosa*, and *S. aureus*	Plastic and wooden kitchen cutting boards	Deza et al. (2007)
	Eurostel EZ-90 Unit (Ecanet, Palamòs, Gìrona, Catalonia, Spain)	1% NaCl	30 ± 5	8.74 ± 0.18	721 ± 12	280	*E. coli*, *L. innocua*, *Salmonella*, and *E. carotovora*	Lettuce, fresh-cut iceberg lettuce, shredded carrot, endive, corn salad, "Four seasons" salad	Abadias et al. (2008)

(continued)

Table 1.1 (continued)

Types of EW generators	Country and model	Substrate used	Current used (A)	pH	ORP (mV)	ACC (mg/L)	Target bacteria	Subject	Reference
SAEW	Apia60 (HOKUTY Co., Kanagawa, Japan)	2% HCl	3	5.6 ± 0.2	940 ± 11	23.7 ± 2.1	*E.* O157:H7, *Salmonella*, and *S. aureus*	Pure culture	Issa-Zacharia et al. 2010, 2011
	DIPS KI/KII/F, *e*-suenc Co., Ltd., Seoul, Korea)	0.1% NaCl	17.1	5.9 ± 0.1	798.0 ± 11.0	35.0 ± 0.8	*Vibrio vulnificus*, *Vibrio parahaemolyticus*	Pure culture	Quan et al. (2010)
	CEW equipment developed by Ding et al. (2015a, b)	0.6% NaCl and 0.15% HCl	13 ± 3	6–6.5	850 ± 50	200 ± 10	*Listeria monocytogenes*, total bacteria	Lettuce, pork, pure culture	Xuan et al. (2016, 2017)
Slightly alkaline electrolyzed water (SAlEW)	Model D-7, sl No. 001171, Dolki Co., Ltd., Wonju, Korea	0.9% NaCl	1.15	6.2	500–520	5	Total bacteria, yeasts and molds, *E. coli* O157:H7, and *L. monocytogenes*	Fresh-cut spinach	Rahman et al. (2010a, b)

adverse effects on human health and the environment owing to its chemical composition and near-neutral pH (Kim et al. 2000; Ding et al. 2015a, b). The strongest chlorine form is HOCl, which has an inactivation efficacy that is 80-fold greater than that of ^{-}OCl in an equivalent concentration when its pH range is from 5.0 to 6.5 (Cao et al. 2009). HOCl will change to ^{-}Ocl in alkaline pH, whereas it will dissociate to Cl_2 at low pH values. The facile escape of Cl_2 from solution decreases its antimicrobial efficacy. Hence, the pH of EW determines the relative fractions of chlorine species in the solution (Park et al. 2004). In addition, it has been proved that the chlorine compound is one of the most important factors responsible for the inactivation efficacy of EW (Hao et al. 2012). Moreover, a few reports have attributed the inactivation action to the ORP of EW (Kim et al. 2000; Liao et al. 2007; Ding et al. 2016; Tkhawkho et al. 2017). Ding et al. (2016) determined the disinfection efficacy of SAEW on *S. aureus* in comparison with that of sodium hypochlorite (NaClO) and hydrochloric acid (HCl). The results showed that a high ORP affects the certain intracellular enzyme systems by changing the electron flow in the cells. Another report showed that a high ORP of EW results in the destruction of layers of bacteria, disturbing the metabolic pathways and oxidation of sulfhydryl mixtures of cells. The result could accelerate the inactivation of bacterial cells (Liao et al. 2007). Therefore, the basic properties of EW including the ACC (Cl_2, ^{-}OCl and HOCl), pH and ORP directly influence its sanitizing efficacy, whereas various electrolytic parameters such as the current, flow rate, salt concentration, electrolyte, electrode materials, water temperature, hardness, and storage environments have been reported to directly affect the properties of EW.

1.5.1 Electrolysis Parameters (Current, Water Flow Rate, Salt/Acid Concentration)

The abovementioned basic properties of EW may be affected by the current, water flow rate, and salt/acid concentration, which has been reported by several researchers (Rahman et al. 2012; Hsu 2003). Rahman et al. (2012) suggested that an increase in the current (1.15–1.45 A) results in an increase in the pH, ACC, and ORP, which eventually enhances the antimicrobial ability toward *E. coli* O157:H7 and *L. monocytogenes*. A higher water flow rate has been reported to cause a larger electric current because a greater amount of salt solution is electrolyzed per unit time, whereas lower ACC and ORP of EW are induced (Hsu 2003, 2005). This phenomenon suggests that the ORP level can be influenced by its machine control action and the ACC can be affected by the amperage parameter. In addition, there is a positive correlation between the conductivity of EW and the salt/acid concentration in the electrolyte solution (Hsu 2005). Furthermore, the salinity is linearly correlated with the concentration of HOCl in EW during the electrolysis process (Al-Haq et al. 2002; Kiura et al. 2002; Rahman et al. 2016). Kiura et al. (2002) demonstrated that the salt concentration and electrolysis time have positive correlations with the free chlorine concentration, which might be explained by considering that the electrolysis efficacy of the electrolysis cell and the separation efficacy of the ion exchange membrane are greatly decreased with increasing flow rate and salt concentration.

1.5.2 Electrolyte and Electrode

The basic properties of EW are also influenced by the type and flow rate of electrolyte, as well as by the electrode settings and materials. EW studies traditionally use NaCl as the electrolyte (Forghani et al. 2015; Al-Haq et al. 2005; Deza et al. 2007; Hsu 2005; Hricova et al. 2008). It has been stated that $MgCl_2$, KCl, and HCl could replace NaCl as the electrolytes (Hricova et al. 2008; Pangloli and Hung 2013). There is a positive correlation between the ACC and the concentration of the electrolyte (Hsu 2005). A higher electrolyte concentration can increase the conductivity, which might increase the chlorine production and enhance its bactericidal ability. Moreover, an increase in pH was induced by increasing the electrolyte concentration (Forghani et al. 2015). Several researchers have been found that the electrode settings and its materials can greatly influence the properties of EW (Hsu et al. 2015; Martínez-Huitle and Brillas 2008; Jeong et al. 2009). Hsu et al. (2015) found that stirring or immersing the electrodes deep under the electrolyte remarkably increased the current density without changing the electric efficiency and current efficiency. An additional change of the electrode size or electrode gap significantly affected the chlorine production and electric current without affecting its electric efficiency and current efficiency. The electrode materials also play an important role in the production of oxidants. Traditionally, platinum is used as the anode in the EW generator. For the production ability of free chlorine, various electrode materials were be ordered by Rahman et al. (2016) as follows: $Ti/IrO_2 > Ti/RuO_2 > Ti/Pt–IrO_2 >$ BDD > Pt. Martínez-Huitle and Brillas (2008) considered that the electrode material governs the production of oxidants and other species in comparison with the current, temperature, and type of electrolysis.

1.5.3 Water Temperature and Hardness

The influence of water temperature and hardness on the basic properties of EW has been reported by a few researchers (Pangloli and Hung 2013; Forghani et al. 2015). Forghani et al. (2015) found that heating (40 °C) EW after production might have a negative effect on its inactivation efficiency due to the partial loss of free chlorine during the heating period. Nevertheless, preheated EW presented a higher ACC and inactivation efficiency on *L. monocytogenes* and *E. coli* O157:H7. The authors also evaluated the effect of water hardness on the properties of SAEW and found that a decrease in pH accompanied increasing water hardness. From this perspective, water hardness is another crucial factor that has a positive effect on the basic properties of EW and its inactivation efficiency. Pangloli and Hung (2013) also reported similar results that water hardness tended to increase the ACC and ORP levels while decreasing the pH of EW. The increase in ACC was probably due to the increase in the concentration of electrolytes and the conductivity which in turn was due to the increase in water hardness.

1.5.4 Storage Environments

The antimicrobial efficacy of EW is indirectly influenced by the storage environment because the breakdown of the main antimicrobial component (HOCl) and evaporation of Cl_2 can be greatly affected by storage conditions, particularly in open/light conditions. In our research, SAEW with an ACC of 20 mg/L and CEW with an ACC of 200 mg/L and 20 mg/L were evaluated for changes in basic properties (pH, ORP, and ACC) during storage in open and closed glass bottles under light and dark conditions at room temperature. The results suggested that a closed-dark container was a more conducive condition for EW storage (Xuan et al. 2016). Len et al. (2002) reported that the loss of chlorine in EW under an open condition was due to chlorine evaporation along with the exposure to the atmosphere, which was also proven in our study. When stored under closed conditions, the chlorine was lost by self-decomposition, which is much higher under open conditions (White 1998). The basic properties of EW are dramatically influenced under closed and open conditions during storage. Rahman et al. (2012) determined the changes in ACC of low-concentration electrolyzed water (LcSAEW, 10 mg/L, pH of 6.8–7.4) under closed and open conditions. They reported that the ACC of LcSAEW gradually decreased from 10 to 0 mg/L in 7 days under the open-dark condition compared to 10–0 mg/L in 21 days under the closed-dark condition. However, the loss of chlorine by lighting is not significant during storage. In addition, a lower storage temperature (4 °C) made these basic properties of EW more stable than that stored at 25 °C and maintained its bactericidal efficiency over 12 months (Nagamatsu et al. 2002; Fabrizio and Cutter 2003; Robinson et al. 2012). Forghani et al. (2015) also found that the preheating method increased the ACC level of EW and enhanced its inactivation efficacy. Furthermore, different types of EW showed different storage characteristics. For instance, it has been reported that NEW is more stable than AEW during the storage period (Nagamatsu et al. 2002; Cui et al. 2009).

1.6 Advantages and Disadvantages

As mentioned previously, EW as a novel cleaning and inactivation technology is generated in an environmentally friendly method from NaCl and distilled water (Hricova et al. 2008). It is potentially applicable to nonthermal food and processing. Its remarkable advantages include (i) the environment friendly type, which poses no threat to humans after used; (ii) the ability for on-site generation, which avoids the chlorination problems during transport, storage, and handling (Jeong et al. 2007); (iii) the broad-spectrum inactivation ability with nonselective properties, which circumvents the growth of bacterial resistance (Hricova et al. 2008); and (iv) no negative influence on the sensory and quality of food by the using of AEW, AlEW, NEW, and SAEW.

However, attention must be paid to the disadvantages and possible downsides as well. First, even though EW is generated by the electrolysis of NaCl solution, it is still composed of chemical compounds. It is allowed by legislation to conduct the surface of the food products and its processing equipment during the cleaning and inactivation procedure, but it is inapplicable to some food with a high porosity and its processing equipment. Second, the main inactivation component, HOCl, is lost with the increment of solution temperature and storage time, which reduces the inactivation activity of EW. It is suggested that a relatively lower solution temperature and closed storage may favor the storage of EW (Hsu and Kao 2004). Third, the leakage of chlorine and hydrogen gas during the EW produces discomfort to the operator and poses a potential threat to the surrounding environment. Hence, better ventilation is needed for on-site generation. Additionally, the relatively high initial cost of the equipment, exhaust system, and the installation greatly limit the wide application of EW by companies and individual users. Another major obstacle for industrial application is the generation rate of EW solution. Finally, the presence of free chlorine, a high ORP, and a low pH of AEW may cause irritation of hands and pitting or minor corrosion of the equipment (Huang et al. 2008).

References

Abadias M, Usall J, Oliveira M et al (2008) Efficacy of neutral electrolyzed water (NEW) for reducing microbial contamination on minimally-processed vegetables. Int J Food Microbiol 123(1):151–158

Al-Haq MI, Seo Y, Oshita S et al (2002) Disinfection effects of electrolyzed oxidizing water on suppressing fruit rot of pear caused by *Botryosphaeria berengeriana*. Food Res Int 35(35):657–664

Al-Haq MI, Sugiyama J, Isobe S (2005) Applications of electrolyzed water in agriculture and food industries. Food Sci Technol Res 11(2):135–150

Al-Holy MA, Rasco BA (2015) The bactericidal activity of acidic electrolyzed oxidizing water against *Escherichia coli* O157:H7, *Salmonella Typhimurium*, and *Listeria monocytogenes* on raw fish, chicken and beef surfaces. Food Control 54:317–321

Anonymous (1997) Principle of formation of electrolytic water. Hoshizaki Electic Co., Ltd., Sakae, Toyoake, Aichi, Japan

Bari ML, Sabina Y, Isobe S et al (2003) Effectiveness of electrolyzed acidic water in killing *Escherichia coli* O 157:H7, *Salmonella Enteritidis*, and *Listeria monocytogenes* on the surfaces of tomatoes. J Food Protect 66(4):542–548

Campus M (2010) High pressure processing of meat, meat products and seafood. Food Eng Rev 2(4):256–273

Cao W, Zhu ZW, Shi ZX et al (2009) Efficiency of slightly acidic electrolyzed water for inactivation of *Salmonella enteritidis* and its contaminated shell eggs. Int J Food Microbiol 130(2):88–93

Cui X, Shang Y, Shi Z et al (2009) Physicochemical properties and bactericidal efficiency of neutral and acidic electrolyzed water under different storage conditions. J Food Eng 91(4):582–586

Cui Y, Lin XD, Kang ML et al (2016) Advances in application of ultra high pressure for preservation and processing of aquatic products. J Food Sci 37(21):291–299

Deza M, Araujo M, Garrido M (2007) Efficacy of neutral electrolyzed water to inactivate *Escherichia coli*, *Listeria monocytogenes*, *Pseudomonas aeruginosa*, and *Staphylococcus aureus* on plastic and wooden kitchen cutting boards. J Food Protect 70(1):102–108

Ding T, Rahman SME, Purev U et al (2010) Modelling of *Escherichia coli* O157:H7 growth at various storage temperatures on beef treated with electrolyzed oxidizing water. J Food Eng 97(4):497–503

Ding T, Xuan XT, Liu DH et al (2015a) Electrolyzed water generated using a circulating reactor. Int J Food Eng 11(1):79–84

Ding T, Ge Z, Shi J et al (2015b) Impact of slightly acidic electrolyzed water (SAEW) and ultrasound on microbial loads and quality of fresh fruits. LWT-Food Sci Technol 60(2):1195–1199

Ding T, Xuan XT, Li J et al (2016) Disinfection efficacy and mechanism of slightly acidic electrolyzed water on *Staphylococcus aureus* in pure culture. Food Control 60:505–510

Fabrizio K, Cutter C (2003) Stability of electrolyzed oxidizing water and its efficacy against cell suspensions of *Salmonella typhimurium* and *Listeria monocytogenes*. J Food Protect 66(8):1379–1384

Fabrizio KA, Sharma RR, Demirci A et al (2002) Comparison of electrolyzed oxidizing water with various antimicrobial interventions to reduce *Salmonella* on poultry. Poultry Sci 81:1598–1605

Forghani F, Park JH, Oh DH (2015) Effect of water hardness on the production and microbicidal efficacy of slightly acidic electrolyzed water. Food Microbiol 48:28–34

Graça A, Abadias M, Salazar M et al (2011) The use of electrolyzed water as a disinfectant for minimally processed apples. Postharvest Biol Technol 61(2–3):172–177

Hao J, Qiu S, Li H et al (2012) Roles of hydroxyl radicals in electrolyzed oxidizing water (EOW) for the inactivation of *Escherichia coli*. Int J Food Microbiol 155(3):99–104

Hao XX, Li BM, Zhang Q et al (2013) Disinfection effectiveness of slightly acidic electrolysed water in swine barns. J Appl Microbiol 115:703–710

Hricova D, Stephan R, Zweifel C (2008) Electrolyzed water and its application in the food industry. J Food Protect 71(9):1934–1937

Hsu SY (2003) Effect of water flow rate, salt concentration and water temperature on efficiency of an electrolyzed oxidizing water generator. J Food Eng 60:460–473

Hsu SY (2005) Effects of flow rate, temperature and salt concentration on chemical and physical properties of electrolyzed oxidizing water. J Food Eng 66:171–176

Hsu S, Kao H (2004) Effects of storage conditions on chemical and physical properties of electrolyzed oxidizing water. J Food Eng 65:465–471

Hsu GSW, Hsia CW, Hsu SY (2015) Effects of electrode settings on chlorine generation efficiency of electrolyzing seawater. J Food Drug Anal 23(4):729–734

Huang YR, Hung YC, Hsu SY et al (2008) Application of electrolyzed water in the food industry. Food Control 19(4):329–345

Issa-Zacharia A, Kamitani Y, Tiisekwa A et al (2010) In vitro inactivation of *Escherichia coli*, *Staphylococcus aureus* and *Salmonella* spp. using slightly acidic electrolyzed water. J Biosci Bioeng 110(3):308–313

Issa-Zacharia A, Kamitani Y, Miwa N et al (2011) Application of slightly acidic electrolyzed water as a potential non-thermal food sanitizer for decontamination of fresh ready-to-eat vegetables and sprouts. Food Control 22:601–607

Jadeja R, Hung YC (2014) Efficacy of near neutral and alkaline pH electrolyzed oxidizing water to control *Eschericchia coli* O 157:H7 and *Salmonella typhimurium* DT 104 from beef hides. Food Control 41:17–20

Jeong J, Kim JY, Cho M et al (2007) Inactivation of *Escherichia coli* in the electrochemical disinfection process using a Pt anode. Chemosphere 67(4):652–659

Jeong J, Kim C, Yoon J (2009) The effect of electrode material on the generation of oxidants and microbial inactivation in the electrochemical disinfection processes. Water Res 43(4):895–901

Keskinen L, Burke A, Annous BA (2009) Efficacy of chlorine, acidic electrolyzed water and aqueous chlorine dioxide solutions to decontaminate *Escherichia coli* O157:H7 from lettuce leaves. Int J Food Microbiol 132:134–140

Kim C, Hung YC (2012) Inactivation of *E. coli* O157:H7 on blueberries by electrolyzed water, ultraviolet light, and ozone. J Food Sci 77(4):M206–M211

Kim C, Hung YC, Brackett RE (2000) Roles of oxidation–reduction potential in electrolyzed oxidizing and chemically modified water for the inactivation of food-related pathogens. J Food Protect 63(1):19–24

Kiura H, Sano K, Morimatsu S et al (2002) Bactericidal activity of electrolyzed acid water from solution containing sodium chloride at low concentration, in comparison with that at high concentration. J Microbiol Method 49(3):285–293

Koide S, Takeda JI, Shi J et al (2009) Disinfection efficacy of slightly acidic electrolyzed water on fresh cut cabbage. Food Control 20:294–297

Koide SJ, Shitanda D, Note M et al (2011) Effects of mildly heated, slightly acidic electrolyzed water on the disinfection and physicochemical properties of sliced carrot. Food Control 22(3):452–456

Len SV, Hung YC, Erickson M et al (2000) Ultraviolet spectrophotometric characterization and bactericidal properties of electrolyzed oxidizing water as influenced by amperage and pH. J Food Protect 63:1534–1537

Len SV, Hung YC, Chung D (2002) Effects of storage conditions and pH on chlorine loss on electrolyzed oxidizing (EO) water. J Agr Food Chem 50:209–212

Li J, Suo YJ, Liao XY et al (2017a) Analysis of *Staphylococcus aureus* cell viability, sublethal injury and death induced by synergistic combination of ultrasound and mild heat. Ultrason Sonoche 39:101–110

Li J, Ding T, Liao XY et al (2017b) Synergetic effects of ultrasound and slightly acidic electrolyzed water against *Staphylococcus aureus* evaluated by flow cytometry and electron microscopy. Ultrason Sonoche 38:711–719

Liao LB, Chen WM, Xiao XM (2007) The generation and inactivation mechanism of oxidation-reduction potential of electrolyzed oxidizing water. J Food Eng 78(4):1326–1332

Martínez-Huitle CA, Brillas E (2008) Electrochemical alternatives for drinking water disinfection. Angew Chem Int Edit 47(11):1998–2005

McCarthy S, Burkhardt W (2012) Efficacy of electrolyzed oxidizing water against *Listeria monocytogenes* and *Morganella morganii* on conveyor belt and raw fish surfaces. Food Control 24(1):214–219

Moreau M, Orange N, Feuilloley MGJ (2008) Non-thermal plasma technologies: new tools for bio-decontamination. Biotechnol Adv 26(6):610–617

Nagamatsu Y, Chen KK, Tajima K et al (2002) Durability of bactericidal activity in electrolyzed neutral water by storage. Dent Mater J 21(2):93–104

Niemira BA (2012) Cold plasma decontamination of foods. Annu Rev Food Sci Technol 3:125–142

Pangloli P, Hung YC (2013) Effects of water hardness and pH on efficacy of chlorine-based sanitizers for inactivating *Escherichia coli* O157:H7 and *Listeria monocytogenes*. Food Control 32(2):626–631

Park H, Hung YC, Chung D (2004) Effects of chlorine and pH on efficacy of electrolyzed water for inactivating *Escherichia coli* O157:H7 and *Listeria monocytogenes*. Int J Food Microbiol 91:13–18

Park EJ, Alexander E, Taylor GA et al (2009) The decontaminative effects of acidic electrolyzed water for *Escherichia coli* O157:H7, *Salmonella typhimurium*, and *Listeria monocytogenes* on green onions and tomatoes with differing organic demands. Food Microbiol 26(4):386–390

Quan Y, Choi KD, Chung D et al (2010) Evaluation of bactericidal activity of weakly acidic electrolyzed water (WAEW) against *Vibrio vulnificus* and *Vibrio parahaemolyticus*. Int J Food Microbiol 136(3):255–260

Rahman SME, Ding T, Oh DW (2010a) Inactivation effect of newly developed low concentration electrolyzed water and other sanitizers against microorganisms on spinach. Food Control 21(10):1383–1387

Rahman SME, Jin YG, Oh DW (2010b) Combined effects of alkaline electrolyzed water and citric acid with mild heat to control microorganisms on cabbage. J Food Sci 75:M111–M115
Rahman SME, Park JY, Song KY et al (2011) Effects of slightly acidic low concentration electrolyzed water on microbiological, physicochemical, and sensory quality of fresh chicken breast meat. J Food Sci 71(1):M35–M41
Rahman SME, Park JH, Wang J et al (2012) Stability of low concentration electrolyzed water and its sanitization potential against foodborne pathogens. J Food Eng 113(4):548–553
Rahman SME, Wang J, Oh DH (2013) Synergistic effect of low concentration electrolyzed water and calcium lactate to ensure microbial safety, shelf life and sensory quality of fresh pork. Food Control 30(1):176–183
Rahman SME, Khan I, Oh DH (2016) Electrolyzed water as a novel sanitizer in the food industry: current trends and future perspectives. Compr Rev Food Sci Food Saf 15:471–490
Ramos B, Miller FA, Brandao TRS et al (2013) Fresh fruits and vegetables—an overview on applied methodologies to improve its quality and safety. Innov Food Sci Emerg 20:1–15
Robinson G, Thorn R, Reynolds D (2012) The effect of long-term storage on the physiochemical and bactericidal properties of electrochemically activated solutions. Int J Mol Sci 14(1):457–469
Sharma RR, Demirci A (2003) Treatment of *Escherichia coli* O157:H7 inoculated alfalfa seeds and sprouts with electrolyzed oxidizing water. Int J Food Microbiol 86:231–237
Sun JL, Zhang SK, Chen JY et al (2012) Efficacy of acidic and basic electrolyzed water in eradicating *Staphylococcus aureus* biofilm. Can J Microbiol 58(4):448–454
Tkhawkho L, Jackson K, Nitzan O et al (2017) Destruction of clostridium difficile spores colitis using acidic electrolyzed water. Am J Infect Control 45(1):1053
Toepfl S, Mathys A, Heinz V et al (2006) Review: potential of high hydrostatic pressure and pulsed electric fields for energy efficient and environmentally friendly food processing. Food Rev Int 22(4):405–423
Walker SP, Demirci A, Graves RE et al (2005) Cleaning milking systems using electrolyzed oxidizing water. Trans ASAE 48(5):1827–1833
Wan J, Coventry J, Swiergon P et al (2009) Advances in innovative processing technologies for microbial inactivation and enhancement of food safety-pulsed electric field and low-temperature plasma. Trends Food Sci Technol 20(9):414–424
Wang JJ, Sun WS, Jin MT et al (2014a) Fate of *Vibrio parahaemolyticus* on shrimp after acidic electrolyzed water treatment. Int J Food Microbiol 179:50–56
Wang JJ, Zhang ZH, Li JB et al (2014b) Modeling *Vibrio parahaemolyticus* inactivation by acidic electrolyzed water on cooked shrimp using response surface methodology. Food Control 36(1):273–279
White GC (1998) Chemistry of chlorination. In: Handbook of chlorination and alternative disinfectants. Wiley, New York
Xuan XT, Wang MM, Ahn J et al (2016) Storage stability of slightly acidic electrolyzed water and circulating electrolyzed water and their property changes after application. J Food Sci 81(3):E610–E617
Xuan XT, Fan YF, Ling JG et al (2017) Preservation of squid by slightly acidic electrolyzed water ice. Food Control 73:1483–14893
Zhang YQ, Wu QP, Zhang JM et al (2011a) Effects of ozone on membrance permeability and ultrastructure in *Pseudomonas aeruginosa*. J Appl Microbiol 111(4):1006–1015
Zhang CL, Lu ZH, Li YY et al (2011b) Reduction of *Escherichia coli* O157:H7 and *Salmonella enteritidis* on mung bean seeds and sprouts by slightly acidic electrolyzed water. Food Control 22(5):792–796

Chapter 2
Decontamination Efficacy and Principles of Electrolyzed Water

Tian Ding and Xinyu Liao

2.1 Introduction

In recent years, EW as a new sanitizer has gained increasing interest. To date, numerous studies have demonstrated the strong bactericidal efficacy of EW on various microorganisms, including bacteria, molds, viruses, and so on (Afari and Hung 2018b; Ding et al. 2016; Huang et al. 2008; Hricova et al. 2008; Rahman et al. 2016). In addition, EW has been applied for the microbial decontamination of food products such as vegetables, fruits, poultry, meat, seafood and fish. Apart from food, EW has been proposed as a novel surface sanitizer for food contact surfaces (e.g., cutting boards, cutlery, plates, etc.) (Ayebah and Hung 2005). The main advantages of EW are environmental safety and a high antimicrobial activity. Therefore, EW has the potential to become an alternative to conventional Pasteurization technologies. An adequate understanding of the effect of inactivation agents on microbial cells is important for optimizing and developing novel Pasteurization technologies in the food industry (Manas and Pagán 2005). Generally, the free chlorine species (e.g., ClO^-, Cl_2, HClO) in EW are responsible for its antimicrobial capacity (Huang et al. 2008). For instance, the HOCl is the major active chlorine species in SAEW and has an 80-fold stronger bactericidal efficacy than ClO^- at an equivalent concentration does (Cao et al. 2009). However, the exact mechanisms underlying microbial inactivation by EW still remain unclear. In particular, the primary targets of microbial

T. Ding (✉) · X. Liao
Department of Food Science and Nutrition, National Engineering Laboratory of Intelligent Food Technology and Equipment, Zhejiang University, Hangzhou 310058, Zhejiang, China
e-mail: tding@zju.edu.cn

T. Ding · X. Liao
Key Laboratory for Agro-Products Postharvest Handling of Ministry of Agriculture, Zhejiang Key Laboratory for Agro-Food Processing, Hangzhou 310058, Zhejiang, China

T. Ding · X. Liao
Fuli Institute of Food Science, Zhejiang University, Hangzhou 310058, Zhejiang, China

T. Ding et al. (eds.), *Electrolyzed Water in Food: Fundamentals and Applications*,
https://doi.org/10.1007/978-981-13-3807-6_2

cells to EW have not been determined.

In this chapter, factors related to the bacterial activity of EW are summarized in detail. Additionally, the proposed mechanisms for different types of EW on microbial inactivation are exhibited. This chapter also discusses various physiological states of microorganisms induced by EW. Finally, recent progress in EW used for spore inactivation and degradation of microbial toxins is also presented in this chapter.

2.2 Factors Influencing Bactericidal Activity of EW

2.2.1 *EW Parameters*

EW properties including the pH, ORP, and ACC have an effect on the antimicrobial efficacy of EW (Table 2.1). The pH values determine the types of chloride compounds present in EW. HOCl is the main form of EW under a pH of 5.0–6.5, but it changes into OCl^- under a higher pH and Cl_2 under a lower pH (Rahman et al. 2010b). Rahman et al. (2010b) found that EW with a pH of 2.5–6.0 resulted in approximately a 5-log reduction of *Listeria monocytogenes*, whereas only a 2.2-log reduction was achieved for EW with a pH of 9.0. Pangloli and Hung (2013) reported a pH ranging from 5 to 8 did not change significantly the inactivation level of *Escherichia coli* O157:H7 (5.68–6.08 $\log_{10}$ CFU/mL) by EW treatment. They explained that the high ORP in this study was not affected by increasing the pH. Additionally, the ORP is another property affecting the inactivation ability of EW (Kim et al. 2000a; Liao et al. 2007). The results from Stevenson et al. (2004) showed that efficient inactivation of *E. coli* O157:H7 strain H4420 required an ORP value higher than 850 mV. They explained that the higher ORP was related to a higher concentration of HOCl, contributing to the stronger bactericidal effect. Park et al. (2004) demonstrated that higher ACC led to a higher reduction of *E. coli* O157:H7 and *L. monocytogenes*. For instance, EW with an ACC over 1.0 mg/L resulted in a 7.94-log inactivated subpopulation of both *E. coli* O157:H7 and *L. monocytogenes*, whereas EW with an ACC of 0.1 mg/L only resulted in a 5.0- and 3.5-log inactivated *E. coli* O157:H7 and *L. monocytogenes*, respectively. In addition, some other parameters of EW were also found to have an effect on bactericidal activity. For example, Forghani et al. (2015) compared the microbial inactivation ability of EW with heating before and after production. It was found that the reduction level of *L. monocytogenes* was 7.51 $\log_{10}$ CFU/mL by EW and only 7.31 $\log_{10}$ CFU/mL by EW with heating before and after production, respectively. The result might be attributed to the loss of ACC during the heating after EW production. Water hardness was observed to affect antimicrobial inactivation of EW in several studies (Forghani et al. 2015; Pangloli and Hung 2013). Pangloli and Hung (2013) demonstrated that when the water hardness increased from 0 to

Table 2.1 The effect of EW parameters on bactericidal efficacy of EW

EW parameters	Microorganisms	Conclusions	References
pH (3.0, 5.0, and 7.0)	*Escherichia coli O157:H7*, *Listeria monocytogenes*	Higher microbial reduction with lower pH	Park et al. (2004)
pH (2.5, 4.0, 5.0, 6.0 and 9.0)	*Escherichia coli* O157:H7, *Listeria monocytogenes*, *Staphylococcus aureus*, and *Salmonella typhimurium*	Higher microbial reduction with lower pH	Rahman et al. (2010a)
pH (5, 6, 7, 8 and 9)	*Escherichia coli O157:H7*	No significant effect ($p > 0.05$)	Pangloli and Hung (2013)
pH (<3, 4–5, 6–7)	*Staphylococcus aureus*	No significant effect ($p > 0.05$)	Zhang et al. (2016a)
OPR (256, 1122 mV)	*Escherichia coli* O157:H7	Higher microbial reduction with higher ORP	Kim et al. (2000a)
OPR	*Escherichia coli* O157:H7	Loss of bactericidal activity at ORP ≤ 848 mV	Stevenson et al. (2004)
ACC (0.1–5.0 mg/L)	*Escherichia coli O157:H7*, *Listeria monocytogenes*	Higher microbial reduction with higher ACC	Park et al. (2004)
Heating before and after production of EW	*Escherichia coli O157:H7*, *Listeria monocytogenes*	Higher reduction with heating before EW production	Forghani et al. (2015)
Water hardness (0, 50, 100, and 200 mg/L as $CaCO_3$)	*Escherichia coli O157:H7*, *Listeria monocytogenes*	Higher microbial reduction with increasing water hardness	Pangloli and Hung (2013)

50 mg/L ($CaCO_3$), the reduction level of *E. coli* O157:H7 increased from 5.8 to 6.4 $\log_{10}$ CFU/mL. However, water hardness with of higher than 50 mg/L was observed to inhibit the inactivation of *E. coli* O157:H7 by EW. The exact mechanism of how water hardness changes the bactericidal efficacy of EW still remains unclear and requires more investigations.

2.2.2 Environmental Factors

Environmental factors such as the temperature, the presence of organic compounds, and matrix properties also affect the antimicrobial effect of EW (Table 2.2). Most studies showed that a higher temperature helped EW to more efficiently kill microorganisms (Ding et al. 2011; Fabrizio and Cutter 2003; Rahman et al. 2010b; Xie et al. 2012). For instance, Ding et al. (2011) found that the antimicrobial activity of AEW on *E. coli O157:H7*, *L. monocytogenes*, *S. typhimurium* and *B. cereus* increased by 100-fold when the temperature increased from 4 to 35 °C. Koide et al. (2011) demonstrated that the survival subpopulation of total aerobic bacteria was 1.3 $\log_{10}$ CFU/g and 2.9 $\log_{10}$ CFU/g at 45 and 18 °C, respectively, after a 2-min SAEW (pH 5.5, ACC 23 mg/L) treatment. The presence of organic compounds tends to have a negative effect on the bactericidal effect of EW. For instance, Li et al. (1996) observed that the reduction rate of *Bacillus subtilis* var. *niger* by 20-min EW exposure decreased from 100 to 19.5% after adding 10% bovine serum albumin (BSA) to AEW (pH 2.6, ORP 1130 mV). Generally, organic matter (e.g., amino acids, proteins, etc.) can react with the active chlorine species in EW, thereby weakening the antimicrobial capacity of EW. On the other hand, organic matters might wrap target microorganisms and protect the outer structures of microbial cells from the attack of EW (Park et al. 2009a; Virto et al. 2005). Similar results were found in other studies (Ayebah et al. 2005; Park et al. 2008a). Park et al. (2009a) demonstrated that the addition of 5 mg/L bovine serum to AEW (pH 2.06, ACC 37.5 mg/L) decreased the reduction level of *E. coli* O157:H7, *S. typhimurium*, and *L. monocytogenes* by 1.2, 1.1 and 0.8 $\log_{10}$ CFU/mL, respectively. In addition, the bactericidal efficacy of EW might be different on different matrices. Chen et al. (2014) compared the survivals of *S. aureus*, *Shigella flexneri*, *Vibrio parahaemolyticus*, *E. coli* and *S. typhimurium* on various surfaces with EW treatment for the same period of time. It was found that for all bacteria the survival rate was higher on the stainless steel than that on a plastic cutting board, an operating floor or gloves. In addition, the bactericidal effect of EW is thought to be better on smooth surfaces than on rough ones (Koseki et al. 2004; Park et al. 2009a). For instance, Park et al. (2009a) observed that the reduction of *E. coli* O157:H7, *S. typhimurium*, and *L. monocytogenes* on the surface of a tomato by AEW exposure was higher than that on the surface of green onions. The authors explained that the smoother surface of tomatoes accelerated the chlorine species in EW to better contact with microorganisms.

2.2.3 Microbial Properties

Microbial properties are important factors influencing the inactivation efficacy of EW (Table 2.3). Most studies have found that Gram-positive bacteria was more resistant to EW exposure than Gram-negative bacteria does (Guentzel et al. 2008; Kim et al. 2000b; Park et al. 2004). Kim et al. (2000b) observed that a 30-s AEW treatment (pH

Table 2.2 Effect of environmental factors on bactericidal activity of EW

Environmental factors	EW types	Microorganisms	Conclusions	References
Temperature (4 and 25 °C)	AEW (pH 2.6, ORP 1150 mV, ACC 50 mg/L)	*Salmonella Typhimurium*, *Listeria monocytogenes*	Highest microbial reduction (>8 $\log_{10}$ CFU/mL) at 25 °C	Fabrizio and Cutter (2003)
Temperature (4, 15, 23, 35, and 50 °C)	SAEW (pH 2.6, ORP 1100 mV, ACC 50 mg/L)	*Listeria monocytogenes*, *Salmonella typhimurium*, *Escherichia coli O157:H7*, and *Staphylococcus aureus*	Higher microbial reduction with higher temperature	Rahman et al. (2010a)
Temperature (4, 15, 23, 30, and 35 °C)	LcEW (pH 6.2, ORP 500–520 mV, ACC 5 mg/L) SAEW (pH 2.54, ORP 1100–1120 mV, ACC 50 mg/L)	*Escherichia coli O157:H7*, *Listeria monocytogenes*, *Salmonella typhimurium*, and *Bacillus cereus*	Higher microbial reduction with higher temperature	Ding et al. (2011)
Temperature (4, 20, and 50 °C)	AEW (pH 2.4, ORP 1163 mV, ACC 51 mg/L)	*Vibrio parahaemolyticus*	Higher microbial reduction with higher temperature	Xie et al. (2012)
Temperature (18 and 40 °C)	SAEW (pH 6.45, ORP 746–805 mV, ACC 30 mg/L)	*Escherichia coli O157:H7*, *Listeria monocytogenes*	Higher microbial reduction with higher temperature	Forghani et al. (2015)
Matrix (phosphate buffered saline, PBS; brain heart infusion, and BHI)	SAEW (pH 5.9, ORP 798 mV, ACC 35.0 mg/L)	*Vibrio vulnificus*, *Vibrio parahaemolyticus*	Higher microbial reduction in PBS than in BHI	Quan et al. (2010)
Matrix (tomatoes and potatoes)	AEW (pH 2.06, ACC 37.5 mg/L)	*Escherichia coli O157:H7*, *Salmonella typhimurium*, and *Listeria monocytogenes*	Higher reduction on smooth surface than rough surface	Park et al. (2009a)

(continued)

Table 2.2 (continued)

Environmental factors	EW types	Microorganisms	Conclusions	References
Matrix (plastic and wooden cutting boards)	NEW (pH 8.27, ORP 774 mV, ACC 60 mg/L)	*Escherichia coli*, *Pseudomonas aeruginosa*, *Staphylococcus aureus*, and *Listeria monocytogenes*	Higher reduction on plastic cutting board than wooden cutting board	Deza et al. (2007)
Organic matters (0, 5, 10, 15, and 20 mg/L bovine serum, BSA)	AEW (pH 2.06, ACC 37.5 mg/L)	*Escherichia coli* O157:H7, *Salmonella typhimurium*, and *Listeria monocytogenes*	Lower microbial reduction as higher BSA level	Park et al. (2009a)
Organic matters (2, 4, 6, 8, 10, 15, and 20 mg/L bovine serum, BSA)	AEW (pH 2.06, ACC 37.5 mg/L)	*Escherichia coli O157:H7*, *Salmonella typhimurium*, and *Listeria monocytogenes*	Lower microbial reduction as higher BSA level	Park et al. (2008a)
Organic matters (0.3%, tryptone, bovine serum albumin, and bovine mucin)	AEW (pH 2.62–2.90, ORP 1118–1188 mV, ACC 60–120 mg/L) SAEW (pH 5.81–5.99, ORP 920–956 mV, ACC 60–120 mg/L)	*Bacillus subtilis* spores	Lower microbial reduction with addition of organic matters	Zhang et al. (2016c)
Organic matters (glucose, sodium citrate, and sucrose)	NEW (pH 8.06–8.34)	*Artemia salina*	Lower microbial reduction with addition of organic matters	Duan et al. (2016)

2.5, ORP 1123 mV, ACC 10 mg/L) resulted in an undetectable survival of Gram-negative *E. coli* but still a 2-log survival of Gram-positive *L. monocytogenes*. In addition, vegetative bacteria cells are found to be more susceptible to EW treatment than spores (Kim et al. 2000b). Kim et al. (2000b) reported that the reduction level of *B. cereus* vegetative cells and *B. cereus* spore cells was 2.6 $\log_{10}$ CFU/mL and less than 1 $\log_{10}$ CFU/mL, respectively, for a 60-s AEW (pH 2.5, 1123 mV, ACC 10 mg/L) exposure. The thicker and more rigid outer structures of Gram-positive or spore bacterial cells make them less vulnerable to EW. In a study by Zhang et al. (2016a), AEW with 30 mg/L ACC only resulted in an 8.5-log reduction for *E. coli* O157:H7 but only a 4.1-log reduction of *B. subtilis*. *B. subtilis* is a Gram-positive and endospore-forming bacteria, which might explain the different inactivation efficacies of *B. subtilis* and *E. coli* O157:H7 by EW treatment to the same extent. In addition, the initial microbial concentration has also been reported to affect the bactericidal activity of EW (Koseki et al. 2003; Zhang et al. 2016a). Zhang et al. (2016a) found that the maximum inactivation level of *E. coli* O157:H7 by AEW (ACC 2 mg/L) was 6.44 $\log_{10}$ CFU/mL when the initial inoculum was 7 $\log_{10}$ CFU/mL, whereas the reduction levels were 4.46 and 1.04 $\log_{10}$ CFU/mL for 5- and 9-log initial inoculum of *E. coli* O157:H7, respectively. The authors explained that the higher bacterial concentration led to a higher level of organic compounds, which might shield bacteria from EW and neutralize the solution with the available chlorides. However, the effect of the initial inoculation population size on the inactivation efficacy remains controversial (Abadias et al. 2008). Abadias et al. (2008) revealed that a high (7 $\log_{10}$ CFU/mL) and low (5 $\log_{10}$ CFU/mL) inoculation population had no significant ($P > 0.05$) effect on reduction of *E. coli*, *L. innocua*, and *Salmonella* by AEW treatment (ACC 40 mg/L). Recently, a study by Zhang et al. (2016a) proposed that the centrifugation time was another factor influencing the inactivation efficacy of EW. They demonstrated that cell surfaces could be injured with the pretreatment of centrifugation, making microbes more easily damaged by the following EW exposure.

2.3 Principles of Microbial Decontamination Using EW

2.3.1 AEW

AEW is generated by the electrolysis of chloride solution through a membrane electrolytic cell (Koseki et al. 2004; Venkitanarayanan et al. 1999). The pH of AEW is determined by the time of electrolysis. AEW has been reported to have a strong antimicrobial effect on various microbes, such as *L. monocytogenes*, *E. coli O157:H7*, *B. cereus*, and *Salmonella* (Hao et al. 2017; Koseki et al. 2003; Ovissipour et al. 2016; Park et al. 2009a; Xiong et al. 2010). A low pH (<2.7) and a high ORP (>1100 mV) values are thought to cause microbial inactivation. Generally, microbial cells can survive under a pH of 4–9 (Gao et al. 2008). A low pH tends to destroy cell wall compounds (e.g., polysaccharides) and increase the permeability, resulting in the

Table 2.3 Effect of microbial properties on bactericidal activity of EW

Microbial properties	EW types	Microorganisms	Conclusions	References
Gram-positive and Gram-negative bacteria	AEW (pH 2.5, ORP 1123 mV, ACC 10 mg/L)	*Escherichia coli* O157:H7, *Listeria monocytogenes*	Higher resistance of Gram-positive bacteria than Gram-negative bacteria	Kim et al. (2000b)
Vegetative and spore cells		*Bacillus cereus* vegetative cells and spore cells	Higher resistance of spore cells than vegetative cells	
Gram-positive and Gram-negative bacteria	SAEW (pH 6.1, ORP 863.5 mV, ACC 30 mg/L)	*Escherichia coli*, *Staphylococcus aureus*	Higher resistance of Gram-positive bacteria than Gram-negative bacteria	Liao et al. (2017b)
Initial microbial concentration	AEW (pH 2.62–2.90, ORP 1118–1188 mV, ACC 60–120 mg/L) SAEW (pH 5.81–5.99, ORP 920–956 mV, ACC 60–120 mg/L)	*Escherichia coli* O157:H7	Higher microbial reduction with lower initial bacterial concentration	Zhang et al. (2016b)
Centrifugation time			Higher microbial reduction with increasing centrifugation times	

death of microbes. Nevertheless, a low pH might not be sufficient to kill microbes, especially spores. Li et al. (1996) reported that the reduction level of *Bacillus subtilis* var. *niger* can reach 100% after a 10-min AEW (pH 2.6, ORP 1130 mV) treatment, whereas it was only 1.06% for an HCl solution with the same pH. In addition, microbial cells maintain a membrane potential of −400 to 900 mV under normal physiological conditions (Quan et al. 2010). A higher surrounding ORP can disturb the distribution of ions on the inner and outer cell surfaces, which lead to the disruption of cell envelopes and a loss of intracellular components (Liao et al. 2007). However, Koseki et al. (2001) observed that the bactericidal effect of AEW with an ORP of 1140 mV was higher than that of ozonated water with an ORP of 1256 mV. The authors demonstrated that the available chlorine, primarily HOCl, was the main contributor to the bactericidal activity of AEW. Therefore, the active chlorine species (e.g., HClO, Cl_2, ClO_2, ClO^-, etc.) in EW were also thought to be another cause of microbial inactivation.

2.3.2 SAEW

SAEW has a near-neutral pH of 5.0–6.5 and a low OPR value (<1000 mV) produced by a nonmembrane cell (Ding et al. 2016; Koide et al. 2009). In SAEW, HClO (~95%) is the main form of active chlorine compounds and thought to be responsible for microbial inactivation. Ding et al. (2016) found that SAEW decreased TCC-dehydrogenase activity in *S. aureus*, which was explained by the fact that the HClO in SAEW reacted with enzymes to form N–Cl bonds. They also observed the leakage of proteins and the destruction of bacterial ultrastructure in *S. aureus* after SAEW treatment. A study by Liao et al. (2017b) from the same research group demonstrated that SAEW showed ROS-independent behavior on inactivation of both *E. coli* and *S. aureus* through multiple cell targets, including the integrity and potentials of cell membranes and intracellular esterase activity. The authors concluded that the oxidative agents in SAEW, especially HOCl, were primarily responsible for the bactericidal activity of SAEW. Similar results were obtained in the study of Hao et al. (2017), who also observed that multiple targets of microbes were damaged by SAEW, including cell membranes and nucleic acids (RNA and DNA), whereas the AEW only resulted in disruption of cell membranes without damages to nucleic acids.

2.3.3 AlEW

AlEW, also called basic electrolyzed water (BEW), is a by-product of AEW production through the electrolysis of diluted chloride solution in the electrolyzed cells with a membrane (Hricova et al. 2008). Generally, AlEW has a pH value of 10–13 and an ORP value of −800 to −900 mV. To date, only limited studies have investigated the bactericidal effect of AlEW (Ayebah et al. 2005; Ovissipour et al. 2016; Rahman et al. 2011). The microbial inactivation mechanisms of AlEW have also received little attention. Ovissipour et al. (2016) found that AlEW resulted in a 1- to 3-log reduction of *E. coli* O104:H4, *L. monocytogenes*, *Campylobacter jejuni*, *Aeromonas hydrophila*, and *Vibrio parahaemolyticus* in suspension. The authors proposed that AlEW can assist the penetration of active chlorine agents and damage the cell walls by destabilizing the extracellular polymeric compounds around bacterial cell envelopes. In addition, the extremely low ORP of AlEW might disturb the cell membranes, leading to the death of microbes. Individual AlEW treatment has been reported to have an inefficient antimicrobial effect in several studies (Ayebah et al. 2005; Park et al. 2008b; Rahman et al. 2011). More investigators have paid attention to the hurdle technologies combining AlEW with other methods, such as mild heat, AEW, and so on, to improve the bactericidal efficacy (Ayebah et al. 2005; Park et al. 2008b; Rahman et al. 2011).

2.3.4 NEW

The products at the anode during the AEW production are redirected into the cathode chamber and form into a new type of EW, which is called NEW (Abadias et al. 2008; Guentzel et al. 2008). NEW has a neutral pH of 7–8, resulting in less corrosion of contact surfaces and greater stability of chlorine agents (Len et al. 2002). Many studies have investigated the application of NEW for microbial inactivation (Abadias et al. 2008; Deza et al. 2003, 2005; Pangloli and Hung 2011; Torlak 2014). The main active agents in NEW are sodium hypochlorite, chlorine, bromide, hypobromous acid, hydrogen peroxide, oxygen free radical, hydroxyl radical, and other disinfection by-products (DBPS), which are thought to be responsible for microbial inactivation (Duan et al. 2016). Xiong et al. (2010) applied UV spectroscopy, ^{17}O-NMR spectroscopy and electron spin resonance (ESR) analysis to detect the radicals in NEW (pH 5.07, 862.2 mV) and AEW (pH 2.69, 1125.7 mV) with same ACC of 30 mg/L. The authors found that NEW had a higher level of OH. than AEW did, which was the main contributor to the inactivation of *Aspergillus flavus* by NEW. These oxidative compounds in NEW tend to damage enzymes and nucleic acids in microbial cells, which result in the final death.

2.4 Sublethal Injury, VBNC, and Apoptosis States of Microorganisms Induced by EW

Apart from the lethal effect, EW treatment can induce other types of physiological states in microorganisms, including sublethal injury, a viable but non-culturable (VBNC) state and apoptosis (Table 2.4).

Sublethal injury is a physiological state caused by a chemical or physical treatment that damages but does not kill microorganisms (Mackey 2000; Wesche et al. 2009). The sublethal injury produces microorganisms that can only grow into colonies on nonselective media but fail to form colonies on selective media. When the conditions become suitable, these sublethally injured microbial cells can repair themselves and outgrow again. Therefore, the underestimation of the sublethally injured pathogens could pose a potential risk to human health and food safety. Limited studies have confirmed that EW can induce the sublethal injury of microorganisms (Han et al. 2018b; Liao et al. 2017b; Xuan et al. 2017). For example, Xuan et al. (2017) found that 20-s SAEW (pH 6.3, ORP 867.4 mV, ACC 30 mg/L) treatment resulted in approximately 90% sublethally injured *L. monocytogenes*. Subsequently, another study from the same group reported that a 15-s SAEW (pH 6.1, ORP 863.5 mV, ACC 30 mg/L) treatment could induce 49.9 and 39.7% of *E. coli* and *S. aureus* into sublethal injury, respectively (Liao et al. 2017b). In the more recent study by Han et al. (2018b), 38.12% of *Yersinia enterocolitica* and 52% of *Salmonella Enteritidis* were

Table 2.4 Sublethal injury, VBNC, and apoptosis states of microorganisms induced by electrolyzed water

Microorganism states	EW properties	Microorganisms	Treatment time	References
Sublethal injury	SAEW (pH 6.3, ORP 867.4 mV, ACC 30 mg/L)	*Listeria monocytogenes*	20–40 s	Xuan et al. (2017)
	SAW (pH 2.7, ACC 25 mg/L)	Coliform bacteria	3 min	Izumi and Inoue (2018)
	SAEW (pH 6.1, ORP 863.5 mV, ACC 30 mg/L)	*Escherichia coli*, *Staphylococcus aureus*	0–30 s	Liao et al. (2017b)
	NEO (pH 7.35, ORP 829.7 mV, ACC 59.1 mg/L)	*Salmonella enteritidis*, *Yersinia enterocolitica*	5 min	Han et al. (2018b)
VBNC	NEO (pH 7.64, ORP 818 mV, ACC 74 mg/L)	*Escherichia coli* O157:H7, *Salmonella enteritidis* and *Yersinia enterocolitica*	5 min	Han et al. (2018a)
	EO (pH 2.5; ACC 0.625–20 mg/L)	*Escherichia* O157:H7 and *L. monocytogenes*	1 or 5 min	Afari and Hung (2018a)
	SAEW (pH 6.1, ORP 893.5 mV, ACC 30 mg/L)	*Staphylococcus aureus*	1–10 min	Li et al. (2017)
Apoptosis	SAEW (pH 6.5, ORP 890 mV, ACC 28 mg/L)	*Aeromonas hydrophila*	5 or 10 min	Ye et al. (2018)
	SAEW (pH 6.40, ORP 910 mV, ACC 60 mg/L)	*Escherichia coli*	1–10 min	Ye et al. (2017)

sublethally damaged after NEW (pH 7.35, ORP value 829.7 mV, ACC 59.1 mg/L) treatment for 5 min, whereas no sublethal injury of *E. coli* O157:H7 was caused under NEW exposure. The authors found that additional treatments of various common stress (temperature, pH, NaCl, and bile salt) in the food industry could achieve synergy on EW-stress microbial cells, which provided possible hurdle technologies to control the occurrence of sublethal injury.

The VBNC state is a type of physiological state, in which microbes fail to grow on the nonselective or selective media but maintain their metabolic activities (Zhao et al. 2017). Pathogens in the VBNC state might still possess their virulent capacity and bring latent risks to public health. In recent years, only limited studies have paid attention to the induction of VBNC by EW (Afari and Hung 2018a, b; Han et al. 2018a, b; Li et al. 2017). For example, Li et al. (2017) demonstrated that 10.42% of *S. aureus* cells was induced into the VBNC state by a 1-min SAEW (pH 6.1, ORP 893.5 mV, ACC 30 mg/L) exposure through the conventional plate method and the flow cytometer technique. Han et al. (2018a) reported that a 3% NEW (pH 7.64, ORP 818 mV, ACC 2.22 mg/L) exposure induced around 68.31% of *Salmonella Enteritidis* cells to enter the VBNC state, and 6% NEW (pH 7.64, ORP 818 mV, ACC 4.44 mg/L) achieved approximate 56% VBNC in *E. coli* O157:H7 cells and 58% VBNC in *Yersinia enterocolitica* cells. No VBNC microbial cells were detected under the exposure of 50% or 100% NEW. Afari and Hung (2018a) found that AEW (pH 2.1) with a low ACC (1.25 and 2.5 mg/L) led both *E. coli* O157:H7 and *L. monocytogenes* into the VBNC state using plate count and flow cytometer methods. In addition, the study further investigated the resuscitation of these VBNC microorganisms. However, only VBNC *E. coli* O157:H7 was observed to resuscitate successfully under suitable conditions. The molecular mechanisms underlying the VBNC physiological state induced by EW should be further investigated in depth.

Apoptosis, or programmed cell death (PCD), is defined as the genetically encoded cellular death (Bayles 2014). Some studies have observed the bacterial apoptosis caused by EW (Ye et al. 2017, 2018). Ye et al. (2018) demonstrated that the 10-min SAEW (pH 6.5, ORP 890 mV, ACC 28 mg/L) treatment induced a proportion of *Aeromonas hydrophila* cells into apoptosis states according to the results from Annexin V-FITC/PI staining and the DNA ladder detection experiment. Ye et al. (2017) observed that a 10-min SAEW (pH 6.40, ORP 910 mV, ACC 60 mg/L) treatment caused around 9% of the *E. coli* stained by an Annexin V fluorescence probe. This probe is usually used for detection of an apoptosis marker-externally exposed phosphatidylserine (PS).

2.5 Inactivation of Spores by EW

Spores are the dormant forms of vegetative bacterial cells, which exhibit extreme resistance towards various stresses, including high temperature, high pressure, radiation, and chemical agents (Wells-Bennik et al. 2016). Spores can return to life rapidly when exposed to appropriate nutrients and lead to foodborne diseases or food spoilage after germinating and resuming vegetative forms (Setlow 2006). Effective methods for the control of spores are required for the food industry. The lethal effect of EW on bacterial spores has been demonstrated in some studies (Table 2.5; Kim et al. 2000b; Okull et al. 2006; Park et al. 2009b; Tang et al. 2011; Vorobjeva et al. 2004). For example, Vorobjeva et al. (2004) demonstrated that 5-min EW (pH 2.84, ORP 1125 mV, ACC 43 mg/L) exposure was sufficient for completed inactivation of *Bacillus cereus* spores. Okull et al. (2006) observed that a 5-min EW (pH 3.5, ORP 1027 mV, ACC 18 mg/L) exposure achieved over 5-log reduction of *Penicillium expansum* spores. In a study by Zhang et al. (2016c), AEW treatment (pH

Table 2.5 The inactivation of spores by electrolyzed water

Spores	EW properties	Treatment time	Reduction level	References
B. cereus spores	SAEW (pH 5.92–6.04, ORP 938–970 mV, ACC 50–150 mg/L)	30 or 60 min	1.7–3.3 logs	Zhang et al. (2018)
	AEW (pH 2.61–2.87, ORP 1146–1185 mV, ACC 50–50 mg/L)		1.6–3.0 logs	
Clostridium difficile spores	AEW (pH 5)	60 min	6 logs	Tkhawkho et al. (2017)
Bacillus subtilis spores, *Bacillus cereus* spores	AEW and SAEW (60 mg/L ACC)	6 min	7 logs	Zhang et al. (2016c)
Bacillus subtilis var. *niger* spores	EOW (ACC 115.7–220.5, pH 2.47–2.57)	20 min	0.17–7.08 logs	Tang et al. (2011)
Penicillium expansum spores	EW (pH 3.5, ORP 1027 mV, ACC 18 mg/L)	120 s	5 logs	Okull et al. (2006)
Bacillus cereus spores	EO water (pH 2.84, ORP 1125 mV, ACC 43 mg/L)	5 min	7.98 logs	Vorobjeva et al. (2004)
Bacillus cereus spores	EO (ACC 10, pH 2.5, ORP 1123 mV)	120 s	>3 logs	Kim et al. (2000b)

2.62–2.90, ORP 1118–1188 mV, ACC 120 mg/L) for 2 min resulted in a 4.45-log reductions of *B. subtilis* spores, whereas SAEW (pH 5.81–5.99, ORP 920–956 mV, ACC 120 mg/L) at the same treatment condition achieved nondetectable level. Tang et al. (2011) reported that a 20-min EW treatment (pH 2.5, ORP 1187 mV, ACC 191 mg/L) resulted in a 100% reduction of *B. subtilis* var. *niger* spores. The authors found that EW exposure destroyed dehydrogenase activity, intensified membrane permeability, elevated suspension conductivity, and caused the leakage of cellular inclusions (K^+, proteins, and DNA), leading to the final death of *B. subtilis* var. *niger* spores.

2.6 Inactivation of Microbial Biofilm by EW

Apart from the form of spores, some bacteria can form biofilm to fight against the external stresses, such as preservatives, antibiotics. Therefore, efficient controlling of biofilm is important for food safety and public health. EW has been documented as a potential method for eradicating the biofilms (Table 2.6). Han et al. (2017) found that AEW (pH 2.3, ORP 1175 mV, 136 mg/mL) exposure for 5 min resulted in 4.51-, 3.57- and 5.9-log inactivation of *E. coli*, *V. parahaemolyticus*, and *L. monocytogenes* biofilm, respectively (Han et al. 2017). The authors observed severe disruptions of EPS induced by AEW with the use of Raman spectroscopic analysis combined with EPS chemical analysis, which was regarded as the main responsibility for the biofilm inactivation. In the study of Hussain et al. (2018), around 6 $\log_{10}$ CFU/cm^2 of *B. cereus* biofilms was inactivated by a 5-min SAEW (pH 5.71, ORP 818–886 mV, ACC 75 mg/L) treatment (Hussain et al. 2018). With the aid of field emission scanning electron microscopy (FESEM) and confocal laser scanning microscopy (CLSM), the authors demonstrated that SAEW disrupted the plasma membrane and caused the leakage of the intracellular components, leading to the final death of the biofilm.

2.7 Degradation of Microbial Toxin by EW

Apart from microorganisms, microbial toxins are also challenges for food safety. During the past few decades, some researchers have demonstrated the capacity of EW in the removal of microbial toxins (Table 2.7; Escobedo-González et al. 2016; Suzuki et al. 2002a, b; Xiong et al. 2012). Suzuki et al. (2002a) reported that staphylococcal enterotoxin A (SEA) can be eliminated by 100% with a 30-min AEW (pH 2.5–2.8, ORP 1180 mV) treatment. The authors explained that EW degraded SEA via an oxidative reaction involving OH radicals and reactive chlorine. Xiong et al. (2012) found that 10-min NEW (pH 5.6, ACC 60 mg/L) and 15-min AEW (pH 2.5, ACC 80 mg/L) can efficiently degrade aflatoxin B1 (AFB_1) by over 90%. In addition, these authors identified the major degradation product—8-chloro-9-hydroxy aflatoxin B1 ($C_{17}H_{13}ClO_7$), which did not have mutagenic activity or cytotoxic effects. Two possible degradation pathways of AFB_1 were revealed in the study by Escobedo-González et al. (2016). To begin with, the double bond of AFB_1 was attacked by electrophilic reagents (chloronium ion for pathway A, hypochlorous acid for pathway B) to reach the ionic activated state or the molecular activated state (TSi1 for pathway A and TSm1 for pathway B). The chloronium ion reactive intermediate (1a) was produced in the next step. Finally, a nucleophilic attack was conducted by the hydroxide ion on the chloronium ion reactive intermediate, producing the second activated state (TS2).

Table 2.6 The inactivation of microbial biofilm by electrolyzed water

Microbial biofilm	EW properties	Treatment time	Reduction level	References
Listeria monocytogenes biofilm	AEW (pH 2.6, ORP 1160 mV, ACC 56 mg/L)	300 s	1.9×10^{10} CFU/82.5 cm^2	Kim et al. (2001)
Listeria monocytogenes biofilm	AEW (pH 2.4, ORP 1163 mV, 47.12 mg/L)	30–120 s	4.33–5.21 logs	Ayebah et al. (2005)
Listeria monocytogenes biofilm	AEW (pH 2.29, ORP 1163 mV, ACC 85 mg/L) AlEW (pH 11.2, ORP −885 mV, ACC 0 mg/L)	30–60 s	1.07–1.19 logs 4.37–4.53 logs	Ayebah et al. (2006)
Listeria monocytogenes biofilm	NEW (pH 7.0, ORP 1110 mV, ACC 65 mg/L)	10 min	6.5 logs	Arevalos-Sánchez et al. (2012)
Staphylococcus aureus biofilm	AlEW (pH 10.8 and 11.6, ORP −856 and −774 mV)	120 s	42 and 78%	Sun et al. (2012)
	AEW (pH 2.5, ORP 1146 mV, ACC 40 mg/L and pH 2.5, ORP 939 mV, ACC 0 mg/L)		13 and 89%	
Candida albicans and *Streptococcus mutans* biofilm	NEW (pH 7.5, ACC 30–200 mg/L)	1 min	100%	Ozaki et al. (2012)
L. monocytogenes biofilm	NEW1 (pH 6.5–7.5, ORP 800–1000 mV, ACC 234 mg/L) NEW2 (pH 6.5–8.5, ORP 900–1000 mV, ACC 4219 mg/L)	3 or 10 min	100%	Arevalos-Sánchez et al. (2013)

(continued)

Table 2.6 (continued)

Microbial biofilm	EW properties	Treatment time	Reduction level	References
Staphylococcus aureus biofilm	AEW (pH 2.99, ORP 1172 mV, ACC 565 mg/L) NEW (pH 6.06, ORP 967 mV, ACC 856 mg/L) AlEW (pH 7.95, ORP 843 mV, ACC 541 mg/L)	30 min	~4 logs	Vázquez-Sánchez et al. (2014)
E. coli biofilm	AEW (pH 2.30, ORP 1175 mV, ACC 136 mg/L)	5 min	65%	Han et al. (2017)
L. monocytogenes biofilm			82%	
V. parahaemolyticus biofilm			52%	
E. coli biofilm	NEW (ACC 50 mg/L)	20 min	3.26 logs	Meireles et al. (2017)
Bacillus cereus biofilm	AEW (pH 2.73, ORP 1177 mV, ACC 50 mg/L)	15 min	100%	Li et al. (2017)
B. cereus (KCTC 13153) biofilm	SAEW (pH of 5.74, ORP 832–855 mV, ACC 150 mg/L)	20 min	3.8 logs	Hussain et al. (2018)

2.8 Conclusions

EW is a novel decontamination technology in the food industry. In this chapter, we analyzed the factors affecting the bactericidal activity of EW in order to help the optimization of EW processing in practice. An overview of the inactivation mechanisms among AEW, SAEW, SlEW, and NEW on microbes is provided (Fig. 2.1). However, the exact mode of EW on microbial cells still remains unclear and requires more investigations to elucidate in future. The knowledge of EW's bactericidal modes is necessary for the optimization and development of EW processing in practice. Reference microbial strains for assessments of EW Pasteurization should be determined. Finally, the establishment of mathematical models based on processing parameters, environmental factors, and microbial physiological states should be improved to accurately predict the inactivation rate of microorganisms by EW. In addition to the microbial inactivation, EW has been demonstrated to induce microbiological physio-

Table 2.7 The degradation of microbial toxin by electrolyzed water

Toxins	EW properties	Treatment times	Removal efficacy (%)	References
Ochratoxin A	AEW (pH 2.46, ORP 1157 mV, ACC 68.89 mg/L)	5 min	~40	Tu et al. (2015)
	SAEW (pH 6.05, ORP 897.5 mV, ACC 49.69 mg/L)		~85	
	AlEW (pH 10.4, ORP 839.15 mV, ACC <1 mg/L)		~1	
Aflatoxins	NEW (pH 7.01, ORP 766 mV, ACC 60 mg/L)	15 min	No	Jardon-Xicotencatl et al. (2015)
Aflatoxin B_1	AEW (pH 2.5, ORP 1117.3 mV, ACC 80.2 mg/L)	5–30 min	20–100	Xiong et al. (2012)
	NEW (pH 5.6, ORP 836.4 mV, ACC 83.7 mg/L)		30–100	
Staphylococcal enterotoxin A	AEW (pH 2.5–2.8, ACC ~36.3 mg/L)	30 min	100	Suzuki et al. (2002a)
Aflatoxin	AEW (pH 2.5, ACC 39.4 mg/L, ORP 1164 mV)	15 min	100	Suzuki et al. (2002b)
Aflatoxin B_1	AlEW (pH 11.7, ORP – 662 mV)	5 min	~100	Fan et al. (2013)

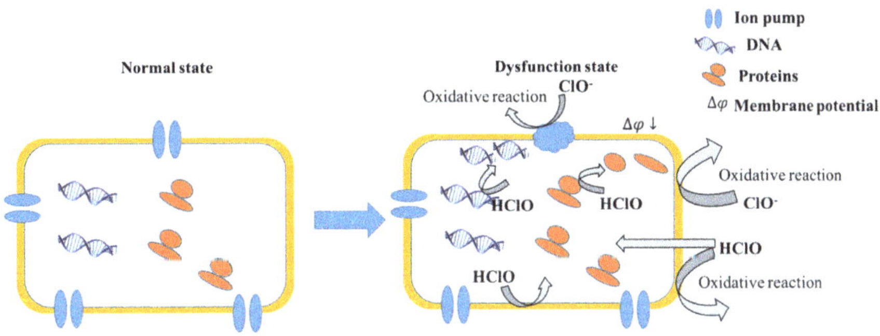

Fig. 2.1 The diagram of proposed microbial inactivation mechanisms under EW treatment

logical changes. More investigations are required to grasp a deeper understanding of the response of microbial cells to EW stress to guarantee the safety of EW. The potential application of EW to microbial spores, biofilm, and microbial toxins degradation by EW should be further explored extensively.

References

Abadias M, Usall J, Oliveira M et al (2008) Efficacy of neutral electrolyzed water (NEW) for reducing microbial contamination on minimally-processed vegetables. Int J Food Microbiol 123(1–2):151–158

Afari GK, Hung YC (2018a) Detection and verification of the viable but nonculturable (VBNC) state of *Escherichia coli* O157:H7 and *Listeria monocytogenes* using flow cytometry and standard plating. J Food Sci 83(7):1913–1920

Afari GK, Hung YC (2018b) A meta-analysis on the effectiveness of electrolyzed water treatments in reducing foodborne pathogens on different foods. Food Control 93:150–164

Arevalos-Sánchez M, Regalado C, Martin SE, Domínguez-Domínguez J, García-Almendárez BE (2012) Effect of neutral electrolyzed water and nisin on Listeria monocytogenes biofilms, and on listeriolysin O activity. Food Control 24(1–2):116–122

Arevalos-Sánchez M, Regalado C, Martin SE et al (2013) Effect of neutral electrolyzed water on lux-tagged *Listeria monocytogenes* EGDe biofilms adhered to stainless steel and visualization with destructive and non-destructive microscopy techniques. Food Control 34(2):472–477

Ayebah B, Hung YC (2005) Electrolyzed water and its corrosiveness on various surface materials commonly found in food processing facilities. J Food Process Eng 28(3):247–264

Ayebah B, Hung YC, Frank JF (2005) Enhancing the bactericidal effect of electrolyzed water on Listeria monocytogenes biofilms formed on stainless steel. J Food Protect 68(7):1375–1380

Ayebah B, Hung YC, Kim C et al (2006) Efficacy of electrolyzed water in the inactivation of planktonic and biofilm *Listeria monocytogenes* in the presence of organic matter. J Food Prot 69(9):2143–2150

Bayles KW (2014) Bacterial programmed cell death: making sense of a paradox. Nat Rev Microbiol 12(1):63

Cao W, Zhu ZW, Shi ZX et al (2009) Efficiency of slightly acidic electrolyzed water for inactivation of *Salmonella enteritidis* and its contaminated shell eggs. Int J Food Microbiol 130(2):88–93

Chen JJ, Deng SG, Lin X et al (2014) Germicidal efficacy of neutral electrolyzed water against the apparatus for the production of aquatic products. Sci Tech Food Ind 1(35):160–163

Deza MA, Araujo M, Garrido MJ (2003) Inactivation of *Escherichia coli* O157:H7, *Salmonella enteritidis* and *Listeria monocytogenes* on the surface of tomatoes by neutral electrolyzed water. Lett Appl Microbiol 37(6):482–487

Deza MA, Araujo M, Garrido MJ (2005) Inactivation of *Escherichia coli*, *Listeria monocytogenes*, *Pseudomonas aeruginosa* and *Staphylococcus aureus* on stainless steel and glass surfaces by neutral electrolysed water. Lett Appl Microbiol 40(5):341–346

Deza MA, Araujo M, Garrido MJ (2007) Efficacy of neutral electrolyzed water to inactivate *Escherichia coli*, *Listeria monocytogenes*, *Pseudomonas aeruginosa*, and *Staphylococcus aureus* on plastic and wooden kitchen cutting boards. J Food Protect 70(5):1070

Ding T, Rahman SME, Oh D (2011) Inhibitory effects of low concentration electrolyzed water and other sanitizers against foodborne pathogens on oyster mushroom. Food Control 22(2):318–322

Ding T, Xuan X, Li J et al (2016) Disinfection efficacy and mechanism of slightly acidic electrolyzed water on *Staphylococcus aureus* in pure culture. Food Control 60:505–510

Duan D, Liu G, Yao P et al (2016) The effects of organic compounds on inactivation efficacy of *Artemia salina* by neutral electrolyzed water. Ocean Eng 125:31–37

Escobedo-González R, Méndez-Albores A, Villarreal-Barajas T et al (2016) A theoretical study of 8-chloro-9-hydroxy-aflatoxin B1, the conversion product of aflatoxin B1 by neutral electrolyzed water. Toxins 8(7):225

Fabrizio KA, Cutter CN (2003) Stability of electrolyzed oxidizing water and its efficacy against cell suspensions of *Salmonella typhimurium* and *Listeria monocytogenes*. J Food Protect 66(8):1379–1384

Fan S, Zhang F, Liu S et al (2013) Removal of aflatoxin B1 in edible plant oils by oscillating treatment with alkaline electrolysed water. Food Chem 141(3):3118–3123

Forghani F, Park J, Oh D (2015) Effect of water hardness on the production and microbicidal efficacy of slightly acidic electrolyzed water. Food Microbiol 48:28–34
Gao X, Liu Z, Li X et al (2008) Sterilization mechanism and application of strong acidic electrolyzed water. Chin Agricult Sci Bull 24(7):393–399
Guentzel JL, Lam KL, Callan MA et al (2008) Reduction of bacteria on spinach, lettuce, and surfaces in food service areas using neutral electrolyzed oxidizing water. Food Microbiol 25(1):36–41
Han Q, Song X, Zhang Z et al (2017) Removal of foodborne pathogen biofilms by acidic electrolyzed water. Front Microbiol 8
Han D, Hung YC, Wang L (2018a) Evaluation of the antimicrobial efficacy of neutral electrolyzed water on pork products and the formation of viable but nonculturable (VBNC) pathogens. Food Microbiol 73:227–236
Han D, Hung YC, Bratcher CL et al (2018b) Formation of sublethally injured *Yersinia enterocolitica*, *Escherichia coli* O157:H7, and *Salmonella enterica* serovar enteritidis cells after neutral electrolyzed oxidizing water treatments. Appl Environ Microbiol 84(17):e01066–18
Hao J, Wu T, Li H et al (2017) Differences of bactericidal efficacy on *Escherichia coli*, *Staphylococcus aureus*, and *Bacillus subtilis* of slightly and strongly acidic electrolyzed water. Food Bioprocess Technol 10(1):155–164
Hricova D, Stephan R, Zweifel C (2008) Electrolyzed water and its application in the food industry. J Food Protect 71(9):1934–1947
Huang Y, Hung Y, Hsu S et al (2008) Application of electrolyzed water in the food industry. Food Control 19(4):329–345
Hussain MS, Kwon M et al (2018) Effect of electrolyzed water on the disinfection of bacillus cereus biofilms: the mechanism of enhanced resistance of sessile cells in the biofilm matrix. J Food Protect 81(5):860–869
Izumi H, Inoue A (2018) Viability of sublethally injured coliform bacteria on fresh-cut cabbage stored in high CO_2 atmospheres following rinsing with electrolyzed water. Int J Food Microbiol 266:207–212
Jardon-Xicotencatl S, Díaz-Torres R, Marroquín-Cardona A et al (2015) Detoxification of aflatoxin-contaminated maize by neutral electrolyzed oxidizing water. Toxins 7(10):4294–4314
Kim C, Hung YC, Bracett RE (2000a) Roles of oxidation-reduction potential in electrolyzed oxidizing and chemically modified water for the inactivation of food-related pathogens. J Food Protect 63(1):19–24
Kim C, Hung YC, Brackett RE (2000b) Efficacy of electrolyzed oxidizing (EO) and chemically modified water on different types of foodborne pathogens. Int J Food Microbiol 61(2–3):199–207
Kim C, Hung YC, Brackett, RE et al (2001) Inactivation of *Listeria monocytogenes* biofilms by electrolyzed oxidizing water. J Food Process Pres 25(2):91–100
Koide S, Takeda JI, Shi J et al (2009) Disinfection efficacy of slightly acidic electrolyzed water on fresh cut cabbage. Food Control 20(3):294–297
Koide S, Shitanda D, Note M et al (2011) Effects of mildly heated, slightly acidic electrolyzed water on the disinfection and physicochemical properties of sliced carrot. Food Control 22(3–4):452–456
Koseki S, Yoshida K, Isobe S et al (2001) Decontamination of lettuce using acidic electrolyzed water. J Food Protect 64(5):652–658
Koseki S, Yoshida K, Kamitani Y et al (2003) Influence of inoculation method, spot inoculation site, and inoculation size on the efficacy of acidic electrolyzed water against pathogens on lettuce. J Food Protect 66(11):2010–2016
Koseki S, Yoshida K, Isobe S et al (2004) Efficacy of acidic electrolyzed water for microbial decontamination of cucumbers and strawberries. J Food Protect 67(6):1247–1251
Len SV, Hung YC, Chung D et al (2002) Effects of storage conditions and pH on chlorine loss in electrolyzed oxidizing (EO) water. J Agr Food Chem 50(1):209–212
Li XW, Sun SH, Li T (1996) Preliminary study of microbiocide effect and its mechanism of electrolyzed oxidizing water. Zhonghua liu xing bing xue za zhi = Zhonghua liuxingbingxue zazhi 17(2):95

Li J, Ding T, Liao X et al (2017) Synergetic effects of ultrasound and slightly acidic electrolyzed water against *Staphylococcus aureus* evaluated by flow cytometry and electron microscopy. Ultrason Sonochem 38:711–719

Liao LB, Chen WM, Xiao XM (2007) The generation and inactivation mechanism of oxidation-reduction potential of electrolyzed oxidizing water. J Food Eng 78(4):1326–1332

Liao X, Liu D, Xiang Q et al (2017a) Inactivation mechanisms of non-thermal plasma on microbes: a review. Food Control 75:83–91

Liao X, Xuan X, Li J et al (2017b) Bactericidal action of slightly acidic electrolyzed water against *Escherichia coli* and *Staphylococcus aureus* via multiple cell targets. Food Control 79:380–385

Mackey BM (2000) Injured bacteria In the microbiological safety and quality of foods, vol I (ed Lund BM, Baird-Parker TC, Gould GW) Aspen Publishers, Inc., Gaithersburg, pp 315–341

Manas P, Pagán R (2005) Microbial inactivation by new technologies of food preservation. J Appl Microbiol 98(6):1387–1399

Meireles A, Ferreira C, Melo L et al (2017) Comparative stability and efficacy of selected chlorine-based biocides against Escherichia coli in planktonic and biofilm states. Food Res Int 102:511–518

Okull DO, Demirci A, Rosenberger D et al (2006) Susceptibility of *Penicillium expansum* spores to sodium hypochlorite, electrolyzed oxidizing water, and chlorine dioxide solutions modified with nonionic surfactants. J Food Prot 69(8):1944–1948

Ovissipour M, Al-Qadiri HM, Sablani SS et al (2016) Efficacy of acidic and alkaline electrolyzed water for inactivating *Escherichia coli* O104:H4, *Listeria monocytogenes*, *Campylobacter jejuni*, *Aeromonas hydrophila*, and *Vibrio parahaemolyticus* in cell suspensions. Food Control 62:405

Ozaki M, Ohshima T, Mukumoto M et al (2012) A study for biofilm removing and antimicrobial effects by microbubbled tap water and other functional water, electrolyzed hypochlorite water and ozonated water. Dent Mater J 31(4):662–668

Pangloli P, Hung Y (2011) Efficacy of slightly acidic electrolyzed water in killing or reducing *Escherichia coli* O157:H7 on iceberg lettuce and tomatoes under simulated food service operation conditions. J Food Sci 76(6):M361–M366

Pangloli P, Hung Y (2013) Effects of water hardness and pH on efficacy of chlorine-based sanitizers for inactivating *Escherichia coli* O157:H7 and *Listeria monocytogenes*. Food Control 32(2):626–631

Park H, Hung YC, Chung D (2004) Effects of chlorine and pH on efficacy of electrolyzed water for inactivating *Escherichia coli* O157:H7 and *Listeria monocytogenes*. Int J Food Microbiol 91(1):13–18

Park EJ, Alexander E, Taylor GA et al (2008a) Effects of organic matter on acidic electrolyzed water for reduction of foodborne pathogens on lettuce and spinach. J Appl Microbiol 105(6):1802–1809

Park EJ, Alexander E, Taylor GA et al (2008b) Effect of electrolyzed water for reduction of foodborne pathogens on lettuce and spinach. J Food Sci 73(6):M268–M272

Park EJ, Alexander E, Taylor GA et al (2009a) The decontaminative effects of acidic electrolyzed water for *Escherichia coli* O157:H7, *Salmonella typhimurium*, and *Listeria monocytogenes* on green onions and tomatoes with differing organic demands. Food Microbiol 26(4):386–390

Park YB, Guo JY, Rahman SME et al (2009b) Synergistic effect of electrolyzed water and citric acid against Bacillus cereus cells and spores on cereal grains. J Food Sci 74(4):M185–M189

Quan Y, Choi K, Chung D et al (2010) Evaluation of bactericidal activity of weakly acidic electrolyzed water (WAEW) against *Vibrio vulnificus* and *Vibrio parahaemolyticus*. Int J Food Microbiol 136(3):255–260

Rahman SME, Ding T, Oh D (2010a) Effectiveness of low concentration electrolyzed water to inactivate foodborne pathogens under different environmental conditions. Int J Food Microbiol 139(3):147–153

Rahman SME, Ding T, Oh D (2010b) Inactivation effect of newly developed low concentration electrolyzed water and other sanitizers against microorganisms on spinach. Food Control 21(10):1383–1387

Rahman SME, Jin Y, Oh D (2011) Combination treatment of alkaline electrolyzed water and citric acid with mild heat to ensure microbial safety, shelf-life and sensory quality of shredded carrots. Food Microbiol 28(3):484–491

Rahman SME, Khan I, Oh D (2016) Electrolyzed water as a novel sanitizer in the food industry: current trends and future perspectives. Compr Rev Food Sci F 15(3):471–490

Setlow P (2006) Spores of Bacillus subtilis: their resistance to and killing by radiation, heat and chemicals. J Appl Microbiol 101(3):514–525

Stevenson S, Cook SR, Bach SJ et al (2004) Effects of water source, dilution, storage, and bacterial and fecal loads on the efficacy of electrolyzed oxidizing water for the control of *Escherichia coli* O157:H7. J Food Protect 67(7):1377–1383

Sun JL, Zhang SK, Chen JY et al (2012) Efficacy of acidic and basic electrolyzed water in eradicating Staphylococcus aureus biofilm. Can J Microbiol 58(4):448–454

Suzuki T, Itakura J, Watanabe M et al (2002a) Inactivation of Staphylococcal enterotoxin-A with an electrolyzed anodic solution. J Agr Food Chem 50(1):230–234

Suzuki T, Noro T, Kawamura Y et al (2002b) Decontamination of aflatoxin-forming fungus and elimination of aflatoxin mutagenicity with electrolyzed NaCl anode solution. J Agr Food Chem 50(3):633–641

Tang W, Zeng X, Zhao Y et al (2011) Disinfection effect and its mechanism of electrolyzed oxidizing water on spores of *Bacillus subtilis* var niger. Food Sci Biotechnol 20(4):889

Tkhawkho L, Jackson K, Nitzan O et al (2017) Destruction of *Clostridium difficile* spores colitis using acidic electrolyzed water. Am J Infect Control 45(9):1053

Torlak E (2014) Inactivation of *Alicyclobacillus acidoterrestris* spores in aqueous suspension and on apples by neutral electrolyzed water. Int J Food Microbiol 185:69–72

Tu R, Ding C, Yin L et al (2015) Study on removal of ochratoxin a using slightly acidic electrolyzed water. J Chin Inst Food Sci Technol 15(11):128–134

Vázquez-Sánchez D, Cabo ML, Rodríguez-Herrera JJ (2014) Single and sequential application of electrolyzed water with benzalkonium chloride or peracetic acid for removal of *staphylococcus aureus* biofilms. J Food Safety 34(3):199–210

Venkitanarayanan KS, Ezeike GO, Hung YC et al (1999) Efficacy of electrolyzed oxidizing water for inactivating *Escherichia coli* O157:H7, *Salmonella enteritidis*, and *Listeria monocytogenes*. Appl Environ Microb 65(9):4276–4279

Virto R, Manas P, Alvarez I et al (2005) Membrane damage and microbial inactivation by chlorine in the absence and presence of a chlorine-demanding substrate. Appl Environ Microb 71(9):5022–5028

Vorobjeva NV, Vorobjeva LI, Khodjaev EY (2004) The bactericidal effects of electrolyzed oxidizing water on bacterial strains involved in hospital infections. Artif Organs 28(6):590–592

Wells-Bennik MH, Eijlander RT, Den Besten HM et al (2016) Bacterial spores in food: survival, emergence, and outgrowth. Annu Rev Food Sci Technol 7:457–482

Wesche AM, Gurtler JB, Marks BP et al (2009) Stress, sublethal injury, resuscitation, and virulence of bacterial foodborne pathogens. J Food Protect 72(5):1121–1138

Xie J, Sun X, Pan Y et al (2012) Combining basic electrolyzed water pretreatment and mild heat greatly enhanced the efficacy of acidic electrolyzed water against Vibrio parahaemolyticus on shrimp. Food Control 23(2):320–324

Xiong K, Liu H, Liu R et al (2010) Differences in fungicidal efficiency against *Aspergillus flavus* for neutralized and acidic electrolyzed oxidizing waters. Int J Food Microbiol 137(1):67–75

Xiong K, Liu HJ, Li LT (2012) Product identification and safety evaluation of aflatoxin B1 decontaminated by electrolyzed oxidizing water. J Agr Food Chem 60(38):9770–9778

Xuan XT, Ding T, Li J et al (2017) Estimation of growth parameters of *Listeria monocytogenes* after sublethal heat and slightly acidic electrolyzed water (SAEW) treatment. Food Control 71:17–25

Ye Z, Wang S, Chen T et al (2017) Inactivation mechanism of *Escherichia coli* induced by slightly acidic electrolyzed water. SCI REP-UK 7(1):6279

Ye Z, Wang S, Gao W et al (2018) Inactivation in *Aeromonas hydrophila* induced by slightly acidic electrolyzed water in freshwater. Trans ASABE 61(1):305–314

Zhang XN, Li WP, Xu XL et al (2016a) Effects of electrolyzed oxidizing water of different ph values on the biofilm of methicillin-resistant *staphylococcus aureus* in vitro. Chin J Clin Res 29:169–172

Zhang C, Li B, Jadeja R, Fang J et al (2016b) Effects of bacterial concentrations and centrifugations on susceptibility of *Bacillus subtilis* vegetative cells and *Escherichia coli* O157:H7 to various electrolyzed oxidizing water treatments. Food Control 60:440–446

Zhang C, Li B, Jadeja R, Hung Y (2016c) Effects of electrolyzed oxidizing water on inactivation of *Bacillus subtilis* and *Bacillus cereus* spores in suspension and on carriers. J Food Sci 81(1):M144–M149

Zhang C, Xia X, Li B et al (2018) Disinfection efficacy of electrolyzed oxidizing water on brown rice soaking and germination. Food Control 89:38–45

Zhao X, Zhong J, Wei C et al (2017) Current perspectives on viable but non-culturable state in foodborne pathogens. Front Microbiol 8:580

Chapter 3
Removal of Pesticide on Food by Electrolyzed Water

Jun Wang and Rongwei Han

3.1 Introduction

3.1.1 Background of Pesticides

Pesticides are widely used in agriculture, forestry, animal husbandry, and other fields to removal of pests, rodents, and bacteria. The Food and Agriculture Organization (FAO) has defined a pesticide as any substance or mixture of substances intended to prevent, destroy, or control any pest, including vectors of human or animal disease or unwanted species of plants or animals; a pesticide causes harm during or otherwise interfering with the production, processing, storage, transport, or marketing of food, agricultural commodities, wood and wood products or animal feedstuffs, as well as substances that may be administered to animals for the control of insects, arachnids, or other pests in or on their bodies. A wide range of pesticides, according to the object of control, is divided into sterilization, killing mites, insecticide, defoliation, killing rats, herbicides, and so on. Based on the origin of the pesticide, they are classified into biopesticides and chemical pesticides. Chemical pesticides such as insecticides, herbicides, and fungicides are commonly used in agriculture. According to the lethal, dose can be divided into low toxicity, moderately high toxicity, and acute toxicity. Although low-toxic pesticides show a slight effect and cannot cause immediate death, pesticides such as thiram have strong accumulation toxicity, which will still pose a serious threat to the environment and human health.

As a special type of chemical, pesticides can not only effectively prevent diseases and pests in agriculture and animal husbandry but also cause harm to humans, animals, and the environment. Therefore, the use of pesticides benefits humans and the environment in which they live. In the process of using pesticides, approximately,

J. Wang (✉) · R. Han
College of Food Science and Engineering,
Qingdao Agricultural University, Qingdao 266109, Shandong, China
e-mail: faithmate@gmail.com

T. Ding et al. (eds.), *Electrolyzed Water in Food: Fundamentals and Applications*,
https://doi.org/10.1007/978-981-13-3807-6_3

70–80% of the pesticides remain in the environment, and a large amount of residual pesticides flow into the water, evaporate in the air, and deposit in the soil, causing pollution of water, air, and soil, and contaminating farm animals and fisheries. Fruit products, through the enrichment and transfer of the food chain, eventually enter the human body. When humans and animals consume foods with pesticide residues, it will cause an accumulation of toxicity and produces cumulative poisoning. These hazards are accumulated through the food chain. After a long period of accumulation, they are often not valued by humans and cause harm to the human body indirectly.

The use of pesticides in agriculture undoubtedly stabilized the pattern of crop production in the last century. On the other hand, pesticides are often applied to fruits and vegetables during cultivation in fields as well as postharvest storage, aimed at controlling the interference of weeds and preventing the damage caused by insects or fungi (Bakirci et al. 2014; Rani et al. 2017). Despite the benefits to agricultural production, their adverse effect on the environment and biology remains a major issue. These problems will be significantly expanded in the coming decades because of the growing demand for food from growing populations, and the sustainability of agricultural production and food processing sectors must be maintained. Pesticides play a crucial role in the stable development of agricultural production. At present, heavily used pesticides can be roughly divided into four categories. The first category is organophosphorus pesticides, which show neurological symptoms and include dichlorvos, trichlorfon, monocrotophos, cotton amphoras, ammonium phosphate, parathion, pine, methyl parathion, fenthion, phoxim, isocarbophos, diaziphos, methamidophos, disulfides, phosphorus in fruits and vegetables, dimethoate, phorate, malathion, etc. The second category is pyrethroid pesticides, such as fenvalerate, permethrin, deltamethrin, cypermethrin, dicyandiamide Chrysanthemum, and other pesticides, which can result in neurological symptoms and skin irritation symptoms. The third category is organochlorines, and the fourth category is commonly used fungicides, such as carbendazim, metalaxyl, triadimefon, triadimenol, tricyclazole, zecorin, dandan, captan, optimum, schizophylline, triamcinolone, thiram, trimethoprim, thiophanate-methyl, thiophanate-methyl, chlorpheniramine, pyridoxine, chlorothalonil, pentachloronitrobenzene, etc. To ensure the normal output of fruits and vegetables, a variety of large quantities of pesticides must be applied to protect agricultural production and grain output. Meanwhile, pesticides residue also poses a threat to the ecological environment and human health.

3.1.2 Removal of Pesticides

Residues and accumulations of pesticides pose a potential threat to human health and cause damage to the environment. The control of pesticide production and the development of methods for degrading pesticides are constructive solutions. With a rising standard of living, people attach more and more importance to pesticide residues. While improving the scientific use of pesticides to avoid pesticide residues, the removal of pesticide residues has become a research hotspot at home and abroad.

The disadvantages of using pesticides make it very urgent to develop and design effective methods for removing different harmful pesticides. Several physical, chemical, and biological methods including adsorption, oxidation, catalytic degradation, membrane filtration, and biological treatment have been developed. At present, there are physical methods for removing pesticide residues: ultrasonics, washing, etc.; chemical methods: oxidation, electrolysis, etc.; and biological methods: microbial degradation, enzyme degradation, etc.

(I) Microbial degradation

The use of microbial degradation technology began in the 1940s. The main principle of the degradation of pesticide residues is the use of microbial decomposition, transformation, and reduction to convert organic substances in pesticides into simple inorganic substances with no secondary pollution. The tiny lives of fungi, bacteria, actinomycetes, algae, etc. have played an indispensable role in the degradation of pesticide residues in water and soil.

Various characteristics of microorganisms influence the degradation and transformation of pesticides, and the degradation of the same target by different types of microorganisms is different. The degradation by microorganisms can only be aimed at specific pesticide structures. Therefore, in the process of degradation of pesticides, researchers need to fully understand the structure of pesticides for effective degradation.

Biodegradation is susceptible to interference from environmental factors, such as the temperature, pH, impurities, etc., so it is difficult to achieve the desired effect.

(II) Physical removal

(i) Common Methods

People currently use various physical methods to degrade pesticides that remain on vegetable plants, such as storage, peeling, heating, washing, and adsorption. The principle of the storage method is that although the agricultural products are harvested, they still have metabolic functions. The longer the storage time, the greater the metabolic amount of pesticides that is metabolized. However, the lower temperature does not easily cause degradation, and at the same time, attention must be paid to the generation of nitrite.

The heating method is due to the thermal instability of some pesticides and the accelerated removal of pesticides due to increasing temperature. Therefore, heating plants by cooking or other means promotes the removal of pesticide residues. However, for some pesticides with high thermal stability, the effect of heating is not obvious.

The washing method is commonly used to remove pesticide residues. Most water-soluble pesticides can be removed by washing. However, for fat-soluble pesticides, washing with pure water is not effective. The addition of detergents such as salt, alkalis, acids, and the like can have partial effects. Some pesticides increase in toxicity when they encounter alkali, so when detergents are used, attention must be paid to the detergent acidity and the toxicity of the detergent itself.

The principle of the adsorption and entrainment methods principle is to use the adsorption of activated carbon. The effectiveness of the use of activated carbon depends on be affected by the temperature, the concentration of pesticides, the pH value, and other organic competition adsorption factors.

(ii) Ultrasound

Ultrasound is a type of sound wave that can reach frequencies above 20,000 Hz per second. It is characterized by a large oscillating intensity and a high frequency, which accelerates the movement of pesticide molecules and increases the dissolution rate. It is applied to pesticide residues and solves the problems of the small dissolution rate and long lifetime of pesticides. Ultrasound has the advantages of simplicity, high efficiency, a short time, and no chemical residue; thus, it has significant advantages in the removal of pesticide residues (Wang et al. 2015).

(iii) Ultrahigh pressure

Ultrahigh pressure is the use of ultrahigh static pressure to destroy the covalent bonds of polymer ions in the target organic compound, thereby achieving the effect of removal of pesticide residues.

Physical methods are inefficient, costly, and some methods can even shorten the shelf life, cause nutritional loss, and have an effect on quality. Biodegradation has the advantages of high efficiency, thoroughness, and no secondary pollution. However, its application scope still has some limitations. It is primarily used in the field of water pollution and environmental control and has not been applied in the field of agricultural products.

(III) Chemical Technology Degradation

The degradation of pesticide residues by chemical technology methods performs well with a high degradation rate and low cost. There is no secondary pollution, and it is easy to implement and popularize. It has a beneficial effect on the quality of agricultural products.

(i) Ozone Degradation

Ozone is extremely oxidizing, and its strong oxidizing ability can be used to decompose the pesticide residues on the surface of fruits and vegetables into small molecule compounds that are easily dissolved in water, and then rinse with water to remove it. Using ozone as a strong oxidizer can avoid the problem of secondary pollution. At present, this method is widely used in food processing, health care, storage, and other industries.

(ii) Hydrogen peroxide degradation

The hydrogen peroxide degradation method generates hydrogen radicals and oxygen radicals by hydrogen peroxide through the catalytic action of divalent iron ions and accelerates organic compounds in pesticides through the propagation of these two radicals and the strong oxidizing property of hydrogen peroxide.

(iii) Ozone method

Ozone is a strong oxidant that dissolves in water and can destroy carbon chains such as those of olefins, alkynes, etc. in the organic molecular structure of methamidophos, monocrotophos, dimethoate, parathion, etc. and strongly oxidize the radicals. These two ways of breaking the bond and the oxidation of the radicals result in a radical change in the molecular structure of the abovementioned substances, thereby acting to detoxify and degrade pesticide residues.

Although the degradation of pesticides by chemical methods is a major type of research and application in the agricultural field with a fast degradation rate, thorough degradation, strong pertinence, low operation requirements, and easy to control, it may cause secondary pollution. With their increasing use, the pollution of pesticides is also becoming increasingly severe, posing a serious threat to the environment and human health. Therefore, development of methods and treatments for the removal of these residues are of great importance. Washing is a very straightforward and common step in household eating or industrial processing, but most pesticides are hydrophobic so that regular wash of tap water cannot reduce the residue effectively (Iizuka and Shimizu 2014). To date, several oxidizing solutions have shown strong pesticide residue removal efficacy, such as chlorine dioxide and ozone (Chen et al. 2014; Souza et al. 2018). In addition, some studies have examined the effectiveness of electrolyzed water on eliminating these hazardous chemicals. The results obtained in recent studies revealed that electrolyzed water might be a suitable environmentally friendly option for this purpose in addition to its other well-known capabilities.

In 1987, electrolyzed water developed by a Japanese scientist was initially used to inactivate clinically resistant strains such as methicillin-resistant Staphylococcus aureus (MRSA). After continuous research and practice, electrolyzed water has been widely used in the medical field and food industry due to its high efficiency in sterilization, safe and simple production and use methods, low residue level, and nontoxic side effects (Issa-Zacharia et al. 2010, Koide et al. 2009, Koseki et al. 2004, 2007, Rahman et al. 2010). The electrolyzed water is AEW, SAEW, and AlEW. At present, alkaline electrolyzed water is generally not used as a sterilizer due to its limited sterilizing capacity but is generally used to remove oil stains; strongly acidic electrolyzed water is used most widely used, but due to its low pH, high redox potential, etc. It usually causes harm to metal and skin, and there are certain hidden dangers in safety; while slightly acidic electrolyzed water not only has a strong bactericidal capacity but also its near-neutral pH and other characteristics that make its use more secure, with a broad application prospect. At present, electrolyzed water is considered to be an effective reagent for microbial inactivation and the removal of pesticides on foods. As introduced in previous chapters, electrolyzed water is produced by electrolyzing a diluted salt (usually NaCl or HCl) solution in an electrolytic chamber with an anode and cathode. AEW is produced on the anode side, while AlEW is generated on the cathode side. AlEW, usually with a high pH (approximately 11.6), is not a good sanitizer but is used to provide an antioxidative effect on highly unsaturated fats and oils. AEW has been widely used in the food industry and food processing in Korea and Japan. Numerous studies have been investigated its bactericidal activity in food systems. Only a few studies have focused on the removal effect of pesticides on foods, especially in vegetables and fruits. The following sections describe the application in detail.

3.2 Pesticide Removal Efficacy of EW

3.2.1 Removal Efficacy of Different Pesticides

The ACC, pH, and ORP are the three main factors that make EW water a strong antimicrobial reagent and are possibly related to pesticide removal. Cl_2, HOCl, and ClO^- are the three forms of chlorine present in EW water (Huang et al. 2008).

Many studies reported that EW showed a strong ability to eliminate the foodborne pathogens from different food matrices such as meat, eggs, and vegetables. Even though electrolyzed water has been tested for its potential for reducing pesticide residues on fresh produce, the published research is very limited. Qi et al. (2018) studied the efficacy of electrolyzed water in degrading and removing three pesticide residues on fresh produce. The effect of the pH, ACC, and the treatment time of AEW and AlEW on degrading diazinon, cyprodinil, and phosmet in aqueous solution and on fresh produce were examined, as well as AlEW. Hao et al. (2011) investigated the efficacy of the EO and ER water treatments in removing the pesticides acephate, omethoate, and dimethyl dichlorovinyl phosphate on fresh vegetables. The results showed that AEW (pH 2.3, available chlorine 70 ppm, ORP 1170 mV) and AlEW (pH 11.6, ORP −860 mV) can remove pesticides (acephate, omethoate, and dichlorovinyl phosphate) effectively on fresh spinach without loss of Vitamin C. The results also showed that electrolyzed water can reduce the pesticide residues effectively. Electrolyzed water reduced the pesticides by 46–86% on different fresh vegetable surfaces, which was better than that of detergents and tap water. The results also indicated that the type of vegetable did not affect the elimination efficacy of EW, in contrast to the results reported by Qi et al. (2018). Lin et al. (2006) reported that AEW and AlEW can effectively degrade methamidophos and dimethoate on leafy cabbage and green pepper. In their study, EO water was shown to effectively remove acephate, omethoate, and DDVP on fresh spinach in 30 min with reductions ranging from 59 to 74%. In addition, EW soaking combined with mechanical washing ensured a 70.5–98.1% reduction of various pesticides residue, which was proven to be most effective for *yuja* (Citrus junos Sieb ex Tanaka) without destroying its antioxidant compounds (Sung et al. 2011). The higher removal rate was primarily due to the combined operations in which there was mechanical washing after soaking in strong acidic electrolyzed water. However, Qi et al. (2018) also evaluated the potential effect of the pH and ACC of EW water in degrading pesticides. More recently, as shown in Table 3.1, some pesticides including diazinon, cyprodinil, phosmet, acephate, omethoate, and DDVP residues were partly removed from spinach, snap beans, and grape after EW washing. Qi et al. (2018) reported the effectiveness of an EOW treatment in removing pesticide (diazinon, cyprodinil, and phosmet) residues and its effect on produce quality. Spinach, snap beans, and grapes were treated with electrolyzed water in this study. Pesticide removals of 8.8–85.7% were observed among different products and types of pesticide (Table 3.1). Electrolyzed oxidizing water was more effective than electrolyzed reduced water, bleach, VegWash, and water in removing pesticide residues. In addition, no significant color or texture deterioration was found on the produce samples treated with electrolyzed water. Thus, Qi et al. concluded that electrolyzed water can be used as an effective tool for the removal of pesticide residues from fresh produce without affecting the produce quality. In addition, the removal effect of pesticide residue on fruit and vegetables studied and published in Chinese is also reviewed, and the results are listed in Table 3.1.

As for alkaline electrolyzed water, Han et al. (2017) investigated its efficiency of electrolyzed water for removing six pesticide residues (isoprocarb, chlorpyrifos, bifenthrin, beta-cypermethrin, difenoconazole, and azoxystrobin) in cowpea. Their results revealed that the removal effect of alkaline electrolyzed water with a pH of 12.2 was superior to that of a tap water, 5% sodium chloride, 5% sodium carbonate, and 5% acetic acid solution. Therefore, it was demonstrated that alkaline electrolyzed water was a potential tool for pesticide removal in cowpea. In addition, the reduction of pesticide residues increased gradually with the washing time. Meanwhile, the status of oxidative compounds such as HClO in AEW could not be denied because of the fact that hydrochloric acid had no marked effect on methamidophos removal (Lin et al. 2006).

3.2.2 *Effect of Different Factors on the Removal Efficacy*

The main factors affecting the pesticide residues removal efficacy are the pH, ACC, and treatment time of electrolyzed water. Qi et al. (2018) investigated the pesticides removal efficacy of EW affected by the pH, ACC, and treatment time in aqueous solution. EW with a pH of 2.8, 5.0, and 6.0, an ACC of 20, 10, and 5 mg/L, and a treatment time of 1, 3, and 5 min was applied to diazinon, cyprodinil, and phosmet, respectively. The results showed that a higher ACC, the lower of pH, and the longer treatment time would lead to higher reductions of pesticides. The ACC has a significant ($P < 0.05$) effect on pesticide removal. The effect of treatment time on the removal efficacy of pesticides primarily depends on the structural differences of different pesticides. It is reported that the P=S, P–O and P–S bonds of pesticides are easier for chlorine to attack, whereas the –NH– bond of pesticides is more difficult for an oxidizing reagent to attack (Deborde and Gunten 2008). The effect of pH on pesticides removal was significant ($P < 0.05$). When the pH was changed from 6.0 to 2.8, the reduction percentage was increased from nearly 98.5 to 99.9%, 77.1 to 99.0%, and 99.1 to 99.9% for diazinon, cyprodinil, and phosmet, respectively.

Qi et al. (2018) investigated the effect of the treatment time and the ACC of EW on the removal of pesticides including diazinon, cyprodinil, and phosmet on produce.

Table 3.1 Efficacy of pesticide removal of food by electrolyzed water with different conditions

EW types	Treatment time (min)	pH	ACC (mg/L)	ORP (mV)	Target samples	Target pesticides	Percentage reduction (%)	References
EO water	30	2.3	70	1170	Fresh spinach	Acephate	74	Hao et al. (2011)
EO water	30	2.3	70	1170	Fresh spinach	Omethoate	62	Hao et al. (2011)
EO water	30	2.3	70	1170	Fresh spinach	DDVP	59	Hao et al. (2011)
ER water	30	11.6	–	−860	Fresh spinach	Acephate	86	Hao et al. (2011)
ER water	30	11.6	–	−860	Fresh spinach	Omethoate	75	Hao et al. (2011)
ER water	30	11.6	–	−860	Fresh spinach	DDVP	46	Hao et al. (2011)
TW	30	–	–	–	Cabbage	DDVP	17.5	Hao et al. (2011)
Detergent	30	–	–	–	Cabbage	DDVP	39.6	Hao et al. (2011)
EO water	30	2.3	70	1170	Cabbage	DDVP	76.8	Hao et al. (2011)
ER water	30	11.6	–	−860	Cabbage	DDVP	82.3	Hao et al. (2011)
TW wash	30	–	–	–	Leek	DDVP	15.8	Hao et al. (2011)
Detergent wash	30	–	–	–	Leek	DDVP	27.7	Hao et al. (2011)
EO water	30	2.3	70	1170	Leek	DDVP	63.5	Hao et al. (2011)
ER water	30	11.6	–	−860	Leek	DDVP	65.7	Hao et al. (2011)
EO water	1	2.8	20	1126 ± 7.0	Grapes	Diazinon	10.2 ± 1.8	Qi et al. (2018)
EO water	1	2.8	70	1144 ± 10.0	Grapes	Diazinon	18.4 ± 7.4	Qi et al. (2018)
EO water	1	2.8	120	1151.5 ± 11.7	Grapes	Diazinon	22.3 ± 4.2	Qi et al. (2018)
EO water	8	2.8	20	1126 ± 7.0	Grapes	Diazinon	21.7 ± 1.8	Qi et al. (2018)
EO water	8	2.8	70	1144 ± 10.0	Grapes	Diazinon	25.5 ± 5.7	Qi et al. (2018)
EO water	8	2.8	120	1151.5 ± 11.7	Grapes	Diazinon	29.5 ± 9.9	Qi et al. (2018)

(continued)

Table 3.1 (continued)

EW types	Treatment time (min)	pH	ACC (mg/L)	ORP (mV)	Target samples	Target pesticides	Percentage reduction (%)	References
EO water	15	2.8	20	1126 ± 7.0	Grapes	Diazinon	27.4 ± 5.3	Qi et al. (2018)
EO water	15	2.8	70	1144 ± 10.0	Grapes	Diazinon	31.9 ± 5.3	Qi et al. (2018)
EO water	15	2.8	120	1151.5 ± 11.7	Grapes	Diazinon	37.1 ± 4.1	Qi et al. (2018)
EO water	1	2.8	20	1126 ± 7.0	Snap beans	Diazinon	12.0 ± 5.2	Qi et al. (2018)
EO water	1	2.8	70	1144 ± 10.0	Snap beans	Diazinon	11.2 ± 7.0	Qi et al. (2018)
EO water	1	2.8	120	1151.5 ± 11.7	Snap beans	Diazinon	18.5 ± 8.2	Qi et al. (2018)
EO water	8	2.8	20	1126 ± 7.0	Snap beans	Diazinon	20.4 ± 2.5	Qi et al. (2018)
EO water	8	2.8	70	1144 ± 10.0	Snap beans	Diazinon	35.7 ± 6.1	Qi et al. (2018)
EO water	8	2.8	120	1151.5 ± 11.7	Snap beans	Diazinon	55.8 ± 5.5	Qi et al. (2018)
EO water	15	2.8	20	1126 ± 7.0	Snap beans	Diazinon	43.9 ± 7.9	Qi et al. (2018)
EO water	15	2.8	70	1144 ± 10.0	Snap beans	Diazinon	64.0 ± 3.1	Qi et al. (2018)
EO water	15	2.8	120	1151.5 ± 11.7	Snap beans	Diazinon	66.5 ± 2.5	Qi et al. (2018)
EO water	1	2.8	20	1126 ± 7.0	Spinach	Diazinon	8.8 ± 5.9	Qi et al. (2018)
EO water	1	2.8	70	1144 ± 10.0	Spinach	Diazinon	19.3 ± 5.4	Qi et al. (2018)
EO water	1	2.8	120	1151.5 ± 11.7	Spinach	Diazinon	29.8 ± 8.4	Qi et al. (2018)
EO water	8	2.8	20	1126 ± 7.0	Spinach	Diazinon	33.2 ± 3.5	Qi et al. (2018)
EO water	8	2.8	70	1144 ± 10.0	Spinach	Diazinon	40.3 ± 8.5	Qi et al. (2018)
EO water	8	2.8	120	1151.5 ± 11.7	Spinach	Diazinon	47.0 ± 6.1	Qi et al. (2018)
EO water	15	2.8	20	1126 ± 7.0	Spinach	Diazinon	33.0 ± 6.7	Qi et al. (2018)
EO water	15	2.8	70	1144 ± 10.0	Spinach	Diazinon	49.5 ± 6.7	Qi et al. (2018)

(continued)

Table 3.1 (continued)

EW types	Treatment time (min)	pH	ACC (mg/L)	ORP (mV)	Target samples	Target pesticides	Percentage reduction (%)	References
EO water	15	2.8	120	1151.5 ± 11.7	Spinach	Diazinon	59.2 ± 5.5	Qi et al. (2018)
EO water	1	2.8	20	1126 ± 7.0	Grapes	Cyprodinil	9.2 ± 3.6	Qi et al. (2018)
EO water	1	2.8	70	1144 ± 10.0	Grapes	Cyprodinil	14.9 ± 2.1	Qi et al. (2018)
EO water	1	2.8	120	1151.5 ± 11.7	Grapes	Cyprodinil	17.8 ± 1.2	Qi et al. (2018)
EO water	8	2.8	20	1126 ± 7.0	Grapes	Cyprodinil	21.5 ± 3.4	Qi et al. (2018)
EO water	8	2.8	70	1144 ± 10.0	Grapes	Cyprodinil	22.2 ± 3.8	Qi et al. (2018)
EO water	8	2.8	120	1151.5 ± 11.7	Grapes	Cyprodinil	26.3 ± 6.3	Qi et al. (2018)
EO water	15	2.8	20	1126 ± 7.0	Grapes	Cyprodinil	21.0 ± 4.9	Qi et al. (2018)
EO water	15	2.8	70	1144 ± 10.0	Grapes	Cyprodinil	25.2 ± 3.6	Qi et al. (2018)
EO water	15	2.8	120	1151.5 ± 11.7	Grapes	Cyprodinil	31.5 ± 2.6	Qi et al. (2018)
EO water	1	2.8	20	1126 ± 7.0	Snap beans	Cyprodinil	12.0 ± 2.9	Qi et al. (2018)
EO water	1	2.8	70	1144 ± 10.0	Snap beans	Cyprodinil	12.9 ± 1.0	Qi et al. (2018)
EO water	1	2.8	120	1151.5 ± 11.7	Snap beans	Cyprodinil	18.0 ± 4.0	Qi et al. (2018)
EO water	8	2.8	20	1126 ± 7.0	Snap beans	Cyprodinil	25.8 ± 4.7	Qi et al. (2018)
EO water	8	2.8	70	1144 ± 10.0	Snap beans	Cyprodinil	26.9 ± 1.0	Qi et al. (2018)
EO water	8	2.8	120	1151.5 ± 11.7	Snap beans	Cyprodinil	40.4 ± 6.9	Qi et al. (2018)
EO water	15	2.8	20	1126 ± 7.0	Snap beans	Cyprodinil	34.7 ± 2.3	Qi et al. (2018)
EO water	15	2.8	70	1144 ± 10.0	Snap beans	Cyprodinil	49.6 ± 6.5	Qi et al. (2018)
EO water	15	2.8	120	1151.5 ± 11.7	Snap beans	Cyprodinil	50.0 ± 8.2	Qi et al. (2018)
EO water	1	2.8	20	1126 ± 7.0	Spinach	Cyprodinil	9.7 ± 3.1	Qi et al. (2018)

(continued)

Table 3.1 (continued)

EW types	Treatment time (min)	pH	ACC (mg/L)	ORP (mV)	Target samples	Target pesticides	Percentage reduction (%)	References
EO water	1	2.8	70	1144 ± 10.0	Spinach	Cyprodinil	17.2 ± 2.7	Qi et al. (2018)
EO water	1	2.8	120	1151.5 ± 11.7	Spinach	Cyprodinil	19.4 ± 0.3	Qi et al. (2018)
EO water	8	2.8	20	1126 ± 7.0	Spinach	Cyprodinil	21.9 ± 6.4	Qi et al. (2018)
EO water	8	2.8	70	1144 ± 10.0	Spinach	Cyprodinil	32.3 ± 2.9	Qi et al. (2018)
EO water	8	2.8	120	1151.5 ± 11.7	Spinach	Cyprodinil	34.3 ± 0.3	Qi et al. (2018)
EO water	15	2.8	20	1126 ± 7.0	Spinach	Cyprodinil	30.5 ± 6.3	Qi et al. (2018)
EO water	15	2.8	70	1144 ± 10.0	Spinach	Cyprodinil	38.3 ± 1.8	Qi et al. (2018)
EO water	15	2.8	120	1151.5 ± 11.7	Spinach	Cyprodinil	43.8 ± 2.9	Qi et al. (2018)
EO water	1	2.8	20	1126 ± 7.0	Grapes	Phosmet	21.4 ± 9.7	Qi et al. (2018)
EO water	1	2.8	70	1144 ± 10.0	Grapes	Phosmet	28.3 ± 4.6	Qi et al. (2018)
EO water	1	2.8	120	1151.5 ± 11.7	Grapes	Phosmet	34.2 ± 5.1	Qi et al. (2018)
EO water	8	2.8	20	1126 ± 7.0	Grapes	Phosmet	28.7 ± 7.1	Qi et al. (2018)
EO water	8	2.8	70	1144 ± 10.0	Grapes	Phosmet	38.5 ± 5.6	Qi et al. (2018)
EO water	8	2.8	120	1151.5 ± 11.7	Grapes	Phosmet	44.1 ± 3.2	Qi et al. (2018)
EO water	15	2.8	20	1126 ± 7.0	Grapes	Phosmet	35.2 ± 3.8	Qi et al. (2018)
EO water	15	2.8	70	1144 ± 10.0	Grapes	Phosmet	45.3 ± 1.4	Qi et al. (2018)
EO water	15	2.8	120	1151.5 ± 11.7	Grapes	Phosmet	49.4 ± 1.3	Qi et al. (2018)
EO water	1	2.8	20	1126 ± 7.0	Snap beans	Phosmet	37.7 ± 6.7	Qi et al. (2018)
EO water	1	2.8	70	1144 ± 10.0	Snap beans	Phosmet	46.0 ± 4.4	Qi et al. (2018)
EO water	1	2.8	120	1151.5 ± 11.7	Snap beans	Phosmet	45.1 ± 8.9	Qi et al. (2018)

(continued)

Table 3.1 (continued)

EW types	Treatment time (min)	pH	ACC (mg/L)	ORP (mV)	Target samples	Target pesticides	Percentage reduction (%)	References
EO water	8	2.8	20	1126 ± 7.0	Snap beans	Phosmet	43.2 ± 2.5	Qi et al. (2018)
EO water	8	2.8	70	1144 ± 10.0	Snap beans	Phosmet	55.9 ± 3.5	Qi et al. (2018)
EO water	8	2.8	120	1151.5 ± 11.7	Snap beans	Phosmet	63.5 ± 2.5	Qi et al. (2018)
EO water	15	2.8	20	1126 ± 7.0	Snap beans	Phosmet	52.1 ± 0.1	Qi et al. (2018)
EO water	15	2.8	70	1144 ± 10.0	Snap beans	Phosmet	66.5 ± 2.8	Qi et al. (2018)
EO water	15	2.8	120	1151.5 ± 11.7	Snap beans	Phosmet	73.0 ± 2.7	Qi et al. (2018)
EO water	1	2.8	20	1126 ± 7.0	Spinach	Phosmet	56.6 ± 2.1	Qi et al. (2018)
EO water	1	2.8	70	1144 ± 10.0	Spinach	Phosmet	57.8 ± 5.5	Qi et al. (2018)
EO water	1	2.8	120	1151.5 ± 11.7	Spinach	Phosmet	64.3 ± 7.9	Qi et al. (2018)
EO water	8	2.8	20	1126 ± 7.0	Spinach	Phosmet	70.9 ± 4.5	Qi et al. (2018)
EO water	8	2.8	70	1144 ± 10.0	Spinach	Phosmet	72.3 ± 6.2	Qi et al. (2018)
EO water	8	2.8	120	1151.5 ± 11.7	Spinach	Phosmet	82.0 ± 4.0	Qi et al. (2018)
EO water	15	2.8	20	1126 ± 7.0	Spinach	Phosmet	74.7 ± 0.3	Qi et al. (2018)
EO water	15	2.8	70	1144 ± 10.0	Spinach	Phosmet	83.5 ± 5.7	Qi et al. (2018)
EO water	15	2.8	120	1151.5 ± 11.7	Spinach	Phosmet	85.7 ± 4.8	Qi et al. (2018)
TW	10	7.34	–	364	Chinese cabbage	Chlorpyrifos	32.28 ± 5.74	Sung et al. (2012)
NaClO-50	10	8.89	52.65	776	Chinese cabbage	Chlorpyrifos	42.47 ± 5.34	Sung et al. (2012)
NaClO-100	10	9.03	90.77	774	Chinese cabbage	Chlorpyrifos	48.07 ± 9.79	Sung et al. (2012)
LAIEW-50	10	8.7	46.98	821	Chinese cabbage	Chlorpyrifos	48.71 ± 7.17	Sung et al. (2012)
LAIEW-100	10	8.58	97.16	814	Chinese cabbage	Chlorpyrifos	63.79 ± 9.01	Sung et al. (2012)

(continued)

Table 3.1 (continued)

EW types	Treatment time (min)	pH	ACC (mg/L)	ORP (mV)	Target samples	Target pesticides	Percentage reduction (%)	References
SAEW-50	10	3.61	43.26	1128	Chinese cabbage	Chlorpyrifos	61.24 ± 1.03	Sung et al. (2012)
SAEW-100	10	3.32	79.43	1141	Chinese cabbage	Chlorpyrifos	68.94 ± 4.10	Sung et al. (2012)
TW	10	7.34	–	364	Chinese cabbage	Prothiofos	34.59 ± 5.26	Sung et al. (2012)
NaClO-50	10	8.89	52.65	776	Chinese cabbage	Prothiofos	46.76 ± 6.35	Sung et al. (2012)
NaClO-100	10	9.03	90.77	774	Chinese cabbage	Prothiofos	48.09 ± 7.68	Sung et al. (2012)
LAIEW-50	10	8.7	46.98	821	Chinese cabbage	Prothiofos	48.85 ± 9.24	Sung et al. (2012)
LAIEW-100	10	8.58	97.16	814	Chinese cabbage	Prothiofos	65.86 ± 11.12	Sung et al. (2012)
SAEW-50	10	3.61	43.26	1128	Chinese cabbage	Prothiofos	61.68 ± 6.95	Sung et al. (2012)
SAEW-100	10	3.32	79.43	1141	Chinese cabbage	Prothiofos	74.06 ± 8.01	Sung et al. (2012)
TW	10	7.34	–	364	Chinese cabbage	Deltamethrin	38.46 ± 9.12	Sung et al. (2012)
NaClO-50	10	8.89	52.65	776	Chinese cabbage	Deltamethrin	36.93 ± 0.10	Sung et al. (2012)
NaClO-100	10	9.03	90.77	774	Chinese cabbage	Deltamethrin	50.13 ± 3.80	Sung et al. (2012)
LAIEW-50	10	8.7	46.98	821	Chinese cabbage	Deltamethrin	41.96 ± 8.16	Sung et al. (2012)
LAIEW-100	10	8.58	97.16	814	Chinese cabbage	Deltamethrin	65.97 ± 1.26	Sung et al. (2012)
SAEW-50	10	3.61	43.26	1128	Chinese cabbage	Deltamethrin	54.30 ± 8.61	Sung et al. (2012)
SAEW-100	10	3.32	79.43	1141	Chinese cabbage	Deltamethrin	78.30 ± 5.19	Sung et al. (2012)
SAEW#	10	2.75	88.63	1057	*yuja*	Chlorpyrifos	70.5 ± 4.8	Sung et al. (2011)
SAEW#	10	2.75	88.63	1057	*yuja*	Prothiofos	70.8 ± 6.9	Sung et al. (2011)
SAEW#	10	2.75	88.63	1057	*yuja*	Spirodiclofen	72.0 ± 5.5	Sung et al. (2011)
SAEW#	10	2.75	88.63	1057	*yuja*	Deltamethrin	81.8 ± 1.7	Sung et al. (2011)

(continued)

Table 3.1 (continued)

EW types	Treatment time (min)	pH	ACC (mg/L)	ORP (mV)	Target samples	Target pesticides	Percentage reduction (%)	References
SAEW[#]	10	2.75	88.63	1057	*yuja*	Benomyl	97.2 ± 0.2	Sung et al. (2011)
SAEW[#]	10	2.75	88.63	1057	*yuja*	Thiophanate-methyl	98.1 ± 0.4	Sung et al. (2011)
SAEW[#]	10	2.75	88.63	1057	*yuja*	Acequinocyl	82.4 ± 1.9	Sung et al. (2011)
AlEW	5	10.1	–	−541	Cowpea	Isoprocarb	16.8	Han et al. (2017)
AlEW	5	10.1	–	−541	Cowpea	Chlorpyrifos	9.5	Han et al. (2017)
AlEW	5	10.1	–	−541	Cowpea	Bifenthrin	8.5	Han et al. (2017)
AlEW	5	10.1	–	−541	Cowpea	Beta-cypermethrin	5.8	Han et al. (2017)
AlEW	5	10.1	–	−541	Cowpea	Difenoconazole	8.9	Han et al. (2017)
AlEW	5	10.1	–	−541	Cowpea	Azoxystrobin	11.2	Han et al. (2017)
AlEW	15	10.1	–	−541	Cowpea	Isoprocarb	28.9	Han et al. (2017)
AlEW	15	10.1	–	−541	Cowpea	Chlorpyrifos	24.1	Han et al. (2017)
AlEW	15	10.1	–	−541	Cowpea	Bifenthrin	16.2	Han et al. (2017)
AlEW	15	10.1	–	−541	Cowpea	Beta-cypermethrin	15.5	Han et al. (2017)
AlEW	15	10.1	–	−541	Cowpea	Difenoconazole	24.0	Han et al. (2017)
AlEW	15	10.1	–	−541	Cowpea	Azoxystrobin	26.3	Han et al. (2017)
AlEW	30	10.1	–	−541	Cowpea	Isoprocarb	46.8	Han et al. (2017)
AlEW	30	10.1	–	−541	Cowpea	Chlorpyrifos	34.8	Han et al. (2017)
AlEW	30	10.1	–	−541	Cowpea	Bifenthrin	36.5	Han et al. (2017)

(continued)

Table 3.1 (continued)

EW types	Treatment time (min)	pH	ACC (mg/L)	ORP (mV)	Target samples	Target pesticides	Percentage reduction (%)	References
AlEW	30	10.1	–	−541	Cowpea	Beta-cypermethrin	33.6	Han et al. (2017)
AlEW	30	10.1	–	−541	Cowpea	Difenoconazole	35.2	Han et al. (2017)
AlEW	30	10.1	–	−541	Cowpea	Azoxystrobin	40.7	Han et al. (2017)
AlEW	45	10.1	–	−541	Cowpea	Isoprocarb	73.5	Han et al. (2017)
AlEW	45	10.1	–	−541	Cowpea	Chlorpyrifos	50.1	Han et al. (2017)
AlEW	45	10.1	–	−541	Cowpea	Bifenthrin	40.2	Han et al. (2017)
AlEW	45	10.1	–	−541	Cowpea	Beta-cypermethrin	52.8	Han et al. (2017)
AlEW	45	10.1	–	−541	Cowpea	Difenoconazole	52.9	Han et al. (2017)
AlEW	45	10.1	–	−541	Cowpea	Azoxystrobin	56.4	Han et al. (2017)
AlEW	5	11.3	–	−639	Cowpea	Isoprocarb	17.4	Han et al. (2017)
AlEW	5	11.3	–	−639	Cowpea	Chlorpyrifos	10	Han et al. (2017)
AlEW	5	11.3	–	−639	Cowpea	Bifenthrin	9.1	Han et al. (2017)
AlEW	5	11.3	–	−639	Cowpea	Beta-cypermethrin	8.5	Han et al. (2017)
AlEW	5	11.3	–	−639	Cowpea	Difenoconazole	10.6	Han et al. (2017)
AlEW	5	11.3	–	−639	Cowpea	Azoxystrobin	8.3	Han et al. (2017)
AlEW	15	11.3	–	−639	Cowpea	Isoprocarb	36.1	Han et al. (2017)
AlEW	15	11.3	–	−639	Cowpea	Chlorpyrifos	24.4	Han et al. (2017)
AlEW	15	11.3	–	−639	Cowpea	Bifenthrin	23.4	Han et al. (2017)

(continued)

Table 3.1 (continued)

EW types	Treatment time (min)	pH	ACC (mg/L)	ORP (mV)	Target samples	Target pesticides	Percentage reduction (%)	References
AlEW	15	11.3	–	−639	Cowpea	Beta-cypermethrin	22.3	Han et al. (2017)
AlEW	15	11.3	–	−639	Cowpea	Difenoconazole	28.7	Han et al. (2017)
AlEW	15	11.3	–	−639	Cowpea	Azoxystrobin	26.9	Han et al. (2017)
AlEW	30	11.3	–	−639	Cowpea	Isoprocarb	55.8	Han et al. (2017)
AlEW	30	11.3	–	−639	Cowpea	Chlorpyrifos	36.4	Han et al. (2017)
AlEW	30	11.3	–	−639	Cowpea	Bifenthrin	30.6	Han et al. (2017)
AlEW	30	11.3	–	−639	Cowpea	Beta-cypermethrin	30.1	Han et al. (2017)
AlEW	30	11.3	–	−639	Cowpea	Difenoconazole	38.8	Han et al. (2017)
AlEW	30	11.3	–	−639	Cowpea	Azoxystrobin	44.5	Han et al. (2017)
AlEW	45	11.3	–	−639	Cowpea	Isoprocarb	74.9	Han et al. (2017)
AlEW	45	11.3	–	−639	Cowpea	Chlorpyrifos	53.5	Han et al. (2017)
AlEW	45	11.3	–	−639	Cowpea	Bifenthrin	42.9	Han et al. (2017)
AlEW	45	11.3	–	−639	Cowpea	Beta-cypermethrin	51.7	Han et al. (2017)
AlEW	45	11.3	–	−639	Cowpea	Difenoconazole	62.9	Han et al. (2017)
AlEW	45	11.3	–	−639	Cowpea	Azoxystrobin	64.6	Han et al. (2017)
AlEW	5	12.2	–	−861	Cowpea	Isoprocarb	21.4	Han et al. (2017)
AlEW	5	12.2	–	−861	Cowpea	Chlorpyrifos	12.7	Han et al. (2017)
AlEW	5	12.2	–	−861	Cowpea	Bifenthrin	9.8	Han et al. (2017)

(continued)

Table 3.1 (continued)

EW types	Treatment time (min)	pH	ACC (mg/L)	ORP (mV)	Target samples	Target pesticides	Percentage reduction (%)	References
AlEW	5	12.2	–	−861	Cowpea	Beta-cypermethrin	12.3	Han et al. (2017)
AlEW	5	12.2	–	−861	Cowpea	Difenoconazole	11.1	Han et al. (2017)
AlEW	5	12.2	–	−861	Cowpea	Azoxystrobin	14.5	Han et al. (2017)
AlEW	15	12.2	–	−861	Cowpea	Isoprocarb	40	Han et al. (2017)
AlEW	15	12.2	–	−861	Cowpea	Chlorpyrifos	30.9	Han et al. (2017)
AlEW	15	12.2	–	−861	Cowpea	Bifenthrin	21.7	Han et al. (2017)
AlEW	15	12.2	–	−861	Cowpea	Beta-cypermethrin	30.9	Han et al. (2017)
AlEW	15	12.2	–	−861	Cowpea	Difenoconazole	27.8	Han et al. (2017)
AlEW	15	12.2	–	−861	Cowpea	Azoxystrobin	35	Han et al. (2017)
AlEW	30	12.2	–	−861	Cowpea	Isoprocarb	63.1	Han et al. (2017)
AlEW	30	12.2	–	−861	Cowpea	Chlorpyrifos	41.8	Han et al. (2017)
AlEW	30	12.2	–	−861	Cowpea	Bifenthrin	37.8	Han et al. (2017)
AlEW	30	12.2	–	−861	Cowpea	Beta-cypermethrin	47	Han et al. (2017)
AlEW	30	12.2	–	−861	Cowpea	Difenoconazole	41.7	Han et al. (2017)
AlEW	30	12.2	–	−861	Cowpea	Azoxystrobin	61.2	Han et al. (2017)
AlEW	45	12.2	–	−861	Cowpea	Isoprocarb	85.1	Han et al. (2017)
AlEW	45	12.2	–	−861	Cowpea	Chlorpyrifos	58	Han et al. (2017)
AlEW	45	12.2	–	−861	Cowpea	Bifenthrin	48.1	Han et al. (2017)

(continued)

Table 3.1 (continued)

EW types	Treatment time (min)	pH	ACC (mg/L)	ORP (mV)	Target samples	Target pesticides	Percentage reduction (%)	References
AlEW	45	12.2	–	−861	Cowpea	Beta-cypermethrin	55.1	Han et al. (2017)
AlEW	45	12.2	–	−861	Cowpea	Difenoconazole	68.5	Han et al. (2017)
AlEW	45	12.2	–	−861	Cowpea	Azoxystrobin	75.2	Han et al. (2017)
AlEW	5	13.2	–	−962	Cowpea	Isoprocarb	20.3	Han et al. (2017)
AlEW	5	13.2	–	−962	Cowpea	Chlorpyrifos	10.7	Han et al. (2017)
AlEW	5	13.2	–	−962	Cowpea	Bifenthrin	8.9	Han et al. (2017)
AlEW	5	13.2	–	−962	Cowpea	Beta-cypermethrin	11.7	Han et al. (2017)
AlEW	5	13.2	–	−962	Cowpea	Difenoconazole	12.5	Han et al. (2017)
AlEW	5	13.2	–	−962	Cowpea	Azoxystrobin	17.9	Han et al. (2017)
AlEW	15	13.2	–	−962	Cowpea	Isoprocarb	41.2	Han et al. (2017)
AlEW	15	13.2	–	−962	Cowpea	Chlorpyrifos	34.4	Han et al. (2017)
AlEW	15	13.2	–	−962	Cowpea	Bifenthrin	19.4	Han et al. (2017)
AlEW	15	13.2	–	−962	Cowpea	Beta-cypermethrin	32.5	Han et al. (2017)
AlEW	15	13.2	–	−962	Cowpea	Difenoconazole	26.5	Han et al. (2017)
AlEW	15	13.2	–	−962	Cowpea	Azoxystrobin	34.1	Han et al. (2017)
AlEW	30	13.2	–	−962	Cowpea	Isoprocarb	61.5	Han et al. (2017)
AlEW	30	13.2	–	−962	Cowpea	Chlorpyrifos	48	Han et al. (2017)
AlEW	30	13.2	–	−962	Cowpea	Bifenthrin	37.6	Han et al. (2017)

(continued)

Table 3.1 (continued)

EW types	Treatment time (min)	pH	ACC (mg/L)	ORP (mV)	Target samples	Target pesticides	Percentage reduction (%)	References
AlEW	30	13.2	–	−962	Cowpea	Beta-cypermethrin	41.7	Han et al. (2017)
AlEW	30	13.2	–	−962	Cowpea	Difenoconazole	39.3	Han et al. (2017)
AlEW	30	13.2	–	−962	Cowpea	Azoxystrobin	60.5	Han et al. (2017)
AlEW	45	13.2	–	−962	Cowpea	Isoprocarb	82.7	Han et al. (2017)
AlEW	45	13.2	–	−962	Cowpea	Chlorpyrifos	59	Han et al. (2017)
AlEW	45	13.2	–	−962	Cowpea	Bifenthrin	46.9	Han et al. (2017)
AlEW	45	13.2	–	−962	Cowpea	Beta-cypermethrin	56.4	Han et al. (2017)
AlEW	45	13.2	–	−962	Cowpea	Difenoconazole	65.4	Han et al. (2017)
AlEW	45	13.2	–	−962	Cowpea	Azoxystrobin	73.7	Han et al. (2017)
AEW	15	2.87 ± 0.06	10.17 ± 1.5	1110.2 ± 10.2	Apple	Dimethoate	34.8	Wuyun (2011)
AEW	15	2.81 ± 0.07	50.34 ± 3.4	1155.4 ± 9.8	Apple	Dimethoate	72.9	Wuyun (2011)
AEW	15	2.73 ± 0.07	101.4 ± 4.2	1203.5 ± 11.3	Apple	Dimethoate	81.4	Wuyun (2011)
AEW	15	2.87 ± 0.06	10.17 ± 1.5	1110.2 ± 10.2	Apple	Chlorpyrifos	27.1	Wuyun (2011)
AEW	15	2.81 ± 0.07	50.34 ± 3.4	1155.4 ± 9.8	Apple	Chlorpyrifos	56.4	Wuyun (2011)
AEW	15	2.73 ± 0.07	101.4 ± 4.2	1203.5 ± 11.3	Apple	Chlorpyrifos	72.3	Wuyun (2011)
AEW	15	2.87 ± 0.06	10.17 ± 1.5	1110.2 ± 10.2	Apple	Parathion	31.1	Wuyun (2011)
AEW	15	2.81 ± 0.07	50.34 ± 3.4	1155.4 ± 9.8	Apple	Parathion	65.6	Wuyun (2011)
AEW	15	2.73 ± 0.07	101.4 ± 4.2	1203.5 ± 11.3	Apple	Parathion	76.9	Wuyun (2011)

(continued)

Table 3.1 (continued)

EW types	Treatment time (min)	pH	ACC (mg/L)	ORP (mV)	Target samples	Target pesticides	Percentage reduction (%)	References
AEW	5	2.81 ± 0.07	50.34 ± 3.4	1155.4 ± 9.8	Apple	Dimethoate	57.4	Wuyun (2011)
AEW	5	2.81 ± 0.07	50.34 ± 3.4	1155.4 ± 9.8	Apple	Chlorpyrifos	27	Wuyun (2011)
AEW	5	2.81 ± 0.07	50.34 ± 3.4	1155.4 ± 9.8	Apple	Parathion	32.5	Wuyun (2011)
AEW	10	2.81 ± 0.07	50.34 ± 3.4	1155.4 ± 9.8	Apple	Dimethoate	61.7	Wuyun (2011)
AEW	10	2.81 ± 0.07	50.34 ± 3.4	1155.4 ± 9.8	Apple	Chlorpyrifos	31.7	Wuyun (2011)
AEW	10	2.81 ± 0.07	50.34 ± 3.4	1155.4 ± 9.8	Apple	Parathion	45	Wuyun (2011)
AEW	15	2.81 ± 0.07	50.34 ± 3.4	1155.4 ± 9.8	Apple	Dimethoate	72.9	Wuyun (2011)
AEW	15	2.81 ± 0.07	50.34 ± 3.4	1155.4 ± 9.8	Apple	Chlorpyrifos	56.4	Wuyun (2011)
AEW	15	2.81 ± 0.07	50.34 ± 3.4	1155.4 ± 9.8	Apple	Parathion	65.5	Wuyun (2011)
AEW	30	2.81 ± 0.07	50.34 ± 3.4	1155.4 ± 9.8	Apple	Dimethoate	87.6	Wuyun (2011)
AEW	30	2.81 ± 0.07	50.34 ± 3.4	1155.4 ± 9.8	Apple	Chlorpyrifos	68.5	Wuyun (2011)
AEW	30	2.81 ± 0.07	50.34 ± 3.4	1155.4 ± 9.8	Apple	Parathion	73.6	Wuyun (2011)
AEW	15	2.73 ± 0.11	65.28 ± 6.1	1212.4 ± 10.1	Rape	Dimethoate	>70	Wuyun (2011)
SAEW	15	6.03 ± 0.16	66.07 ± 5.3	820 ± 6.5	Rape	Dimethoate	>60	Wuyun (2011)
AEW	15	2.73 ± 0.11	65.28 ± 6.1	1212.4 ± 10.1	Tomato	Dimethoate	>50	Wuyun (2011)
SAEW	15	6.03 ± 0.16	66.07 ± 5.3	820 ± 6.5	Tomato	Dimethoate	>50	Wuyun (2011)
AEW	15	2.73 ± 0.11	65.28 ± 6.1	1212.4 ± 10.1	Beans	Dimethoate	>50	Wuyun (2011)
SAEW	15	6.03 ± 0.16	66.07 ± 5.3	820 ± 6.5	Beans	Dimethoate	>50	Wuyun (2011)
AEW	15	2.73 ± 0.11	65.28 ± 6.1	1212.4 ± 10.1	Rape	Chlorpyrifos	>50	Wuyun (2011)
SAEW	15	6.03 ± 0.16	66.07 ± 5.3	820 ± 6.5	Rape	Chlorpyrifos	>50	Wuyun (2011)

(continued)

Table 3.1 (continued)

EW types	Treatment time (min)	pH	ACC (mg/L)	ORP (mV)	Target samples	Target pesticides	Percentage reduction (%)	References
AEW	15	2.73 ± 0.11	65.28 ± 6.1	1212.4 ± 10.1	Tomato	Chlorpyrifos	>50	Wuyun (2011)
SAEW	15	6.03 ± 0.16	66.07 ± 5.3	820 ± 6.5	Tomato	Chlorpyrifos	>50	Wuyun (2011)
AEW	15	2.73 ± 0.11	65.28 ± 6.1	1212.4 ± 10.1	Beans	Chlorpyrifos	>50	Wuyun (2011)
SAEW	15	6.03 ± 0.16	66.07 ± 5.3	820 ± 6.5	Beans	Chlorpyrifos	>50	Wuyun (2011)
AEW	15	2.73 ± 0.11	65.28 ± 6.1	1212.4 ± 10.1	Rape	Parathion	>60	Wuyun (2011)
SAEW	15	6.03 ± 0.16	66.07 ± 5.3	820 ± 6.5	Rape	Parathion	>60	Wuyun (2011)
AEW	15	2.73 ± 0.11	65.28 ± 6.1	1212.4 ± 10.1	Tomato	Parathion	>60	Wuyun (2011)
SAEW	15	6.03 ± 0.16	66.07 ± 5.3	820 ± 6.5	Tomato	Parathion	>60	Wuyun (2011)
AEW	15	2.73 ± 0.11	65.28 ± 6.1	1212.4 ± 10.1	Beans	Parathion	>50	Wuyun (2011)
SAEW	15	6.03 ± 0.16	66.07 ± 5.3	820 ± 6.5	Beans	Parathion	>50	Wuyun (2011)
AEW	60	2.25 ± 0.1	75 ± 5	1270 ± 10	Rape	Acephate	82	Hao and Li (2006)
AlEW	60	12.15 ± 0.1	–	−890 ± 10	Rape	Acephate	>90	Hao and Li (2006)
AlEW	90	11.0 ± 0.1	–	−830	Apple	Lambda-cyhalothrin	67.9	Liu et al. (2015)
AlEW	90	8.0	–	–	Apple	Lambda-cyhalothrin	40	Liu et al. (2015)
AlEW	90	9.0	–	–	Apple	Lambda-cyhalothrin	67.9	Liu et al. (2015)
AlEW	90	10.0	–	–	Apple	Lambda-cyhalothrin	67.9	Liu et al. (2015)

(continued)

Table 3.1 (continued)

EW types	Treatment time (min)	pH	ACC (mg/L)	ORP (mV)	Target samples	Target pesticides	Percentage reduction (%)	References
AlEW	90	11.0	–	–	Apple	Lambda-cyhalothrin	72.84	Liu et al. (2015)
AlEW	90	12.0	–	–	Apple	Lambda-cyhalothrin	90.66	Liu et al. (2015)
AlEW	30	11.0±0.1	–	−830	Apple	Lambda-cyhalothrin	>30	Liu et al. (2015)
AlEW	60	11.0±0.1	–	−830	Apple	Lambda-cyhalothrin	~50	Liu et al. (2015)
SAEW	15	6.0	79.66	–	Leek	Dimethoate	66.02	Hu et al. (2016)
SAEW	15	6.0	79.66	–	Leek	Chlorpyrifos	72.55	Hu et al. (2016)
SAEW	15	6.0	1.95	–	Leek	Dimethoate	>40	Hu et al. (2016)
SAEW	15	6.25	5.66	–	Leek	Dimethoate	~40	Hu et al. (2016)
SAEW	15	6.32	11.3	–	Leek	Dimethoate	78.59	Hu et al. (2016)
SAEW	15	5.95	29.36	–	Leek	Dimethoate	52	Hu et al. (2016)
SAEW	15	5.98	79.66	–	Leek	Dimethoate	52	Hu et al. (2016)
SAEW	15	6.12	126.59	–	Leek	Dimethoate	>52	Hu et al. (2016)
SAEW	15	6.0	1.95	–	Leek	Chlorpyrifos	>40	Hu et al. (2016)
SAEW	15	6.25	5.66	–	Leek	Chlorpyrifos	~40	Hu et al. (2016)

(continued)

Table 3.1 (continued)

EW types	Treatment time (min)	pH	ACC (mg/L)	ORP (mV)	Target samples	Target pesticides	Percentage reduction (%)	References
SAEW	15	6.32	11.3	–	Leek	Chlorpyrifos	~70	Hu et al. (2016)
SAEW	15	5.95	29.36	–	Leek	Chlorpyrifos	~70	Hu et al. (2016)
SAEW	15	5.98	79.66	–	Leek	Chlorpyrifos	~70	Hu et al. (2016)
SAEW	15	6.12	126.59	–	Leek	Chlorpyrifos	~70	Hu et al. (2016)
SAEW	30	–	–	–	Cabbage	Phoxim	92	Luo et al. (2014)

*TW: Tap water
EO: electrolyzed oxidizing water
ER: alkaline electrolyzed water
NaClO-50: NaClO aqueous solution containing 50 ppm concentration of available chlorine
NaClO-100: NaClO aqueous solution containing 100 ppm concentration of available chlorine
LAlEW-50: low alkaline electrolyzed water containing 50 ppm concentration of available chlorine
LAlEW-100: low alkaline electrolyzed water containing 100 ppm concentration of available chlorine
SAEW-50: strong acidic acid electrolyzed water containing 50 ppm concentration of available chlorine
SAEW-100: strong acidic acid electrolyzed water containing 100 ppm concentration of available chlorine
SAEW[#]: There was operation of mechanical washing after soaking strong acidic electrolyzed water

An ACC of 20, 70, and 120 mg/L of EW and a treatment time of 1, 8, and 15 min was applied to conduct experiments with a total of nine possible treatment combinations. All treatments were conducted in glass beakers and after then, all produces were immediately transferred into 1000 mL of tap water and soaked for approximately 5 s to remove residual chlorine. QuEChERS method was used to extract the pesticides from samples, and UPLC-MS/MS was used to identify pesticides including diazinon, cyprodinil, and phosmet. The results in Table 3.1 show that a higher ACC and a longer EW treatment time lead to greater reductions of pesticides on grapes, spinach, and snap peas, respectively. Although EW with a higher ACC and a longer treatment time could obtain higher reductions of pesticides on different produces, the differences among different experimental conditions were not significant ($P > 0.05$). The presence of chlorine in EW was highly dependent on the pH. Moreover, some pesticides such as methamidophos and methamidophos had reduced sulfur moieties (P=S) that made them easily react with HClO in EW (Qi et al. 2018), whereas other pesticides such as methamidophos and dimethoate were more readily oxidized and hydrolyzed in basic solutions (Lin et al. 2006), which accelerated their removal. Finally, moderately prolonging the washing time improved the efficacy of fruits and vegetables sanitization. In addition, this study also demonstrated that the surface structure of each produce is the main factor affecting the reduction efficacy and the abovementioned sensitivity of each pesticide to EW mentioned above.

In addition, Wuyun (2011) investigated the removal effects of AEW and SAEW on organophosphorus pesticides on fruit and vegetables. The results showed that there was no significant difference ($P > 0.05$) in the removal effect of organophosphorus pesticides on apples between AEW and SAEW at the same ACC. The degradation rates of dimethoate in AEW and SAEW were 72.8 and 69.5%, and those of chlorpyrifos in AEW and SAEW were 66.4 and 63.4%, and those of parathion in AEW and SAEW were 66.6 and 45.0%, respectively. For the removal effect of AEW and SAEW on vegetables, there was no significant difference ($P > 0.05$) between these treatments in their removal effect on dimethoate and parathion on tomato and beans between AEW and SAEW at the same ACC, in contrast to rape, whereas there was no significant difference ($P > 0.05$) in their removal effect on chlorpyrifos on all target vegetables including rape, tomato, and beans.

3.3 Mechanism of Pesticide Removal by EW

A few studies have reported on the mechanism of pesticide removal by EW. Hao and Li (2006) studied the effect of EW on the removal of pesticide residue on rape and hypothesized that EW can be used to eliminate pesticide residues in vegetables primarily because of its physical and chemical properties. Organophosphorus pesticides including acephate primarily contain P=S and C=O double bonds. A nucleophilic reaction occurs under acidic or alkaline conditions and the double bond is broken because AEW has a low pH and a high ORP value, whereas AlEW has a high pH and a good emulsifying property. Wuyun (2011) investigated the degradation of chlorpyrifos and parathion in aqueous solutions using AEW. The degradation products were identified using the GC-MS method. The degradation test of chlorpyrifos with an initial concentration of 2 mg/L was performed using AEW (pH 2.8, ACC 10 mg/L), and samples were taken at 1 min and 10 min for GC-MS analysis. In the total, ion chromatogram at 1 min, 3,5,6-trichloro-2-pyridinol (TCP) and chlorpyrifos oxides (chlorpyrifos oxon, phosphoric acid, diethyl 3,5,6-trichloro-2-pyridyl ester, CPO) were found with chlorpyrifos, whereas only TCP and CPO were found at 10 min. The results showed that after AEW treatment for chlorpyrifos, there is a chromatographic peak of chlorpyrifos at a reaction time of 1 min, but the chlorpyrifos was degraded at 10 min; the peak heights of the degradation products TCP and CPO at 1 min are significantly higher than those at 10 min, indicating that the degradation rate increases with time. The initial ion concentration of 2 mg/L parathion was degraded by AEW at pH 2.8 and an ACC of 10 mg/L and then analyzed via GC-MS. In the total ion chromatogram (TIC) at 1 min, 4-nitrophenol and paraoxon were found with parathion, whereas only 4-nitrophenol and paraoxon were found at 10 min. Based on the above results, Wuyun et al. (2018) proposed a degradation pathway of chlorpyrifos and parathion after AEW treatment. In addition, the degradation of pesticides is a very complicated chemical process. Wuyuandalai found many degradation products in the total ion chromatogram without further investigation because it is difficult to determine their chemical structure.

In addition, Wan (2015) studied the residues of dichlorvos, dimethoate, and chlorpyrifos in electrolyzed water cane, and the degradation rates reached 98.79, 51.70, and 72.55%, respectively, which were better than those of the same concentration of NaClO solution. The increase in chlorine concentration can enhance the removal effect of acidic electrolytic ionized water on pesticide residues. From 20 to 50 °C, the temperature of the electrolytic ionized water has a slight effect on the removal of pesticide residues; a ratio of raw material to electrolytic ionized water of 15:500 (g:mL) is the best process condition. The difference in the removal effect of electrolytic ionized water on the residue of dimethoate and chlorpyrifos after storage for different times indicates that the removal rate of organic ionized water by electrolytic ionized water is related to its ability to generate free radicals.

3.4 Conclusions

In summary, the electrolyzed water has an obvious effect on the removal of pesticide residues on food without a significant decrease in quality. It will have a good application prospect, even though the related research is lacking. However, there are few studies on the mechanism of electrolyzed water removal of pesticides. In addition, the variety of pesticides has made it difficult to study the mechanism of electrolyzed water removal of pesticides. This is also the bottleneck limiting the application of electrolyzed water for the removal of pesticide residues in food area.

References

Bakirci GT, Yaman Acay DB, Bakirci F et al (2014) Pesticide residues in fruits and vegetables from the Aegean region, Turkey. Food Chem 160:379–392
Chen Q, Wang Y, Chen F et al (2014) Chlorine dioxide treatment for the removal of pesticide residues on fresh lettuce and in aqueous solution. Food Control 40:106–112
Deborde M, Gunten UV (2008) Reactions of chlorine with inorganic and organic compounds during water treatment—kinetics and mechanisms: a critical review. Water Res 42(1–2):13–51
Han YT, Song L, An QS et al (2017) Removal of six pesticide residues in cowpea with alkaline electrolysed water. J Sci Food Agric 97:2333–2338
Hao JX, Li LT (2006) Study on the removal of pesticide residue of vegetables by electrolyzed functional water. Sci Technol Food Ind 5:164–166
Hao J, Wuyun DL, Liu HJ et al (2011) Reduction of pesticide residues on fresh vegetables with electrolyzed water treatment. J Food Sci 76(4):C520–C524
Hu ZH, Wu TJ, Wan YF et al (2016) Study on the removal of dimethoate and chlorpyrifos in leek by slightly acidic electrolyzed water. Sci Technol Food Ind 37(01):49–52
Huang YR, Hsieh HS, Lin SY et al (2006) Application of EO water on the reduction of bacterial contamination for seafood. Food Control 17:987–993
Huang YR, Hung YC, Hsu SY et al (2008) Application of electrolyzed water in the food industry. Food Control 19:329–345
Iizuka T, Shimizu A (2014) Removal of pesticide residue from Brussels sprouts by hydrostatic pressure. Innov Food Sci Emerg Technol 22(Supplement C): 70–Ñ75
Issa-Zacharia A, Kamitani Y, Morita K et al (2010) Sanitization potency of slightly acidic electrolyzed water against pure cultures of *Escherichia coli* and *Staphylococcus aureus*, in comparison with that of other food sanitizers. Food Control 21:740–745
Koide S, Takeda J, Shi J et al (2009) Disinfection efficacy of slightly acidic electrolyzed water on fresh cut cabbage. Food Control 20:294–297
Koseki S, Yoshida K, Kamitani Y et al (2004) Effect of mild heat pre-treatment with alkaline electrolyzed water on the efficacy of acidic electrolyzed water against *Escherichia coli* O157:H7 and *Salmonella* on lettuce. Food Microbiol 21:559–566
Koseki M, Tanaka Y, Noguchi H et al (2007) Effect of pH on the taste of alkaline electrolyzed water. J Food Sci 72:298–302
Lin CS, Tsai PJ, Wu C et al (2006) Evaluation of electrolysed water as an agent for reducing methamidophos and dimethoate concentrations in vegetables. Int J Food Sci Technol 41(9):1099–1104
Liu HJ, Li RZ, Su DH et al (2015) Degradation of Lambda-cyhalothrin in fruits and vegetable by alkaline electrolyzed water. Food Sci Technol 40(2):123–127
Luo Q, Zu YH, Shi KQ et al (2014) Study on the reduction effect of pesticide residues of vegetables with slightly acidic electrolyzed water. J Food Saf Qual 5(11):3657–3663
Qi H, Huang Q, Hung YC (2018) Effectiveness of electrolyzed oxidizing water treatment in removing pesticide residues and its effect on produce quality. Food Chem 239:561–568
Rahman SME, Ding T, Oh DH (2010) Effectiveness of low concentration electrolyzed water to inactivate foodborne pathogens under different environmental conditions. Int J Food Microbiol 139:147–153
Rani M, Shanker U, Jassal V (2017) Recent strategies for removal and degradation of persistent & toxic organochlorine pesticides using nanoparticles: a review. J Environ Manage 190:208–222
Souza LP, Faroni L, Heleno FF (2018) Ozone treatment for pesticide removal from carrots: optimization by response surface methodology. Food Chem 243:435–441
Sung JM, Kwon KH, Kim JH et al (2011) Effect of washing treatments on pesticide residues and antioxidant compounds in Yuja (Citrus junos Sieb ex Tanaka). Food Sci Biotechnol 20(3):767–773
Sung JM, Park KJ, Lim JH et al (2012) Removal effects of microorganism and pesticide residues on Chinese cabbages by electrolyzed water washing. Korean J Food Sci Technol 44(5):628–633
Wan YF (2015) Study on degradation mechanism of organophosphorus pesticides in fruits and vegetables by electrolyzed water. Thesis, Hebei University of Science and Technology

Wang B, Zhu C, Gong RH et al (2015) Degradation of acephate using combined ultrasonic and ozonation method. Water Sci Eng 8(3):233–238

Wuyun DL (2011) Mechanism and application of slightly acid electrolyzed functional water degradation of organophosphorus pesticide residues. China Agricultural University

Wuyun DL, Liang D, Hao JX et al (2018) Kinetic model and pathway of parathion degraded by electrolyzed oxidizing water. Sci Technol Food Ind 39(12):63–68

Chapter 4
Application of Electrolyzed Water in Fruits and Vegetables Industry

Jianxiong Hao and Qingfa Wang

4.1 Introduction

Fruits and vegetables, as an indispensable part of the human diet, provide various nutrients and phytochemical components to prevent us from the risk of chronic diseases. And these produce may be minimally processed to gain their unique sensory quality, meeting the increasing demand of consumers pursuing healthy products of natural flavor (Mahajan et al. 2017; Masis et al. 2017; Ramos et al. 2013).

However, there are complex but essential steps during the processing and storage of fruits and vegetables, such as sorting, grading, washing, peeling, cutting, and shredding, providing opportunities for the occurrence of food safety hazards including microbial contamination, pesticide residue, moth infection, and so on (Bempah et al. 2016; Castro-Ibáñez et al. 2017; Rady et al. 2017). For instance, foodborne disease related to pathogens existing in fruits and vegetables have appeared frequently in recent years throughout the world. And, *L. monocytogenes*, *Salmonella* spp. and shigatoxin-producing *Escherichia coli* had been considered as three main pathogens detected in fruits and vegetables (Silva et al. 2017; Vojkovska et al. 2017). There were 128,000 hospitalization and 3,000 deaths every year in the United States as a consequence of intake of food contaminated by microorganisms (Scharff et al. 2016). Furthermore, it is inevitable to use pesticide in order to protect fruits and vegetables from the attack of fungi or insects during their growth. Consequently, the toxic pesticide residues remaining in fresh produce can pose a threat to human health, which violates people's demand for safe and green food (Liu and Niyongira 2017; Malarkodi et al. 2017; Zhang et al. 2015). Additionally, as living entities, fruits and vegetables after harvest still maintain the physiological activities like respiration, transpiration, and maturation. Once treated inappropriately, they will be easily perished or lose better characteristics including reduction of functional components,

J. Hao (✉) · Q. Wang
College of Bio Science and Engineering, Hebei University of Science and Technology, Shijiazhuang 050018, Hebei, China
e-mail: cauhjx@163.com

T. Ding et al. (eds.), *Electrolyzed Water in Food: Fundamentals and Applications*,
https://doi.org/10.1007/978-981-13-3807-6_4

weight loss or shriveling (Khalid et al. 2017; Moggia et al. 2016; Xanthopoulos et al. 2017). And it would virtually reduce the value of fresh produce.

To address the above-mentioned problems, it is necessary to hunt for new methods for the disinfection, removing pesticide residues, and improving the overall quality of fruits and vegetables. So far, a number of physical or chemical techniques have been employed throughout the food chain. For example, ionizing irradiation, ultraviolet light, and chlorine can be used for decontaminating microorganisms (Inatsu et al. 2017; Islam et al. 2016; Meireles et al. 2016); Ozone, chlorine dioxide, and membrane filtration have been reported to remove or reduce the pesticide residues of fruits and vegetables (Chen et al. 2014; Doulia et al. 2016; Souza et al. 2018). But disadvantages of these techniques do exist, such as remaining harmful by-products, having a negative impact on organoleptic quality, low stability, and high cost.

Electrolyzed water (EW), on the other hand, has shown its potential use in fruits and vegetables industry not only as a novel sanitizer and cleaner but also due to its significant role in improving postharvest quality as well as preventing physiological disease (Feliziani et al. 2016; Pinto et al. 2016; Yoon and Lee 2017). Compared to preceding techniques, EW is a more effective method that is easy to produce, consumes lower costs and does less harm to health and environment. Generally recognized as safe, EW could be generated by the electrolysis of a dilute salt solution (commonly sodium chloride) in an electrolytic cell containing an anode and cathode separated by a bipolar membrane (Huang et al. 2008; Joshi et al. 2013). Thus, two types of EW generated simultaneously: AEW at the anode possessing a pH of 2–3, ACC of 10–90 mg/L and a high oxidation–reduction potential (ORP) over 1100 mV; AlEW at the cathode with a high pH of 10–13 and a low ORP of 700–900 mV below zero (Hricova and Zweifel 2008). Besides, NEW can be formed if the electrolytic cell does not have the membrane, and its pH is within the range of 7–8 and ORP is approximately 750–900 mV (Duan et al. 2016; Rahman et al. 2016).

Admittedly, EW has been widely applied to be a disinfectant in dairy, meat, cereal, fruits, and vegetables industry (Al-Holy and Rasco 2015; Audenaert et al. 2012; Jiménez-Pichardo et al. 2016; Joshi et al. 2013). But more researches were focused on the effects of it on the quality of fruits and vegetables. Therefore, the article reviews the application of EW in the processing and storage of fruits and vegetables, mainly in recent years including its disinfection efficacy, the effect on the postharvest quality and reduction of pesticide residues. Owing to the popularity of sprouts vegetables recently, effects of EW on their microbial reduction, functional components accumulation and growth are also summarized. Finally, the review aims to promote the development and utilization of EW in fruits and vegetables industry as well as contribute to eliminating the hazards existing in fresh produce and eventually ensuring their safety.

4.2 Application of EW in the Processing and Storage of Fruits and Vegetables

4.2.1 *Application of EW in the Disinfection of Fruits and Vegetables During Processing*

It is well known that there is a diversity of microorganisms inhabited on the surface of fruits and vegetables that provide an approach for the transmission of foodborne illness (Graça et al. 2017a; Jonathan and Leff 2013). One of the most important points of EW in fruits and vegetables industry is its microbial load reduction efficacy during their processing, which has attracted numerous researchers' concern. The mechanism of disinfection efficacy of EW has not become a consensus. But it is clear that it is highly dependent on the property of EW, namely, pH, ACC and ORP. First, the pH is responsible for the existing form of chlorine. The chlorine gas (Cl_2) as the main form is easy to escape from the solution whose pH is lower than 2.7, thus causing the decrease of its antimicrobial efficacy. Most chlorine existing in EW of higher pH is hypochlorite (ClO^-) whose disinfection efficacy is only about one-eightieth as effective as hypochlorous acid (HClO) in equivalent concentration that is mainly present in EW of lower pH (5.0–6.5). And, HClO could exhibit strong sterilizing power because it produces hydroxyl radicals after penetrating the cell membranes and incurs disruption of key metabolic pathways (Athayde et al. 2017; Cao et al. 2009; Huang et al. 2008). Then, our research group proved that chlorine compounds are an important factor accounting for the disinfection efficacy of EW (Hao et al. 2012). It was indicated that complete inactivation of pathogens could be achieved by EW in a pH range of 2.6–7.0 as long as its ACC was sufficiently high (Park et al. 2004). And, Ni et al. demonstrated that bactericidal activity of EW increased with the increase of ACC (Ni et al. 2016). Nevertheless, there were still some reports attributing it to the ORP of EW. From their perspectives, higher ORP could lead to the destruction of outer and inner membranes of bacteria or change the metabolic process and influence the generation of adenosinetriphosphate, thus accelerating its death (Liao et al. 2007; Tkhawkho et al. 2017).

Additionally, these physiochemical indices of EW may be affected by preparation parameters (such as salt concentration, current, and flow rate) and storage conditions, which eventually has an impact on its disinfection efficacy. For example, previous studies suggested that EW prepared with higher current remained higher pH, ACC and ORP and had stronger antimicrobial ability (Rahman et al. 2012). When AEW had been exposed to light during storage, ACC, ORP, and bactericidal efficacy of it significantly decreased, according to Cui et al. (2009). And, lower storage temperature (4 °C) made these basic properties of EW more stable and maintain its bactericidal efficiency over 12 months (Robinson et al. 2012). Furthermore, preheating EW could increase its ACC value in comparison with postproduction heating and enhance its disinfection efficacy (Forghani et al. 2015).

Undoubtedly, washing raw materials is a required step before their being processed when spoilage microorganisms or pathogens may be accumulated and EW is most frequently acted as a sanitizer (Gil et al. 2015). Based on the common properties, EW can be further classified into AEW, SAEW, NEW, and AlEW. Disinfection efficacy of every type of EW applied to fruits and vegetables has been summarized in the following parts. Besides, EW combined with other newly developed techniques in fruits and vegetables industry has aroused many researchers' concern, which is also discussed.

4.2.1.1 Disinfection Efficacy of AEW on Fruits and Vegetables

Owing to its high oxidizing capacity, AEW, is also called electrolyzed oxidizing water (EOW) and has been acknowledged as a novel disinfection agent recently (Cui et al. 2009; Jadeja et al. 2013). Compared with conventional chlorine sanitizers such as sodium chlorite (NaClO), the use of AEW could avoid the generation of toxic by-products like trihalomethanes (THMs) and be produced on site (Gómez-López et al. 2015). Besides, it is well documented that AEW have shown its strong bactericidal effect against various microorganisms attached to fruits and vegetables including *Escherichia coli* (Graça et al. 2011; Hao et al. 2006, 2015a; Pangloli and Hung 2013), *Listeria innocua* (Graça et al. 2009, 2011), *Salmonella choleraesuis* (Graça et al. 2011), *Salmonella typhimurium* (Ding et al. 2011; Park et al. 2009), *Salmonella enteric* (Fishburn et al. 2012), *Cronobacter sakazakii* (Santo et al. 2016), *Bacillus cereus* (Ding et al. 2011; Ju et al. 2017), *Bacillus subtilis* (Hao et al. 2011a, b), and *Pseudomonas fluorescens* (Ignat et al. 2016).

A number of investigations have been launched for the sterilization mechanism of AEW. Generally, it is involved in the destruction of cell structures, changes in protein, and other essential elements of microbes. For one thing, AEW treatment deformed the carbohydrate C–O–C bond and aromatic rings in tyrosine and phenylalanine which were the components of extracellular polymeric substances, leading to the detachment and inactivation of biofilms of *Vibrio parahaemolyticus* and *Listeria monocytogenes* (Han et al. 2017). And, it could effectively kill both static and flow *Enterococcus faecalis*as well as *Staphylococcus aureus* by destructing their biofilms without eradicating them (Cheng et al. 2016; Sun et al. 2012). In addition, as Xiong et al reported, AEW had strong antifungal efficacy on *Aspergillus flavus* because it destroyed the cellular structures, making K^+ and Na^+ drained and depriving them of normal functions (Xiong et al. 2014). For another, our research group concluded that cell membranes of *Escherichia coli*, *Staphylococcus aureus*, and *Bacillus subtilis* could be damaged when treated by AEW, but it could not make their DNA and RNA degraded (Hao et al. 2016a, b). Aoki et al found that AEW had the ability to kill *Mycobacteria* via denaturing cytoplasmic proteins after damaging cell walls (Yamamoto et al. 2012).

Application of AEW in Disinfection of Fruits

As illustrated in Table 4.1, AEW has been utilized for sterilization of many fruits comprising apple, pear, melon, blueberry, strawberry, sweet basil, and so forth. Basically, AEW treatment could make a reduction of 1–2 log CFU/g of bacteria existing on samples. However, there are also studies revealing that the bactericidal activity of AEW seemed not high enough. For, instance, the study of Udompijitkul et al. achieved only 0.71 log CFU/g reduction of *Escherichia coli O157:H7* attached to strawberry with the AEW treatment (pH 2.6, ACC 47 mg/L and ORP 1293 mV) for 5 min at 22 °C (Udompijitkul et al. 2007). It can be observed from Table 4.1 that washing fruits by immersing them into AEW is the main method for disinfection and the efficacy is governed by the property of AEW and processing parameters (such as rinsing time, temperature, and the frequency of rotary shaker). An example is that the reduction of *Escherichia coli O157:H7* increased with the increase of treating time (from 1to 3 min) when blueberries were disinfected (Pangloli and Hung 2013). Besides, spraying can also be used and the disk orifice size in combination with lasting time affects the disinfection efficacy more or less (Kim and Hung 2012).

Application of AEW in Disinfection of Vegetables

It is observed from Table 4.1 that various microbes including bacteria, yeasts and molds can be removed from edible vegetables. More researches have been focused on the disinfection of lettuce by AEW and most reduction level could reach 2–3 log CFU/g, and varying treatment conditions (washing time or temperature), as well as physiochemical properties of EW, may result in different efficiency, even in terms of the same microbe. When treated by AEW (pH 2.06 and ACC 37.5 mg/L) for 1 min at 22 °C, the number of *Escherichia coli O157:H7*, *Salmonella typhimurium*, and *Listeria monocytogenes* on tomatoes and green onions decreased more than 4 log CFU/g, which exhibited the strong bactericidal activity of AEW (Park et al. 2009). In case of chicory, chive, mushroom, broccoli, spinach, lotus roots, cilantro, and other leafy greens, washing with AEW could generally reduce microbes by 1–2 log CFU/g. Indeed, some reduction level was not significant (<1 log CFU/mL, which could be explained by the initial counts of different microbes and the discrepancy in their resistance to AEW. In summary, AEW would be a substitute for conventional sanitizers when it refers to disinfection.

4.2.1.2 Disinfection Efficacy of SAEW on Fruits and Vegetables

Although having potential antimicrobial activity, AEW has a relatively low pH of 2–3 at which Cl_2 is easy to volatilize into the air which consequently poses threat to health. Moreover, AEW may be corrosive to organic materials due to its high acidity, which ultimately limits its application (Rahman et al. 2010a, b, c, 2012). On the contrary, the acidity of SAEW is slighter (pH 5.0–6.5) and it has an ORP no more than 1000 mV

Table 4.1 Disinfection efficacy of acidic electrolyzed water in fruits and vegetables

Microorganism	Origin	Physiochemical properties of EW			Washing conditions	Reduction (log CFU/g)	Reference
		pH	ACC (mg/L)	ORP (mV)			
Escherichia coli	Apple	3.08	53	1111	5 min, 150 rpm	2.09	Graça et al. (2011)
Listeria innocua	Apple	3.08	53	1111	5 min, 150 rpm	1.00–1.50	Graça et al. (2011)
Listeria innocua	Apple	2.68	200	1111	5 min, 150 rpm	1.6–1.8	Graça et al. (2010)
Salmonella choleraesuis	Apple	3.08	53	1111	5 min, 150 rpm	1.40–1.50	Graça et al. (2011)
Cronobacter sakazakii	Apple, pear and melon	2.82	103	1121	5 min, 150 rpm	1.3–1.8	Santo et al. (2016)
Escherichia coli O157:H7	Blueberry	2.63	31.1	1163	5 min, 150 rpm	4.44	Pangloli and Hung (2013)
Escherichia coli O157:H7	Blueberry	2.67	46.1	1169.6	4 min, 2.311 mm (spraying orifice diameter)	1.10	Kim and Hung (2012)
Escherichia coli O157:H7	Strawberry	2.7	55.5	1067.6	5 min, 4 °C	1.58	Hung et al. (2010)
Escherichia coli O157:H7	Strawberry	2.6	47	1293	5 min, 100 rpm, 22 °C	0.71	Udompijitkul et al. (2007)
Listeria monocytogenes	Strawberry	2.6	47	1293	5 min, 100 rpm, 22 °C	1.65	Udompijitkul et al. (2007)
Aerobicmesophilic bacteria	Strawberry	2.6	32.1	1130	Overflowing washing for 10 min	0.8–0.9	Shigenobu et al. (2004)
Mesophilic bacteria	Sweet basil	2.58	32.4	1144	5 min, 25 °C	About 2.5	Tirawat et al. (2016)

(continued)

Table 4.1 (continued)

Microorganism	Origin	Physiochemical properties of EW			Washing conditions	Reduction (log CFU/g)	Reference
		pH	ACC (mg/L)	ORP (mV)			
Escherichia coli O157:H7	Lettuce	2.6	50	1200	2 min, 70 rpm, 22 °C	<1	Keskinen et al. (2009)
Escherichia coli O157:H7	Lettuce	2.89	13	1097	3 min, 15 °C	3.43	Fishburn et al. (2012)
Listeria monocytogenes	Lettuce	2.89	13	1097	3 min, 15 °C	2.55	Fishburn et al. (2012)
Salmonella	Lettuce	2.89	13	1097	3 min, 15 °C	3.72	Fishburn et al. (2012)
Escherichia coli O157:H7	Tomato and green onion	2.06	37.5	–	1 min, 22 °C	>4	Park et al. (2009)
Salmonella typhimurium	Tomato and green onion	2.06	37.5	–	1 min, 22 °C	>4	Park et al. (2009)
Listeria monocytogenes	Tomato and green onion	2.06	37.5	–	1 min, 22 °C	>4	Park et al. (2009)
Escherichia coli O157:H7	Chicory, chive and onion	–	100	–	5 min	<1	Ju et al. (2017)
Bacillus cereus	Chicory	–	100	–	5 min	1.27	Ju et al. (2017)
Bacillus cereus	Chive	–	100	–	5 min	1.15	Ju et al. (2017)
Bacillus cereus	Onion	–	100	–	5 min	0.00	Ju et al. (2017)
Escherichia coli O157:H7	Leafy greens	2.1	30–35	1100	60 s, 22 °C	2.4	Stopforth et al. (2008)
Salmonella	Leafy greens	2.1	30–35	1100	60 s, 22 °C	2.3	Stopforth et al. (2008)

(continued)

Table 4.1 (continued)

Microorganism	Origin	Physiochemical properties of EW			Washing conditions	Reduction (log CFU/g)	Reference
		pH	ACC (mg/L)	ORP (mV)			
Listeria monocytogenes	Leafy greens	2.1	30–35	1100	60 s, 22 °C	2.5	Stopforth et al. (2008)
Listeria monocytogenes	Mushroom	2.54	50	1100–1120	3 min, 23 °C	2.08	Ding et al. (2011)
Salmonella typhimurium	Mushroom	2.54	30	1100–1120	3 min, 23 °C	2.04	Ding et al. (2011)
Bacillus cereus	Mushroom	2.54	30	1100–1120	3 min, 23 °C	1.95	Ding et al. (2011)
Total aerobic bacteria	Spinach	2.54	50	1130	3 min, 23 °C	1.94	Rahman et al. (2010a, b)
Yeasts and molds	Spinach	2.54	50	1130	3 min, 23 °C	1.57	Rahman et al. (2010a, b, c)
Escherichia coli O157:H7	Broccoli	2.5	54.1	1108.6	1 min, 24 °C	1.28	Hung et al. (2010a, b)
Enterobacteria	Kailan-hybrid broccoli	–	70	–	2 min, 8 °C	1–1.5	Navarro-Rico et al. (2014)
Yeasts and molds	Kailan-hybrid broccoli	–	70	–	2 min, 8 °C	3.5–4	Navarro-Rico et al. (2014)
Yeasts and molds	Lotus roots	2.28	79.35	1134	5 min	1.26	Li et al. (2017)
Total aerobic bacteria	Lotus roots	2.28	79.35	1134	5 min	1.39	Li et al. (2017)
Aerobic mesophilic bacteria	Minizuna baby leaves	3.1	260	>1100	2 min, 5 °C	<1	Tomás-Callejas et al. (2011)

(continued)

Table 4.1 (continued)

Microorganism	Origin	Physiochemical properties of EW			Washing conditions	Reduction (log CFU/g)	Reference
		pH	ACC (mg/L)	ORP (mV)			
Total aerobic bacteria	Cilantro	2.48	79.35	1134	5 min	2.58	Hao et al. (2011a, b)
Yeasts and molds	Cilantro	2.48	79.35	1134	5 min	2.04	Hao et al. (2011a, b)
Coli-forming bacterial	Cilantro	2.48	68.35	1127	5 min, room temperature	1.51	Hao et al. (2015a, b)
Coli-forming bacterial	Cilantro	2.48	79.35	1134	5 min	1.31	Hao et al. (2011a, b)

AEW acidic electrolyzed water; *EW* electrolyzed water; *ACC* available chlorine concentration; *ORP* oxidation—reduction potentisal; "–" means the data wasn't shown in literature and ACC with the range from 10 to 30 mg/L while its bactericidal capacity is pretty strong on account of HClO as the main form of available chlorine (Ding et al. 2016; Ni et al. 2016; Tango et al. 2015). Therefore, the use of SAEW as a means of disinfection has gained popularity with investigators in food processing and preservation (Xuan et al. 2017). To date, SAEW has been found to possess antimicrobial ability against both bacteria and fungi related to fruits and vegetables, such as *Listeria* spp. (Ding et al. 2011; Tango et al. 2017), *Salmonella* spp. (Issa-Zacharia et al. 2011), Coliform populations (Gómez-López et al. 2013; Hao et al. 2015a), *Bacillus* spp. (Hao et al. 2011a, b), *Pseudomonas* spp. (Gómez-López et al. 2013), and so on.

Application of SAEW in Disinfection of Fruits

A few investigations have studied the disinfection efficacy on fruits treated by SAEW alone in literature. As reflected in Table 4.2, SAEW has exhibited strong antimicrobial ability against microorganisms on apple, cherry tomato, and strawberry. Overall, it could reduce microbial population by 2–3 log CFU of bacteria or fungi attached to fruits, which is obviously higher than the reduction level (less than 2 log) caused by conventional chlorine treatment (Pinto et al. 2016). As for the disinfection of SAEW, more researches have been focused on vegetables that may further turn out its potential.

Application of SAEW in Disinfection of Vegetables

From Table 4.2, we are aware that a variety of vegetables can be disinfected by SAEW including celery, cabbage, cilantro, lettuce, mushroom, spinach, daikon sprout, tomato, endive, and kale. In many cases, the reduction level of microbes on vegetables is about 1.5–3 log CFU. Particularly, SAEW could reduce the total aerobic bacteria existing on the surface of celery by 4.33 log CFU (Zhang et al. 2016a, b). These facts are enough to elucidate that SAEW could rival AEW with regard to the disinfection efficacy. As an environmentally friendly disinfection agent, SAEW is bound to be an alternative for AEW and traditional chlorine in vegetables industry.

4.2.1.3 Disinfection Efficacy of NEW on Fruits and Vegetables

To overcome the limitation of AEW of corrosive characteristics in the process of food sterilization, another method besides SAEW treatment is the prevalence of NEW that is also considered as mixed oxidants and could not corrode metal equipments or irritate human skins (Arevalos-Sánchez et al. 2012; Cui et al. 2009). And, the main constituents accounting for sterilization in it are hypochlorous acid, hypochlorite, and hydroperoxyl (Torlak 2014). In comparison with other EWs, physiochemical properties (pH, ACC or ORP), and disinfection efficacy of NEW remains relatively stable in varying storage conditions (such as light exposure) owing to its neutral pH

Table 4.2 Disinfection efficacy of slightly acidic electrolyzed water in fruits and vegetables

Microorganism	Origin	Physiochemical properties of EW			Washing conditions	Reduction (log CFU)	Reference
		pH	ACC (mg/L)	ORP (mV)			
Escherichia coli O157:H7	Apple	5.42	30	818–854	23 °C, 3 min	2.28/fruit	Tango et al. (2017)
Listeria monocytogenes	Apple	5.42	30	818–854	23 °C, 3 min	2.30/fruit	Tango et al. (2017)
Total aerobic bacteria	Apple	5.42	30	818–854	23 °C, 3 min	1.89/fruit	Tango et al. (2017)
Total bacteria	Cherry tomato	5.49	34.33	853.7	–	3.29/g	Ding et al. (2015)
Yeasts and molds	Cherry tomato	5.49	34.33	853.7	–	3.59/g	Ding et al. (2015)
Total bacteria	Strawberry	5.49	34.33	853.7	–	2.32/g	Ding et al. (2015)
Yeasts and molds	Strawberry	5.49	34.33	853.7	–	3.01/g	Ding et al. (2015)
Total aerobic bacteria	Celery	5.76–6.05	25	879.3–923.6	20 °C, 5 min	4.33/g	Zhang et al. (2016a, b)
Yeasts and molds	Celery	5.76–6.05	25	879.3–923.6	20 °C, 5 min	3.86/g	Zhang et al. (2016a, b)
Escherichia coli	Celery	5.6	22.1	931	5 min, 500 rpm	2.74/g	Issa-Zacharia et al. (2011)
Salmonella spp.	Celery	5.6	22.1	931	5 min, 500 rpm	2.87/g	Issa-Zacharia et al. (2011)
Total aerobic bacteria	Cilantro	5.76–6.05	25	879.3–923.6	20 °C, 5 min	4.14/g	Zhang et al. (2016a, b)

(continued)

Table 4.2 (continued)

Microorganism	Origin	Physiochemical properties of EW			Washing conditions	Reduction (log CFU)	Reference
		pH	ACC (mg/L)	ORP (mV)			
Total aerobic bacteria	Cilantro	5.85	19.58	809	5 min, at room temperature	2.52/g	Hao et al. (2015a, b)
Yeasts and molds	Cilantro	5.76–6.05	25	879.3–923.6	20 °C, 5 min	3.75/g	Zhang et al. (2016a, b)
Yeasts and molds	Cilantro	5.85	19.58	809	5 min, at room temperature	1.42/g	Hao et al. (2015a, b)
Coli-forming bacterial	Cilantro	5.85	19.58	809	5 min, at room temperature	1.68/g	Hao et al. (2015a, b)
Escherichia coli O78	Cilantro	5.85	19.46	815	5 min	2.49/g	Hao et al. (2011a, b)
Bacillus subtilis 1.1849	Cilantro	5.85	19.46	815	5 min	1.54/g	Hao et al. (2011a, b)
Total aerobic bacteria	Cabbage	6.1	20	–	10 min, 20–22 °C	1.5/g	Koide et al. (2009)
Yeasts and molds	Cabbage	6.1	20	–	10 min, 20–22 °C	1.3/g	Koide et al. (2009)
Total microbial counts	Lettuce	6.4	30	562	10 min, 20 °C	1.9/g	Park et al. (2017)
Total viable counts	Lettuce	6.2–6.9	20	820–960	1 min, 20 °C	1.9/g	Xuan et al. (2016)
Escherichia coli	Lettuce	5.6	22.1	931	5 min, 500 rpm	2.84/g	Issa-Zacharia et al. (2011)
Escherichia coli O157:H7	Lettuce	5.89	18.2	906.9	30 s, 22 °C	1.91/leaf	Pangloli and Hung (2011)
Salmonella spp.	Lettuce	5.6	22.1	931	5 min, 500 rpm	2.91/g	Issa-Zacharia et al. (2011)

(continued)

Table 4.2 (continued)

Microorganism	Origin	Physiochemical properties of EW			Washing conditions	Reduction (log CFU)	Reference
		pH	ACC (mg/L)	ORP (mV)			
Listeria monocytogenes	Mushroom	6.2	5	500–520	3 min, 23 °C	2.16/g	Ding et al. (2011)
Salmonella typhimurium	Mushroom	6.2	5	500–520	3 min, 23 °C	2.08/g	Ding et al. (2011)
Bacillus cereus	Mushroom	6.2	5	500–520	3 min, 23 °C	2.02/g	Ding et al. (2011)
Total bacteria counts	Spinach	6.3	5	520	3 min, 23 °C	1.93/g	Rahman et al. (2010a, b, c)
Yeasts and molds	Spinach	6.3	5	520	3 min, 23 °C	1.64/g	Rahman et al. (2010a, b, c)
Pseudomonas spp.	Baby spinach	6.5	4.4	814	1 min, 7 °C	<1/g	Gómez-López et al. (2013)
Escherichia coli	Daikon sprout	5.6	22.1	931	5 min, 500 rpm	2.85/g	Issa-Zacharia et al. (2011)
Salmonella spp.	Daikon sprout	5.6	22.1	931	5 min, 500 rpm	2.91/g	Issa-Zacharia et al. (2011)
Escherichia coli O157:H7	Tomato	5.89	18.2	906.9	15 s, 22 °C	6.55/tomato	Pangloli and Hung (2011)
Total microbial count	Endive	6.4	30	562	10 min, 20 °C	1.3/g	Park et al. (2017)
Total microbial count	Kale	6.4	30	562	10 min, 20 °C	1.7/g	Park et al. (2017)

"–" means the data wasn't shown in literature

(Cui et al. 2009; Monnin et al. 2012). For example, the chlorine loss rate of NEW at 5 °C is only 0.013 mg/(L min) (Meireles et al. 2017). Besides, it does not cause loss of organoleptic qualities of fresh produce (Meireles et al. 2017).

Until now, NEW has been reported to effectively inactivate *Alicyclobacillus acidoterrestris* (Torlak 2014), *Salmonella* (Abadias et al. 2008), *Listeria* (Jung et al. 2017a; Navarro-Rico et al. 2014), *Escherichia coli* (Graça et al. 2010; Jung et al. 2017b), *Yersinia enterocolitica* (Lehto et al. 2017), *Cronobacter sakazakii* (Santo et al. 2016), and *Pseudomonas* (Gómez-López et al. 2013; Ignat et al. 2016; Pinto et al. 2015) in fruits and vegetables industry. Furthermore, numerous studies have revealed the disinfection mechanism of NEW. Some concluded that NEW could destroy the cellular structures of *Aspergillus flavus*, causing the leakage of K^+ and Mg^{2+} and eliminating it (Xiong et al. 2014). Meanwhile, others found NEW could reduce viable cells embedded in biofilms and even eradicate them and the disinfection efficacy might be ascribed to its higher ACC or ORP (Arevalos-Sánchez et al. 2012, 2013; Chuang et al. 2013; Machado et al. 2016). Furthermore, NEW could promote the formation of reactive oxygen species which was also a contributing factor to its antimicrobial capacity (Meireles et al. 2017).

Application of NEW in Disinfection of Fruits

Disinfection efficacy of NEW on various fruits against pathogens and spoilage microorganisms has been described in Table 4.3, including apple, pear, melon, orange, cantaloupe, and mango. On average, it could eliminate the number of microbes by approximately 1 log CFU during washing. Besides, few researchers have studied the efficacy of EW on killing spores surviving on fruits, which is a exception for Torlak who had washed apples using NEW (pH of 7.76, ACC of 191 mg/L and ORP of 786 mV) for 3 min at the agitating speed of 50 rpm and it reduced the inoculated *Alicyclobacillus acidoterrestris* spores by more than 4 log CFU/apple (Torlak 2014). And, it further turned out the advantage of NEW of decontamination.

Application of NEW in Disinfection of Vegetables

It can be observed from Table 4.3 that numerous investigations have focused on the sanitizing potential of NEW in vegetables. As a whole, microbial reduction of different vegetables treated with NEW under varying dipping time and temperature ranges from 1 log to 4 log. And when it comes to lettuce, significant difference could be seen in the reduction of *Escherichia coli O157:H7* treated with NEW originated from different literatures, which may be attributed to the critical role of the physiochemical properties of it. Furthermore, more researchers tend to seek out optimized parameters of NEW washing conditions so that microbial safety could be ensured in practical processing line without deteriorating nutritional and sensory quality (Afari et al. 2015; Lee et al. 2014).

Table 4.3 Disinfection efficacy of neutral electrolyzed water on fruits and vegetables

Microorganism	Origin	Physicochemical properties of EW			Washing conditions	Reduction (log CFU)	Reference
		pH	ACC (mg/L)	ORP (mV)			
Escherichia coli O157:H7	Apple	8.18	100	725	5 min, 150 rpm	0.5–0.6/g	Graça et al. (2010)
Listeria innocua	Apple	8.18	100	725	5 min, 150 rpm	1.0–1.1/g	Graça et al. (2010)
*Alicyclobacillus acidoter-restris*spores	Apple	7.76	191	786	3 min, at room temperature, 50 rpm	>4/apple	Torlak (2014)
Cronobacter sakazakii	Apple, pear and melon	8.18	94	762	5 min, 150 rpm	1–1.2/g	Santo et al. (2016)
Escherichia coli O157:H7	Pear	8.18	100	725	5 min, 150 rpm	0.7–0.8/g	Graça et al. (2010)
Listeria innocua	Pear	8.18	100	725	5 min, 150 rpm	1.0–1.1/g	Graca et al. (2010)
Salmonella enterica	Pear	8.20	102	754	5 min, 150 rpm	About 1/g	Graça et al. (2017a, b)
Escherichia coli O157:H7	Orange	8.18	100	725	5 min, 150 rpm	1.3–1.4/g	Graça et al. (2010)
Listeria innocua	Orange	8.18	100	725	5 min, 150 rpm	0.5–0.6/g	Graça et al. (2010)
Listeria monocytogenes	Cantaloupe	6.93	58.3	–	5 min, 21.4 °C	>1.5/g	Jung et al. (2017a, b)
Salmonella	Cantaloupe	6.93	58.3	–	5 min, 21.4 °C	1.2/g	Jung et al. (2017a, b)
Escherichia coli	Mango	7.95	101	797	5 min, 150 rpm	2.2/g	Santo et al. (2018)

(continued)

Table 4.3 (continued)

Microorganism	Origin	Physiochemical properties of EW			Washing conditions	Reduction (log CFU)	Reference
		pH	ACC (mg/L)	ORP (mV)			
Enterobacteriaceae	Lettuce	–	200	–	5 min, 4 °C	1.91/g	Pinto et al. (2015)
Listeria monocytogenes	Lettuce	6.83	43	–	5 min	2.6–3.4/g	Jung et al. (2017a, b)
Escherichia coli	Lettuce	6.5–6.7	100	800–900	10 min, 25 °C	0.25/mL	Guentzel et al. (2008)
Escherichia coli O157:H7	Lettuce	6.83	43	–	5 min	3.4–3.7/g	Jung et al. (2017a, b)
Escherichia coli O157:H7	Lettuce	6.5	50	>450	30 s, 4 °C	1.0/g	Posada-Izquierdo et al. (2014)
Escherichia coli O157:H7	Lettuce	7.52	155	760	5 min, 65 rpm	1.9–2.5/g	Afari et al. (2015)
Salmonella	Lettuce	6.5–6.7	100	800–900	10 min, 25 °C	2.69–3.34/mL	Guentzel et al. (2008)
Salmonella typhimurium	Lettuce	7.52	155	760	5 min, 65 rpm	2.9–4.2/g	Afari et al. (2015)
Salmonella enterica	Lettuce	6.83	43	–	5 min	2.8–3.0/g	Jung et al. (2017a, b)
Pseudomonas	Lettuce	6.5	30	–	2 min, 8 °C	1.0/g	Ignat et al. (2016)
Psychrophilic bacteria	Kailan-hybrid broccoli	7	100	900	2 min, 8 °C	3.5–4/g	Navarro-Rico et al. (2014)
Yeasts and molds	Kailan-hybrid broccoli	7	100	900	2 min, 8 °C	3.5–4/g	Navarro-Rico et al. (2014)
Salmonella enteritidis	Kailan-hybrid broccoli	7	100	900	2 min, 5 °C	2.6/g	Martínez-Hernández et al. (2015)

(continued)

Table 4.3 (continued)

Microorganism	Origin	Physiochemical properties of EW			Washing conditions	Reduction (log CFU)	Reference
		pH	ACC (mg/L)	ORP (mV)			
Escherichia coli	Spinach	6.5–6.7	100	800–900	10 min, 25 °C	>4/mL	Guentzel et al. (2008)
Salmonella	Spinach	6.5-6.7	100	800–900	10 min, 25 °C	3.93–4.62/mL	Guentzel et al. (2008)
Psychrophilic bacteria	Baby spinach	6.7	1.9	521	1 min, 7 °C	about 1/g	Gómez-López et al. 2013
Coliform count	Carrot	7.54	146.53	798.4	5 min, 15 °C	0.97/g	Lee et al. (2014)
Aerobic bacteria count	Carrot	7.54	146.53	798.4	5 min, 15 °C	1.93/g	Lee et al. (2014)
Salmonella typhimurium	Tomato	7.52	155	760	5 min, 65 rpm	8.0/tomato	Afari et al. (2015)
Escherichia coli O157:H7	Tomato	7.52	155	760	5 min, 65 rpm	4.4/tomato	Afari et al. (2015)
Coliform count	Cabbage	7.54	146.53	798.4	5 min, 15 °C	1.51/g	Lee et al. (2014)
Aerobic plate count	Cabbage	7.54	146.53	798.4	5 min, 15 °C	2.17/g	Lee et al. (2014)
Total aerobic bacteria	Leek	7.87	30	–	–	0.43/g	Vandekinderen et al. (2009a, b, c)
Enterobacteriaceae	Chicory	–	200	–	5 min, 4 °C	1.41/g	Pinto et al. (2015)
Aerobic mesophilic bacteria	Minizuna baby leaves	7.0	410	–	2 min, 5 °C	<1/g	Tomás-Callejas et al. (2011)

"–" means the data wasn't shown in literature

Disinfection Efficacy of AlEW Combined with AEW on Fruits and Vegetables

AlEW, also known as electrolyzed reducing water, is usually used as a detergent for washing away substances that are hard to remove like proteins or fats (Hati et al. 2012). As for its application to disinfection, the number of *Listeria monocytogenes* inoculated in lettuce treated by AlEW was markedly lower than untreated ones during their whole lifespan and the maximum reduction reached 1 log CFU/g, according to Ding's study (Ding et al. 2009). Also, AlEW has been reported to effectively remove biofilms of *Staphylococcus aureus* but have little effect on killing cells in biofilms (Sun et al. 2012). The research group of Guilherme's suggested that AlEW (pH 11.02) could inhibit the growth of bacteria only at the initial period, whereas it could not show any eliminating effects in later period (Dilarri et al. 2016). Ovissipour et al. (2015) implied that *Escherichia coli O104:H4* in cell suspensions was more resistant to AlEW in contrast with AEW. Obviously, use of AlEW alone is limited. On the contrary, as mentioned in the preceding part, AEW has strong antimicrobial efficacy. Therefore, to make full utilization of AlEW and AEW, the combined use of them for disinfection has attracted researchers' attention in fruits and vegetables industry.

From Table 4.4, we can see the disinfection efficacy of AlEW alone or combined with AEW in various fruits and vegetables including avocados, lemon, lettuce, cilantro, green onion and tomato. With respect to the application to cilantro in the table, it is easy to notice that reduction of both total aerobic bacteria and yeasts and molds in samples treated with AlEW for 5 min followed by AEW for 5 min is 4–5 log more than that treated with AlEW alone. Moreover, the synergy of both EWs could reduce microbial counts existing in fruits and vegetables by 4–6 log, in general, which proved the strong antimicrobial ability of their combination.

Disinfection Efficacy of EW Combined with Other Techniques on Fruits and Vegetables

Although EW has become a promising disinfection method, sometimes the use of it individually has limited effectiveness on controlling the growth of microbes adhered to fruits and vegetables. To further enhance the disinfection efficacy and broaden antimicrobial action, it is essential to utilize hurdle approach combining EW with other physical, chemical or biological methods so that synergistic effects could be achieved while each of them shows lower intensities during the sequential or simultaneous process (Meireles et al. 2016). Meanwhile, the nutritional and sensory quality, as well as shelf life and economic values of fresh produce, could be improved (Khan et al. 2017; Luo et al. 2016). To date, EW in conjunction with several techniques have been successfully conducted for the sterilization of fruits and vegetables including ultrasound (US), microbubbles, ultraviolet light C (UV-C), organic acids, calcium oxide (CaO), and so on, as is depicted in Table 4.5.

Table 4.4 Disinfection efficacy of alkaline electrolyzed water alone or combined with AEW on fruits and vegetables

Microorganism	Origin	AlEW pretreatment	AEW treatment	Temperature	Reduction/log CFU	Reference
Listeria monocytogenes	Avocados	30 s	90 s	–	4.9/g	Rodriguez-Garcia et al. (2011)
Salmonella	Avocados	30 s	90 s	–	5.8/g	Rodriguez-Garcia et al. (2011)
Escherichia coli O157:H7	Avocados	30 s	90 s	–	4.9/g	Rodriguez-Garcia et al. (2011)
Escherichia coli O157:H7	Lemon	15 s	15 s	–	5.3/lemon	Pangloli (2009)
Total aerobic bacteria	Cilantro	5 min	–	At room temperature	1.06/g	Hao et al. (2015a, b)
Total aerobic bacteria	Cilantro	5 min	5 min	At room temperature	>6/g	Hao et al. (2015a, b)
Yeasts and molds	Cilantro	5 min	–	At room temperature	0.82/g	Hao et al. (2015a, b)
Yeasts and molds	Cilantro	5 min	5 min	At room temperature	>5/g	Hao et al. (2015a, b)
Escherichia coli O157:H7	Lettuce	5 min	–	22 °C	1.34/g	Park et al. (2008a)
Escherichia coli O157:H7	Lettuce	10 s	4.8 min	22 °C	>3.72/g	Park et al. (2008a)
Escherichia coli O157:H7	Lettuce	5 min	5 min	20 °C	1.82/g	Koseki et al. (2004)
Salmonella	Lettuce	5 min	5 min	20 °C	1.65/g	Koseki et al. (2004)
Escherichia coli O157:H7	Green onion	10 s	4.8 min	22 °C	4.69/g	Park et al. (2008b)
Salmonella typhimurium	Tomato	10 s	4.8 min	22 °C	>5.57/g	Park et al. (2008b)

"–" means the data wasn't shown in literature

Table 4.5 Disinfection efficacy of electrolyzed water alone or combined with other techniques on fruits and vegetables

Microorganisms	Origin	Treatments	Processing conditions	Reduction (log CFU)	Reference
Escherichia coli O157:H7	Apple	SAEW	23 °C, 3 min	2.28/apple	Tango et al. (2017)
Escherichia coli O157:H7	Apple	SAEW + fumaric acid	23 °C, 3 min	3.12/apple	Tango et al. (2017)
Listeria monocytogenes	Apple	SAEW	23 °C, 3 min	2.30/apple	Tango et al. (2017)
Listeria monocytogenes	Apple	SAEW + fumaric acid	23 °C, 3 min	3.00/apple	Tango et al. (2017)
Total aerobic bacteria	Cherry tomato	SAEW	–	1.45/g	Ding et al. (2015)
Total aerobic bacteria	Cherry tomato	SAEW + US	40 kHz, 240 W, 10 min	1.77/g	Ding et al. (2015)
Yeasts and molds	Cherry tomato	SAEW	–	1.10/g	Ding et al. (2015)
Yeasts and molds	Cherry tomato	SAEW + US	40 kHz, 240 W, 10 min	1.50/g	Ding et al. (2015)
Bacillus cereus	Potato	SAEW	40 °C, 3 min	2.3/g	Luo et al. (2016)
Bacillus cereus	Potato	SAEW + US	40 kHz, 400 W/L, 40 °C, 3 min	3.0/g	Luo et al. (2016)
Escherichia coli O157:H7	Cabbage	SAEW	23 °C, 3 min	1.22/g	Forghani and Oh (2013)
Escherichia coli O157:H7	Cabbage	SAEW + US	40 kHz, 400 W, 23 °C, 3 min	2.5–3.0/g	Forghani and Oh (2013)
Total bacteria count	Cabbage	AlEW	50 °C, 5 min	3.08/g	Rahman et al. (2010a, b, c)

(continued)

Table 4.5 (continued)

Microorganisms	Origin	Treatments	Processing conditions	Reduction (log CFU)	Reference
Total bacteria count	Cabbage	AlEW + citric acid	50 °C, 5 min	3.98/g	Rahman et al. (2010a, b, c)
Listeria monocytogenes	Lettuce	SAEW	23 °C, 3 min	1–1.5/g	Forghani and Oh (2013)
Listeria monocytogenes	Lettuce	SAEW + US	40 kHz, 400 W, 23 °C, 3 min	1.5–2.0/g	Forghani and Oh (2013)
Escherichia coli O157:H7	Lettuce	NEW	40 °C, 3 min	2.32/g	Forghani et al. (2013)
Escherichia coli O157:H7	Lettuce	NEW—US	NEW washing for 3 min followed by US (40 kHz, 400 W/L) for 3 min, 40 °C	2.47/g	Forghani et al. (2013)
Escherichia coli O157:H7	Lettuce	US—NEW	US (40 kHz, 400 W/L) for 3 min followed by NEW washing for 3 min	3.18/g	Forghani et al. (2013)
Salmonella typhimurium	Romaine lettuce	NEW	5 min, 100 rpm	2.48/g	Afari et al. (2016)
Salmonella typhimurium	Romaine lettuce	NEW + US	5 min, 100 rpm, 210 W	3.25/g	Afari et al. (2016)
Listeria monocytogenes	Spinach	SAEW	23 °C, 3 min	1–1.5/g	Forghani and Oh (2013)
Listeria monocytogenes	Spinach	SAEW + US	40 kHz, 400 W, 23 °C, 3 min	1.5–2.0/g	Forghani and Oh (2013)
Staphylococcus aureus	Spinach	SAEW	23 °C, 3 min	2–3/g	Ngnitcho et al. (2017)
Staphylococcus aureus	Spinach	SAEW + fumaric acid	23 °C, 3 min	3–4/g	Ngnitcho et al. (2017)

(continued)

Table 4.5 (continued)

Microorganisms	Origin	Treatments	Processing conditions	Reduction (log CFU)	Reference
Staphylococcus aureus	Spinach	CaO - SAEW + fumaric acid	3 min CaO washing followed by SAEW simultaneous with famaric acid washing for 3 min, 23 °C	Approximately 4/g	Ngnitcho et al. (2017)
Escherichia coli O157:H7	Tomato	NEW	5 min, 100 rpm	4.76/tomato	Afari et al. (2016)
Escherichia coli O157:H7	Tomato	NEW + US	5 min, 100 rpm, 210 W	8.44/tomato	Afari et al. (2016)
Salmonella typhimurium	Bell pepper	SAEW	23 °C, 3 min	1.19/g	Luo and Oh (2016)
Salmonella typhimurium	Bell pepper	SAEW + US	40 kHz, 400 W/L, 60 °C, 3 min	3–4/g	Luo and Oh (2016)
Escherichia coli	Sweet basil	AEW	5 min	<1/g	Klintham et al. (2017)
Escherichia coli	Sweet basil	AEW + microbubbles	5 min	<1/g	Klintham et al. (2017)
Listeria monocytogenes	Kale	SAEW	40 °C, 3 min	2–3/g	Mansur and Oh (2015)
Listeria monocytogenes	Kale	SAEW + US	40 kHz, 400 W/L, 40 °C, 3 min	2–3/g	Mansur and Oh (2015)
Salmonella enteritidis	Kailan-hybrid broccoli	NEW	8 °C, 2 min	2.60/g	Martínez-Hernández et al. (2015)
Salmonella enteritidis	Kailan-hybrid broccoli	NEW + UV-C	8 °C, 2 min, 7.5 kJ/m^2	About 3/g	Martínez-Hernández et al. (2015)

US ultraosound; *UV–C* ultraviolet light C; "–" means the data wasn't shown in literature

US techniques have bactericidal effects because it could generate many cavitation bubbles which would further motivate mechanical, chemical and heat effects and consequently cause the death of microbes (Zhou et al. 2015; Zou and Wang 2017). Nevertheless, it could not reach complete lethality unless high ultrasonic intensity is exerted. Ultraviolet light (UV) has the ability of disrupting the transcription and translation of DNA, causing its damage and eliminating cells (Birmpa et al. 2013). But it is hard for UV to penetrate opaque fruits or vegetables and kill microbes on total surface when they are stacked together (Alexandre et al. 2012). Although organic acids have been widely used as a preservative in the food industry, their cost is expensive and high acidity is corrosive to equipments (Meireles et al. 2016). CaO has been considered as a natural antimicrobial due to its hydration and the releasing hydroxyl ions when it reacts with water, whereas the disinfection efficacy on pathogens could only reach a reduction of 1–2 log CFU when single CaO treatment was added (Tango et al. 2017). These limitations may be overcome if they are combined with EW.

According to Table 4.5, a number of investigations have shown the significant improvement in disinfection efficacy on various fruits and vegetables treated with hurdle technology. In general, EW combined with other techniques could make an extra reduction of 0.5–2 log of microorganisms in comparison with using it alone. Among them, the most frequently associated technique is ultrasound which could be performed simultaneously with EW treatment and also prior to or followed by it. And, changes in their sequence would lead to the varying of antimicrobial efficacy (Forghani et al. 2013). Appropriate treatment temperature and time should be selected to achieve the maximum microbial reduction in spite of their combination (Luo et al. 2016). Furthermore, more hurdles EW is combined which tend to result in a better disinfection efficacy (Ngnitcho et al. 2017). Finally, although a few studies revealed that little enhancement had been found on eliminating microorganisms when EW is combined with microbubbles, it is still necessary for microbubbles to maintain the freshness of leafy vegetables (Klintham et al. 2017). Anyway, hurdle technology will add brilliance to the present conditions of EW in fruits and vegetables industry.

4.2.2 Application of EW in Postharvest Storage of Fruits and Vegetables

4.2.2.1 Effect of EW Treatment on Quality Attributes of Fruits and Vegetables After Harvest

It is strongly evident that sensory attributes including appearance, flavor, and texture as well as nutritional values are fundamental elements responsible for the acceptability of fresh fruits and vegetables (Brasil and Siddiqui 2018). As metabolically active products, fresh fruits and vegetables during postharvest period could still undergo a series of physiological processes such as ripening (Giovenzana et al. 2015). And quality loss will easily occur if inappropriate treatments are exerted. Although EW could ensure microbial safety of fresh produce to some degree, it would be worthless if the use of it had adverse impact on physical properties or chemical components of fruits and vegetables. Therefore, it is essential for reviewing the effects of EW on these quality attributes to seek out better parameters so that shelf life of them could be extended and consumers' demand for nutritional and fresh-like food would be satisfied.

Effect of EW Treatment on the Physical Properties of Fruits and Vegetables After Harvest

Associated with tissue deterioration, changes of texture can be used to evaluate the freshness or quality decline of fruits and vegetables (Toivonen and Brummell 2008). Firmness or hardness is one of the crucial attributes reflecting the texture of them, the study of which has aroused many researchers' concern in terms of improving the acceptability of products (Hampson et al. 2014). Stored at 4 °C, sweet cherry had an increasing firmness in the first 10 days and then it decreased during storage when they were treated with EW (ACC of 25–400 mg/L) (Eda and Seckin 2015). However, treating shredded carrots with AlEW combined with citric acid, the firmness kept increasing in 15 days at the same temperature (Rahman et al. 2011). Firmness of cilantro under AEW treatment (pH 2.45, ACC 16.8 mg/L) was consistently lower than that treated by tap water during 2 weeks' preservation (Wang et al. 2004). On the contrary, our previous study implied that SAEW treatment (pH 5.85, ACC 19.46 mg/L) could enhance the firmness of cilantro in comparison with control during 14 days storage (Hao et al. 2015b). And, Chen et al. washed blueberries with AEW (pH 2.8, ACC 48 mg/L), indicating that their firmness after 15 days storage at 4 °C remained higher than untreated ones. They also found that the reason why AEW treatment delayed the softening of fruits was that it deactivated polygalacturonase, cellulase, and β-galactosidase contributing to the degradation of cell wall components (Chen et al. 2017). The combination of AlEW (pH 11.28, ACC 1.83 mg/L) and UV-C treatment made the firmness of date palm reach 2.10 N after 30 days preservation at 20 °C, which was above initial value (1.72 N) (Jemni et al. 2014a). Besides, EW treatment (ACC 400 mg/L) could increase the hardness of grape skin by 0.028–0.068 mJ after 24 h storage at 20 °C (Laureano et al. 2016).

Electrolyte leakage (EL) has been considered as an indicator of membrane integrity of fruits and vegetables, which is especially true when they are exposed to environmental stresses (Rolny et al. 2011). Fresh produce during senescence or getting damaged tends to have higher membrane permeability that can be characterized by the increase of EL (Antunes and Sfakiotakis 2008; Jungklang et al. 2017). So EL is also a reflection of plant tissues' quality. Markedly lower EL values (55.32%) of mushroom were achieved when it was treated by EW (ACC 25 mg/L) after 12 days of storage at 4 °C (Aday 2016). Having been stored for 44 days at 2 °C, peaches washed by EW (pH 6.1, ACC 21 mg/L) had an EL value that was 91.3% of control (Zhou et al. 2012). Therefore, washing fruits and vegetables with EW may have a positive effect on reducing the decline of EL. However, Wang et al. concluded that AEW (pH 2.45, ACC 16.8 mg/L) treatment impaired tissues of cilantro and the EL reached 28.33 μs/cm before storage that was apparently higher than control (15.78 μs/cm) (Wang et al. 2004), which was similar to the results of our research group. Moreover, our previous study also showed that treating cilantro with SAEW may lead to lower EL in comparison with AEW (Hao et al. 2015b).

Color is an another primary physical property constituting the sensory quality of fruits and vegetables and it can directly influence their market values and plays a pivotal role in customers' choosing desire for goods (Ma et al. 2017). EW treatments have varying effects on visual quality of fruits and vegetables, and three parameters including hue, chroma, and total color difference (ΔE) are often used to assess the color changes of them. For instance, several studies found no significant difference in these chromatic parameters as to lettuce, broccoli, strawberry, and date palm before and after EW washing (Hung et al. 2010a; Jemni et al. 2014b; Park et al. 2001). However, significant difference was observed in ΔE for sliced carrots: it was 4.0 after washing by tap water at 18 °C, whereas it only reached 2.1 in terms of SAEW treatment (pH 5.5, ACC 23.0 mg/L) at the same temperature (Koide et al. 2011). There were also a few investigations revealing that sensory quality of broccoli sprouts could be improved due to their brighter color after AEW (pH 3.66, ACC 230 mg/L) washing for 20 s (Puligundla et al. 2018). Additionally, browning is a major concern associated with the shortened shelf life of fresh-cut fruits and vegetables. And the appearance of brown color is mainly caused by polyphenol oxidase (PPO) that plays a vital role in catalyzing phenols into colorless quinines subsequently polymerizing a black compound (Sun et al. 2018). EW treatment may lessen the PPO activity and inhibit samples' browning. The study of Zhi reported a decreased PPO activity and browning index of peach presoaked in SAEW (pH 5.91, ORP 904 mV) after 4 days of storage at 25 °C (Zhi et al. 2017), which was in agreement with Zhou's assay suggesting that SAEW treated peaches had a PPO activity 27.9% lower than control at the end of 44 days preservation at 2 °C and its lightness and hue values were 2.10–2.22 times as great as control while chroma was 12–25% of the control value by the 44th day (Zhou et al. 2012). Besides, similar results were also found on apples (Kim and Hung 2014; Wang et al. 2007) and lotus roots (Li et al. 2017) and thus proved that appropriate EW treatment could be considered a means of alleviating the deterioration of the sensory quality of fruits and vegetables.

Effect of EW Treatment on the Chemical Components of Fruits and Vegetables After Harvest

As known to us, fruits and vegetables are integral parts of human diet which supply various nutrients including vitamins, minerals, and dietary fiber as well as other bioactive compounds such as phenolics that have many health promoting benefits (Dufour et al. 2018; Ramsay et al. 2017). However, it has not been summarized in literature that EW treatment has any impact on these chemical components of fruits and vegetables during postharvest storage. So, there is an urgent need for more information about EW treatment conditions favorable for the maintenance or elevation of these components contents.

As a hydrophilic antioxidant, vitamin C present in foods has two forms: one is ascorbic acid (AA), a major compound with vitamin C activity, and the other is dehydroascorbic acid (DHAA) that could convert to AA intracellularly (Phillips et al. 2016). Vitamin C could scavenge various radicals, alleviate oxidative stress (Mock et al. 2017), regulate the expression of cancer epigenome (Gillberg et al. 2018) and prevent DNA damage and apoptosis (Chen et al. 2017). Therefore, it is of vital importance to analyze the changes of it in fruits and vegetables treated by EW. Reductions of 10.8% in AA content were observed when sliced carrots were washed by SAEW (pH 5.5, ACC 23.0 mg/L, 18 °C), obviously lower than that treated by tap water (12.2%) (Koide et al. 2011). Washing Iceberg lettuce with NEW (pH 7.50, ACC 4.5 mg/L) decreased the loss of vitamin C by 0.38 mg/100 g fresh weight compared with tap water (Vandekinderen et al. 2009a). And as to white cabbage treated by NEW (pH 7.87, ACC 31.7 mg/L), the decrease reached 0.43 mg/100 g fresh weight. Although washing white cabbage with NEW (pH 7.50, ACC 4.9 mg/L) caused an additional loss of vitamin C in contrast to that with tap water, there was no significant difference in its contents (Vandekinderen et al. 2009a, b, c). Hence, it seems that EW treatment has no adverse impact on samples' nutritional quality with regard to vitamin C loss.

β-carotene, a member of carotenoids family, exists in many fruits and vegetables, adequate intake of which has numerous physiological functions (Calligaris et al. 2018), such as adjuvant therapy for multidrug-resistant cancer (Teng et al. 2016), antioxidant property, preventing cardiovascular diseases (Gómez-Mascaraque et al. 2017), and so on. Also, it could impart an attractive color to food (Lin et al. 2018). According to literature, no marked difference had been reported in relation to β-carotene in carrots treated by SAEW and tap water at 18 °C or 45 °C (Koide et al. 2011). Furthermore, the research group of Isabelle's indicated that NEW had no special effect on all-trans-β-carotene content in both Iceberg lettuce and white cabbage with NEW washing (Vandekinderen et al. 2009a, b, c).

Phenolic compounds have been confirmed to have various benefits to human health including antioxidant activity, lowering the occurrence of cardiovascular disease, inhibition of neurodegeneration, anticancer activity, and so forth (Del et al. 2013; Shahidi and Ambigaipalan 2015). Originating from secondary metabolites in plant tissues, they can be accumulated through many abiotic stresses, such as water (Qin et al. 2017). And EW is without exception, which could be exemplified by Tomas-Callejas' experiments suggesting that there was a slightly increasing trend for the content of total phenolics in mizuna baby leaves treated by NEW (pH 7.0, ACC 410 mg/L) during 11 days of storage at 5 °C and the eventual content, remarkably higher than that of deionized water, increased by 33% compared to the initial value (Tomás-Callejas et al. 2011). Besides, the total phenolics content in kailan-hybrid broccoli washed with AEW (ACC 100 mg/L) reached its summit (about 800 mg gallic acid equivalent/kg fresh weight) at an earlier period (the 5th day) and it kept quiet stable throughout the storage (19 days) and it constantly prevail over the NaClO-treated ones (Navarro-Rico et al. 2014). However, there were also literature documenting that EW washing could not bring any differential effects on these functional constituents in broccoli sprout and date palm in contrast to other sanitizing solutions (Jemni et al. 2014b; Puligundla et al. 2018).

The accumulated starch during maturation of fruits tends to degrade into low-molecular sugars in the subsequent process of ripening. And, the sugar content can be regarded as an indicator of the maturity of fruits (Prasad et al. 2018). EW treatment decreased total sugars content of date palms from 76.55–78.56% to 62.08–64.61% after 30 days preservation at 20 °C and reducing sugars including glucose and fructose declined by 0.32–5.10% (Jemni et al. 2014b). In comparison with untreated ones, no special effect was found on both sucrose and reducing sugar contents of date fruits washed with NEW (pH 7.15, ORP 835 mV) (Bessi et al. 2014). However, spraying sugarcane with EW could alleviate the inversion of sucrose into reducing sugars, the content of which only increased by 13.84 folds after 168 h storage in comparison with the initial value while that of control was 23.98 times more than that before storage (Solomon and Sugar 2009). Moreover, EW could regulate the utilization for sugars of yeasts during alcoholic fermentation. For instance, AEW treatment (pH 2.7, ACC 102 mg/L, ORP 863 mV) soaking persimmons enhanced the content of reducing sugar in wine to 11.93 g/L, obviously higher than pure water treatment (5.13 g/L), via suppressing the fermentation activity of yeasts (Zhu et al. 2016).

Undoubtedly, acidity is an important attribute determining the flavor of fruits which could influence consumers' preference (Etienne et al. 2013). During the processing of fruits and vegetables, sterilization is an unavoidable procedure to ensure microbial safety and products with higher acidity (pH below 4.6) do not need to be exposed to drastic thermal conditions so that sensory and nutrition loss could be minimized for these foods (Anthon and Barrett 2012). And acidity is often evaluated with titratable acidity (TA) and pH. Basically, the use of EW had no noticeable effect on both TA and pH compared to other treatments in the case of date fruits, date palm, and strawberry (Bessi et al. 2014; Hung et al. 2010a, b; Jemni et al. 2014b). Monia et al. also confirmed that the pH values of date palms treated by EW slightly decreased with the prolonging storing time within a month but no significant changes were found on TA (Jemni et al. 2014b). In contrast, TA value of date palm treated by EW combined with UV-C rose by 0.25–0.5 g citric acid/kg fresh weight through 30 days storage and pH decreased by 0.63–0.72 (Jemni et al. 2014b).

4.2.2.2 Application of EW in Postharvest Physiology and Disease Control of Fruits and Vegetables

Admittedly, physiological activities including respiration, transpiration, maturation, ripening, and senescence could affect the postharvest quality of fruits and vegetables. For example, water in fresh produce tends to escape from its surface via transpiration when the water potential of ambient atmosphere is lower, thus causing the decline of freshness (Favre et al. 2018; Jalali et al. 2017). Another commonly occurring problem is the disease or decay of them, which is mainly responsible for the quality loss throughout shelf life and eventually poses economic loss (Martínez-Hernández et al. 2017; Mditshwa et al. 2017). Hence, significance should be attached to selecting proper treatment conditions for alleviating the appearance of these symptoms. When it comes to EW, more researches are focused on its effect on respiration and disease/decay control of fruits and vegetables, which is elaborated in the following paragraphs.

Effect of EW Treatment on the Respiration of Fruits and Vegetables During Postharvest Storage

Respiration is one of the essential metabolic processes involved in the breakdown of complex substances including sugars and organic acids into simpler molecules along with the generation of energy and medium molecules so that living cells are able to maintain a series of anabolic reactions to keep the integrity of cellular organizations and membranes. Inappropriate control of respiration does not only cause weight loss impairing the appearance of produce, but also induces physiological decay resulting from anaerobic fermentation in conditions of lower O_2 level (Jalali et al. 2017; Xanthopoulos et al. 2017). EW treatment, as an external stress, has diversified effects on respiration of fruits and vegetables. For example, SAEW treatment (pH 5.91, ORP 904 mV) diminished the respiration rate and postponed the occurrence of climacteric peak of peaches during 8 days of ripening at 25 °C (Zhi et al. 2017). Furthermore, Rico et al. implied that the higher ACC (12–120 mg/L) of EW was, the lower respiration rate of lettuce was within 24 h after packaging at 4 °C, which may be explained by the inhibition of microbial spoilage. But little noticeable difference was found on the 7th day of storage (Rico et al. 2008). By contrast, there was no significant difference before the 5th day of storage at 5 °C between AEW and NEW (ACC 40–100 mg/L) treated mizuna baby leaves and control while minor difference appeared on the 7th day that the respiration rate of AEW or NEW treated ones were slightly lower than control (Tomás-Callejas et al. 2011). Meanwhile, some investigations also demonstrated that changes of respiration rate during refrigerated storage could be avoided when EW was used as a pretreatment for disinfection of shredded cabbages and broccoli (Gómez-López et al. 2007; Navarro-Rico et al. 2014).

Application of EW in Postharvest Disease Control of Fruits and Vegetables

It is estimated that losses of fruits and vegetables after harvest have reached more than one-third of the total number, most of which is due to disease or decay particularly caused by fungal infections during the sequential handling before preservation (Romanazzi et al. 2016a; Sardella et al. 2017). The use of fungicide, a traditional strategy for managing the disease, has been eliminated in food market because of its potential risks including induced resistance of the chemicals, the residues of toxic by-products together with increasing health concerns (Romanazzi et al. 2016b). Therefore, efforts have been made to search for alternatives, one of which is EW that has displayed its efficacy on controlling disease associated with *Fusarium* (Khayankarn et al. 2013), *Monilinia*, *Botrytis* (Guentzel et al. 2010), *Penicillium* (Fallanaj et al. 2016; Whangchai et al. 2010), *Botryosphaeria* (Al-Haq et al. 2002), Galactomyces, and *Alternaria* (Vasquez-Lopez et al. 2016).

Specific examples have been listed in Table 4.6 where we can notice that EW treatment has been successfully conducted to lower the incidence of postharvest disease of many products with artificially inoculated microorganisms including pineapple, peach, grape, tangerine, orange, pear, and tomato. Briefly, the reduction varies from 20 to 90% after 7–28 days of storage of samples dipped or sprayed with EW. And, the difference of treatments efficacy might lie in physiochemical properties of EW, dipping or spraying time, the resistance of pathogens to stress ,and so on. Furthermore, as apparently reflected in the table, EW treatment had no inhibiting effect on the lesion formation of wounded apples inoculated with *Penicilium expansum*. But it could avoid cross contamination, specifically, incidence of decayed apples infected with this pathogen from washing water was only 10.2% of that treated with distilled water, and 30% of NaClO (pH 10.2, ACC 198 mg/L, ORP 755 mV) treated one (Okull and Laborde 2004). Thereby, EW has a promising prospect in controlling disease of fruits and vegetables during storage.

4.2.3 Other Applications of EW in Fruits and Vegetables Industry

4.2.3.1 Application of EW to Sprouts Vegetables

Recently, sprouts vegetables have been widely consumed as an important source of many nutrients including free amino acids, minerals, vitamins, dietary fiber as well as polyphenols in Asia, Europe, and the United States (Gan et al. 2017; Nam et al. 2015; Ren and Sun 2014). Researches on sprouts cultivated from various edible seeds have been constantly mentioned in literature, which is also what our research group is focused on. EW treatment, an environmental factor, has shown its effects on both the growth and the final products of sprouts vegetables. Overall, they can be categorized into three parts, namely, inhibitory effect on the microbial growth, influence on the morphological features and accumulation of functional components of sprouts.

Inhibitory Effect of EW on the Microbes Attached to Sprouts

Microbial growth can be easily found in sprouting process due to its warm, moist and nutritious environment (Butscher et al. 2016). Consumed raw to maintain higher nutritional value, sprouts carrying pathogens are potentially hazardous to humans' health because of the frequently occurring foodborne illness (Erdozain et al. 2013; Millan-Sango et al. 2017). Just as discussed in the preceding parts, disinfection efficacy of EW has also been observed in the production of sprouts vegetables. From

Table 4.6 Application of electrolyzed water in postharvest disease control of fruits and vegetables

Symptom	Origin	Related pathogens	Physiochemical properties of EW			Storage conditions	Disease control effect (%)	Reference
			pH	ACC (mg/L)	ORP (mV)			
Decay	Pineapple	*Fusarium* sp.	–	100	–	13 °C, 15 d	About 50	Khayankarn et al. (2013)
Brown rot	Peach	*Monilinia fructicola*	6.3–6.5	250	800–900	25 °C, 16 d	About 80	Guentzel et al. (2010)
Gray mold	Grape	*Botrytis cinerea*	6.3–6.5	250	800–900	25 °C, 28 d	About 21	Guentzel et al. (2010)
Decay	Tangerine	*Penicillium digitatum*	3.9	102	–	5 °C, 18 d	80–90	Whangchai et al. (2010)
Green mold	Orange	*Penicillium digitatum*	9.0	30	–	20 °C, 7 d	92	Fallanaj et al. (2016)
Rot	Pear	*Botryosphaeria berengeriana*	2.6	250	1170	20 °C, 7 d	15.8–52.6	Al-Haq et al. (2002)
Rot	Tomato	*Fusarium oxysporum*	7.0	60	850	22 °C, 8 d	40	Vasquez-Lopez et al. (2016)
Rot	Tomato	*Galactomyces geotrichum*	7.0	60	850	22 °C, 8 d	57.1	Vasquez-Lopez et al. (2016)
Rot	Tomato	*Alternaria* sp.	7.0	60	850	22 °C, 8 d	33.3	Vasquez-Lopez et al. (2016)
Decay	Apple	*Penicilium expansum*	3	80	1154	25 °C, 6 d	0	Okull and Laborde (2004)

"–" means the data wasn't shown in literature. Disease control effect (%) = (disease incidence of samples in control—disease incidence of samples treated with EW)/disease incidence of samples in control

Table 4.7, we are aware that the number of total (aerobic) bacteria or yeast and mold on sprouts treated with EW after about 1 week's germination was 1–2 log lower than that watered by tap water in general. Additionally, EW could eliminate the inoculated pathogens on sprouts vegetables products to an acceptable level, such as *Escherichia coli O157:H7*, *Salmonella enteritidis*, and *Listeria monocytogenes* (Phua et al. 2014; Zhang et al. 2011). Furthermore, our research group have demonstrated that the antimicrobial effect of EW increased with the increase of its ACC when it was applied to the disinfection of brown millet sprout (Li et al. 2015), brown rice sprout (Liu et al. 2013), and broccoli sprout (Li et al. 2018). Thus, the role of ACC of EW in sterilization has been clarified over and over again. Meanwhile, a few authors emphasized the importance of pH with the fact that AEW (pH 4.47) showed the strongest antimicrobial capacity for washing mung bean sprouts in comparison with AEW possessing pH values of 3.02, 558, and 6.51 (Liu and Yu 2017).

Influence of EW on the Morphological Features of Sprouts

Growth conditions of sprouts vegetables could be directly perceived by their morphological features including sprouts length, hypocotyl length as well as hypocotyl diameter. Spraying them with EW during germination has varying influence on these indices for different species. Specifically, SAEW (pH 6.5, ACC of 20 and 30 mg/L) could greatly elevate the length of radish sprouts by 1.21–2.08 cm in comparison with tap water treatment after a week's germination (Zhang et al. 2016a). Similar growth promoting effects were also achieved for brown rice sprouts treated with AEW (pH 3.27, ACC 17.76 mg/L) (Liu et al. 2013) and mung bean sprouts sprayed with SAEW (pH of about 5.8, ACC of 10–30 mg/L) (Liu et al. 2014). However, adverse impact on the growth was found when SAEW was applied to treat brown millet sprouts that had significantly shorter length than control (Li et al. 2015), which was consistent with our latest study showing that SAEW treatment decreased the hypocotyls length of broccoli sprouts by 59.6% in comparison with tap water sprayed ones (Li et al. 2018). Besides, our research group ever concluded that mung bean sprouts watered by EW with pH values of about 2, 7.88, or 10.08 had shortened hypocotyls length while that whose pH was 4–6.5 was favorable for their growth, and moderate ACC (10–30 mg/L) could produce sprouts with longer hypocotyls (Rui et al. 2011). Furthermore, increasing ACC values within the range of different SAEWs (ACC 10–50 mg/L) meant the decline of hypocotyls length of broccoli sprouts (Li et al. 2018). Still, the special effects have not been observed in the above-mentioned three parameters of buckwheat sprouts when SAEW treatment was cxerted (Hao et al. 2016a, b). And changes of pH or ACC of EW did not lead to any difference in hypocotyls diameter of mung bean sprouts (Rui et al. 2011).

Table 4.7 Reduction of the microbes attached to sprouts sprayed with electrolyzed water during germination

Microorganisms	Origin	Physiochemical properties of EW			Conditions of sprouting process	Reduction (log CFU/g)	Reference
		pH	ACC (mg/L)	ORP (mV)			
Total aerobic bacteria	Radish sprouts	6.5	50	–	22–25 °C, 7 d	1.9	Zhang et al. (2016a)
Yeast and mold	Radish sprouts	6.5	50	–	22–25 °C, 7 d	2.19	Zhang et al. (2016a)
Total bacterial count	Buckwheat sprouts	5.95	20.25	–	20 °C, 8 d	1.12	Hao et al. (2016a, b)
Total bacterial count	Brown millet sprouts	5.96	30.35	834	25 °C, 3 d	0.97	Li et al. (2015)
Total aerobic bacteria	Brown rice sprouts	5.95	29.42	873.85	25 °C, 4 d	2.54	Liu et al. (2013)
Total bacterial count	Broccoli sprouts	5.47	40.8	–	25 °C, 8 d	About 1.5	Li et al. (2018)
Total bacterial count	Mung bean sprouts	5.79	18.87	–	23 °C, 6.5 d	1.06	Rui et al. (2011)

"–" means the data wasn't shown in literature

Effect of EW on the Accumulation of Bioactive Compounds in Sprouts

Sprouts commonly have more functional components than their original seeds, such as vitamins, minerals, phenolics, and γ-aminobutyric acid (GABA) (Pongrac et al. 2016; Tsurunaga et al. 2013; Tuan et al. 2013). The synthesis of some secondary metabolites like flavonoids may be modified by EW as an external abiotic elicitor throughout germination. And, it has aroused considerable concerns about whether EW could promote the accumulation of these substances in sprouts due to their numerous health benefits.

GABA, a nonprotein amino acid, has multiple physiological functions on health including lowering blood pressure (Diana et al. 2014; Kimura et al. 2002), improving immunity (Abdou et al. 2006), relieving oxidative stress (Xie et al. 2014), and mediating antiobesity effects (Imam et al. 2014), etc. SAEW (pH 5.83, ACC 20.3 mg/L) spraying made the content of GABA in buckwheat sprouts reach 143.20 mg/100 g after 6 days of germination, approximately 14 times as high as that in seeds (10.34 mg/100 g) and 22.4% more than control (117.04 mg/100 g) (Hao et al. 2016a, b). Similar increasing tendency has also been found in germinated brown millet (Li et al. 2015) and brown rice (Liu et al. 2013) with SAEW treatment, whose GABA content was enhanced by 2–3 times compared with their initial values and 3.61–23.2% higher than tap water treated ones. And, the accumulating mechanism could be attributed to the increased glutamic aciddecarboxylase activity which could effectively catalyze the conversion of glutamic acid into GABA, especially in acidic environment (Hao et al. 2016a, b; Liu et al. 2013).

As a group of dietary phenolics in plants, flavonoids can be also accumulated during the growth of buckwheat and broccoli sprouts, and these compounds have been clarified to improve cognitive function, show anti-inflammatory effects (Jaeger et al. 2017), lower serum cholesterol level (Burkard et al. 2017) and have anticonvulsant activity (Nieoczym et al. 2014). SAEW (pH 5.61, ACC 20.4 mg/L) had been considered as a useful means for increasing the content of total flavonoid content in broccoli sprouts, which reached a climax of 1.22 mg rutin equivalents/g dried weight among SAEW of various ACCs (10–50 mg/L) and tap water treatments (Li et al. 2018). Regarding individual flavonoids, significant higher content of rutin (739.9 mg/100 g) in buckwheat sprouts was revealed on the 8th day of germination when SAEW (pH 5.83, ACC 20.3 mg/L) was exerted, 1.29 times as many as control. Furthermore, our former studies suggested that the rutin accumulation under EW stress should be ascribed to higher activity of phenylalanine ammonialyase that was a crucial enzyme in phenylpropanoid secondary metabolic pathway (Hao et al. 2016a, b).

4.2.3.2 Application of EW to Preventing Moth Infestation of Fruits

Insect infestation is one of the serious problems occurring in the utilization of fruits, which does not only cause the decline of yield, but also lead to economic losses (Ben-Amor et al. 2016). Recently, more investigations have been focused on methods controlling moth of pyrale parasitic on date palms. Methyl bromide, a conventional fumigant, is usually used for managing pests attack on fruits. Nonetheless, the use of it is limited due to the harm to both human health and environment (Jemâa et al. 2013). As a more eco-friendly and safer alternative, EW has exhibited its capacity in effectively inhibiting moth proliferation on date palms. As Monia et al. reported, NEW pretreatment decreased the rate of date palms infested by *Ectomyelois ceratoniae* Zeller after 30 days of shelf life at 20 °C to 9.00%, obviously lower than control (26.67%) and other treatments including AEW, NaClO, and ozone gas (12–15%) (Jemni et al. 2014b). However, more extensive investigations should be launched when it refers to the eliminating efficacy of moth infestation of EW on other fruits and vegetables.

4.3 Developing Prospects and Challenges

This chapter has reviewed recent advance in EW applied to fruits and vegetables industry, elaborating on its application to disinfection and pesticide removal during their processing and disease control along with inhibiting moth infestation throughout storage as well as the effect on physiochemical property, chemical components, and postharvest physiology. Regarding sprouts vegetables, the investigating focus of our research group, the article has discussed the particular application of EW in them to make our achievements clearly understood.

Admittedly, more importance will be attached to the progress of fruits and vegetables industry in the forthcoming years and novel alternatives for traditional methods are in crying needs to eliminate food safety hazards and extend shelf life of fresh produce. EW has been proved to be a promising choice owing to its sufficient obliterating efficacy on spoilage or pathogenic microorganisms, pesticide residues and some insect pests existing on fruits and vegetables without affecting sensory and nutritional quality of them in most occasions. Besides, the employment of EW to sprouts vegetables makes those healthful components such as CABA and flavonoids accumulated, which could promote the development of functional foods and further broaden the application field of EW. Additionally, thanks to its lower producing costs, there will be little difficulty in putting EW into large-scale industrial production.

On the other hand, at some point, EW appropriately combined with other methods will ensure better quality of products. So based on advantages of EW, the addition of other novel techniques such as ultrasound and UV-C should be considered when necessary. Furthermore, more extensive trials should be conducted to hunt for optimal EW processing conditions (such as washing time and temperature) on various fruits and vegetables in order to minimize quality loss, get the utmost out of resources and make the process more efficient. At the same time, more researches are essential to monitor changes in physiochemical properties of EW during their contact with fruits and vegetables, thus further clarifying the role of these characteristic factors (including pH, ACC and ORP). Anyway, the status of EW in fruits and vegetables industry will attract more attention as it has a bright prospect.

References

Abadias M, Usall J, Oliveira M et al (2008) Efficacy of neutral electrolyzed water (NEW) for reducing microbial contamination on minimally-processed vegetables. Int J Food Microbiol 123(1–2):151–158

Abdou A, Higashiguchi S, Horie K et al (2006) Relaxation and immunity enhancement effects of gamma-aminobutyric acid (GABA) administration in humans. BioFactors 26(3):201–208

Aday MS (2016) Application of electrolyzed water for improving postharvest quality of mushroom. Food Sci Technol 68:44–51

Afari GK, Hung Y, King CH (2015) Efficacy of neutral pH electrolyzed water in reducing *Escherichia coli* O157:H7 and *Salmonella typhimurium* DT 104 on fresh produce items using an automated washer at simulated food service conditions. J Food Sci 80(8):M1815–M1822

Afari GK, Hung YC, King CH et al (2016) Reduction of *Escherichia coli* O157:H7 and *Salmonella typhimurium* DT 104 on fresh produce using an automated washer with near neutral electrolyzed (NEO) water and ultrasound. Food Control 63:246–254

Alexandre EMC, Brandão TRS, Silva CLM (2012) Emerging technologies to improve the safety and quality of fruits and vegetables, 261–297

Al-Haq MI, Seo Y, Oshita S et al (2002) Disinfection effects of electrolyzed oxidizing water on suppressing fruit rot of pear caused by *Botryosphaeria berengeriana*. Food Res Int 35(7):657–664

Al-Holy MA, Rasco BA (2015) The bactericidal activity of acidic electrolyzed oxidizing water against *Escherichia coli* O157:H7, *Salmonella typhimurium*, and *Listeria monocytogenes* on raw fish, chicken and beef surfaces. Food Control 54:317–321

Anthon GE, Barrett DM (2012) Pectin methylesterase activity and other factors affecting pH and titratable acidity in processing tomatoes. Food Chem 132(2):915–920

Antunes MD, Sfakiotakis EM (2008) Changes in fatty acid composition and electrolyte leakage of 'Hayward' kiwifruit during storage at different temperatures. Food Chem 110(4):891–896

Arevalos-Sánchez M, Regalado C, Martin SE et al (2012) Effect of neutral electrolyzed water and nisin on *Listeria monocytogenes* biofilms, and on listeriolysin O activity. Food Control 24(1–2):116–122

Arevalos-Sánchez M, Regalado C, Martin SE et al (2013) Effect of neutral electrolyzed water on lux-tagged *Listeria monocytogenes* EGDe biofilms adhered to stainless steel and visualization with destructive and non-destructive microscopy techniques. Food Control 34(2):472–477

Athayde DR, Flores DRM, da Silva JS et al (2017) Application of electrolyzed water for improving pork meat quality. Food Res Int 100(Pt 1):757–763

Audenaert K, Monbaliu S, Deschuyffeleer N et al (2012) Neutralized electrolyzed water efficiently reduces *Fusarium* spp. in vitro and on wheat kernels but can trigger deoxynivalenol (DON) biosynthesis. Food Control 23(2):515–521

Bempah CK, Agyekum AA, Akuamoa F et al (2016) Dietary exposure to chlorinated pesticide residues in fruits and vegetables from Ghanaian markets. J Food Compos Anal 46:103–113

Ben-Amor R, de Miguel-Gómez MD, Martínez-Sánchez A et al (2016) Effect of hot air on Deglet Noor palm quality parameters and on *Ectomyelois ceratoniae*. J Stor Prod Res 68:1–8

Bessi H, Debbabi H, Grissa K et al (2014) Microbial reduction and quality of stored date fruits treated by electrolyzed water. J Food Qua 37(1):42–49

Birmpa A, Sfika V, Vantarakis A (2013) Ultraviolet light and ultrasound as non-thermal treatments for the inactivation of microorganisms in fresh ready-to-eat foods. Int J Food Microbiol 167(1):96–102

Brasil IM, Siddiqui MW (2018) Chapter 1—postharvest quality of fruits and vegetables: an overview. In: Preharvest modulation of postharvest fruit and vegetable quality. Academic Press, pp 1–40

Burkard M, Leischner C, Lauer UM et al (2017) Dietary flavonoids and modulation of natural killer cells: implications in malignant and viral diseases. J Nutr Biochem 46:1–12

Butscher D, Van H, Waskow A et al (2016) Plasma inactivation of microorganisms on sprout seeds in a dielectric barrier discharge. Int J Food Microbiol 238:222–232

Calligaris S, Valoppi F, Barba, L et al (2018) β-Carotene degradation kinetics as affected by fat crystal network and solid/liquid ratio. Food Res Int 105(Supplement C):599–604

Cao W, Zhu ZW, Shi ZX et al (2009) Efficiency of slightly acidic electrolyzed water for inactivation of *Salmonella enteritidis* and its contaminated shell eggs. Int J Food Microbiol 130(2):88–93

Castro-Ibáñez I, Gil MI, Allende A (2017) Ready-to-eat vegetables: current problems and potential solutions to reduce microbial risk in the production chain. Food Sci Technol 85:284–292

Chen Q, Wang Y, Chen F et al (2014) Chlorine dioxide treatment for the removal of pesticide residues on fresh lettuce and in aqueous solution. Food Control 40:106–112

Chen Y, Hung YC, Chen M et al (2017) Effects of acidic electrolyzed oxidizing water on retarding cell wall degradation and delaying softening of blueberries during postharvest storage. Food Sci Technol 84:650–657

Cheng X, Tian Y, Zhao C et al (2016) Bactericidal effect of strong acid electrolyzed water against flow *Enterococcus faecalis* biofilms. J Endod 42(7):1120–1125

Chuang CY, Yang S, Chang MY et al (2013) Inactivation efficiency to *Bacillus subtilis* and *Escherichia colibacterial* aerosols of spraying neutral electrolyzed water. J Air Waste Manag Assoc 63(12):1447–1456

Cui X, Shang Y, Shi Z et al (2009) Physicochemical properties and bactericidal efficiency of neutral and acidic electrolyzed water under different storage conditions. J Food Eng 91(4):582–586

Del D, Rodriguez-Mateos A, Spencer JP et al (2013) Dietary (poly)phenolics in human health: structures, bioavailability, and evidence of protective effects against chronic diseases. Antioxid Redox 18(14):1818–1892

Diana M, Quílez J, Rafecas M (2014) Gamma-aminobutyric acid as a bioactive compound in foods: a review. J Funct Foods 10:407–420

Dilarri G, Silva VLd, Pecora HB et al (2016) Electrolytic treatment and biosurfactants applied to the conservation of *Eugenia uniflora* fruit. Food Sci Technol (Campinas) 36(3):456–460

Ding T, Jin YG, Oh DH (2009) Predictive model for growth of *Listeria monocytogenes* in untreated and treated lettuce with alkaline electrolyzed water. World J Microb Biot 26(5):863–869

Ding T, Rahman SME, Oh DH (2011) Inhibitory effects of low concentration electrolyzed water and other sanitizers against foodborne pathogens on oyster mushroom. Food Control 22(2):318–322

Ding T, Ge Z, Shi J et al (2015) Impact of slightly acidic electrolyzed water (SAEW) and ultrasound on microbial loads and quality of fresh fruits. Food Sci Technol 60(2):1195–1199

Ding T, Xuan XT, Li J et al (2016) Disinfection efficacy and mechanism of slightly acidic electrolyzed water on *Staphylococcus aureus* in pure culture. Food Control 60:505–510

Doulia DS, Anagnos EK, Liapis KS et al (2016) Removal of pesticides from white and red wines by microfiltration. J Hazard Mater 317:135–146

Duan D, Liu G, Yao P et al (2016) The effects of organic compounds on inactivation efficacy of *Artemia salina* by neutral electrolyzed water. Ocean Eng 125:31–37

Dufour C, Loonis M, Delosière M et al (2018) The matrix of fruit and vegetables modulates the gastrointestinal bioaccessibility of polyphenols and their impact on dietary protein digestibility. Food Chem 240(Supplement C):314–322

Eda H, Seckin MSA (2015) The effect of different electrolyzed water treatments on the quality and sensory attributes of sweet cherry during passive atmosphere packaging storage. Postharvest Biol Technol 102:32–41

Erdozain MS, Allen KJ, Morley KA et al (2013) Failures in sprouts-related risk communication. Food Control 30(2):649–656

Etienne A, Génard M, Bancel D et al (2013) A model approach revealed the relationship between banana pulp acidity and composition during growth and post harvest ripening. Sci Hortic 162:125–134

Fallanaj F, Ippolito A, Ligorio A et al (2016) Electrolyzed sodium bicarbonate inhibits *Penicillium digitatum* and induces defence responses against green mould in citrus fruit. Postharvest Biol Technol 115:18–29

Favre N, Bárcena A, Bahima JV et al (2018) Pulses of low intensity light as promising technology to delay postharvest senescence of broccoli. Postharvest Biol Technol 142:107–114

Feliziani E, Lichter A, Smilanick JL et al (2016) Disinfecting agents for controlling fruit and vegetable diseases after harvest. Postharvest Biol Technol 122:53–69

Fishburn JDT, Tang YJ, Frank FJ (2012) Efficacy of various consumer-friendly produce washing technologies in reducing pathogens on fresh produce. Food Prot Trends 32(8):456–466

Forghani F, Oh DH (2013) Hurdle enhancement of slightly acidic electrolyzed water antimicrobial efficacy on Chinese cabbage, lettuce, sesame leaf and spinach using ultrasonication and water wash. Food Microbiol 36(1):40–45

Forghani F, Rahman SME, Park MS et al (2013) Ultrasonication enhanced low concentration electrolyzed water efficacy on bacteria inactivation and shelf life extension on lettuce. Food Sci Biotechnol 22(1):131–136

Forghani F, Park JH, Oh DH (2015) Effect of water hardness on the production and microbicidal efficacy of slightly acidic electrolyzed water. Food Microbiol 48:28–34

Gan RY, Lui WY, Wu K et al (2017) Bioactive compounds and bioactivities of germinated edible seeds and sprouts: an updated review. Trends Food Sci Tech 59:1–14

Gil MI, Gómez-López VM et al (2015) Potential of electrolyzed water as an alternative disinfectant agent in the fresh-cut industry. Food Bioprocess Technol 8(6):1336–1348

Gillberg L, Orskov AD et al (2018) Vitamin C—a new player in regulation of the cancer epigenome. Semin Cancer Biol 51:59–67

Giovenzana V, Beghi R, Civelli R et al (2015) Optical techniques for rapid quality monitoring along minimally processed fruit and vegetable chain. Trends Food Sci Tech 46(2):331–338

Gómez-López VM, Ragaert P et al (2007) Shelf-life of minimally processed cabbage treated with neutral electrolysed oxidising water and stored under equilibrium modified atmosphere. Int J Food Microbiol 117(1):91–98

Gómez-López VM, Marín A et al (2013) Generation of trihalomethanes with chlorine-based sanitizers and impact on microbial, nutritional and sensory quality of baby spinach. Postharvest Biol Technol 85:210–217

Gómez-López VM, Gil MI, Pupunat L et al (2015) Cross-contamination of *Escherichia coli* O157:H7 is inhibited by electrolyzed water combined with salt under dynamic conditions of increasing organic matter. Food Microbiol 46:471–478

Gómez-Mascaraque LG, Perez-Masiá R, González-Barrio R et al (2017) Potential of microencapsulation through emulsion-electrospraying to improve the bioaccesibility of β-carotene. Food Hydrocolloids 73:1–12

Graça A, Nunes C, Centro M et al (2009) Efficacy of neutral and acidic electrolyzed water for reducing microbial contamination on fresh-cut fruits. In: 6th international postharvest symposium

Graça A, Nunes C, Salazar M (2010) Efficacy of neutral and acidic electrolyzed water for reducing microbial contamination on fresh-cut fruits. Acta Hortic 877(Part 1):649–655

Graça A, Abadias M, Salazar M et al (2011) The use of electrolyzed water as a disinfectant for minimally processed apples. Postharvest Biol Tech 61(2–3):172–177

Graça A, Esteves E, Nunes C et al (2017a) Microbiological quality and safety of minimally processed fruits in the marketplace of southern Portugal. Food Control 73:775–783

Graça A, Santo D, Quintas C et al (2017b) Growth of *Escherichia coli*, *Salmonella enterica* and *Listeria* spp., and their inactivation using ultraviolet energy and electrolyzed water, on 'Rocha' fresh-cut pears. Food Control 77:41–49

Guentzel JL, Liang L, Callan K et al (2008) Reduction of bacteria on spinach, lettuce, and surfaces in food service areas using neutral electrolyzed oxidizing water. Food Microbiol 25(1):36–41

Guentzel JL, Lam KL, Callan MA et al (2010) Postharvest management of gray mold and brown rot on surfaces of peaches and grapes using electrolyzed oxidizing water. Int J Food Microbiol 143(1–2):54–60

Hampson CR, Stanich K, McKenzie DL et al (2014) Determining the optimum firmness for sweet cherries using just-about-right sensory methodology. Postharvest Biol Technol 91:104–111

Han Q, Song X, Zhang Z et al (2017) Removal of foodborne pathogen biofilms by acidic electrolyzed water. Front Microbiol 8:988

Hao J, Liu H, Liu R et al (2011a) Efficacy of slightly acidic electrolyzed water (SAEW) for reducing microbial contamination on fresh-cut Cilantro. J Food Saf 31(1):28–34
Hao J, Liu H et al (2011b) Reduction of pesticide residues on fresh vegetables with electrolyzed water treatment. J Food Sci 76(4):C520–C524
Hao J, Qiu S, Li H et al (2012) Roles of hydroxyl radicals in electrolyzed oxidizing water (EOW) for the inactivation of *Escherichia coli*. Int J Food Microbiol 155(3):99–104
Hao J, Li H, Wan Y et al (2015a) Combined effect of acidic electrolyzed water (AEW) and alkaline electrolyzed water (AlEW) on the microbial reduction of fresh-cut cilantro. Food Control 50:699–704
Hao J, Li H, Wan Y et al (2015b) Effect of slightly acidic electrolyzed water (SAEW) treatment on the microbial reduction and storage quality of fresh-cut Cilantro. J Food Process Pres 39(6):559–566
Hao J, Wu T, Li H et al (2016a) Differences of bactericidal efficacy on *Escherichia coli*, *Staphylococcus aureus*, and *Bacillus subtilis* of slightly and strongly acidic electrolyzed water. Food Bioprocess Technol 10(1):155–164
Hao J, Wu T, Li H et al (2016b) Dual effects of slightly acidic electrolyzed water (SAEW) treatment on the accumulation of gamma-aminobutyric acid (GABA) and rutin in germinated buckwheat. Food Chem 201:87–93
Hati S, Mandal S, Minz PS et al (2012) Electrolyzed oxidized water (EOW): non-thermal approach for decontamination of food borne microorganisms in food industry. Food Nutr Sci 03(06):760–768
Hricova D, Zweifel R (2008) Electrolyzed water and its application in the food industry. J Food Protect 71(9):1934–1947
Hua W, Feng H, Luo Y (2006) Dual-phasic inactivation of *Escherichia coli* O157:H7 with peroxyacetic acid, acidic electrolyzed water and chlorine on cantaloupes and fresh-cut apples. J Food Saf 26:335–347
Huang YR, Hung YC, Hsu SY et al (2008) Application of electrolyzed water in the food industry. Food Control 19(4):329–345
Hung YC, Bailly D, Kim C et al (2010a) Effect of electrolyzed oxidizing water and chlorinated water treatments on strawberry and broccoli quality. J Food Qual 33(5):578–598
Hung YC, Tilly P, Kim C (2010b) Efficacy of electrolyzed oxidizing (Eo) water and chlorinated water for inactivation of *Escherichia coli* O157:H7 on strawberries and broccoli. J Food Qual 33(5):559–577
Ignat A, Manzocco L, Maifreni M et al (2016) Decontamination efficacy of neutral and acidic electrolyzed water in fresh-cut salad washing. J Food Process Pres 40(5):874–881
Imam MU, Ishaka A, Ooi DJ et al (2014) Germinated brown rice regulates hepatic cholesterol metabolism and cardiovascular disease risk in hypercholesterolaemic rats. J Funct Foods 8:193–203
Inatsu Y, Weerakkody K, Bari ML et al (2017) The efficacy of combined (NaClO and organic acids) washing treatments in controlling *Escherichia coli* O157:H7, *Listeria monocytogenes* and spoilage bacteria on shredded cabbage and bean sprout. Food Sci Technol 85:1–8
Islam MS, Patras A, Pokharel B et al (2016) UV-C irradiation as an alternative disinfection technique: study of its effect on polyphenols and antioxidant activity of apple juice. Innov Food Sci Emerg Technol 34:344–351
Issa-Zacharia A, Kamitani Y, Miwa N et al (2011) Application of slightly acidic electrolyzed water as a potential non-thermal food sanitizer for decontamination of fresh ready-to-eat vegetables and sprouts. Food Control 22(3–4):601–607
Jadeja R, Hung YC, Bosilevac JM (2013) Resistance of various shiga toxin-producing *Escherichia coli* to electrolyzed oxidizing water. Food Control 30(2):580–584
Jaeger BN, Parylak SL, Gage FH (2017) Mechanisms of dietary flavonoid action in neuronal function and neuroinflammation. Mol Aspects Med 61:50–62
Jalali A, Seiiedlou S, Linke M et al (2017) A comprehensive simulation program for modified atmosphere and humidity packaging (MAHP) of fresh fruits and vegetables. J Food Eng 206:88–97

Jemâa MB, Haouel S, Khouja ML (2013) Efficacy of Eucalyptus essential oils fumigant control against Ectomyelois ceratoniae (Lepidoptera: Pyralidae) under various space occupation conditions. J Stored Prod Res 53:67–71

Jemni M, Gómez P, Souza M et al (2014a) Combined effect of UV-C, ozone and electrolyzed water for keeping overall quality of date palm. Food Sci Technol 59(2):649–655

Jemni M, Otón M, Ramirez JG et al (2014b) Conventional and emergent sanitizers decreased *Ectomyelois ceratoniae* infestation and maintained quality of date palm after shelf-life. Postharvest Biol Tech 87:33–41

Jiménez-Pichardo R, Regalado C, Castaño-Tostado E et al (2016) Evaluation of electrolyzed water as cleaning and disinfection agent on stainless steel as a model surface in the dairy industry. Food Control 60:320–328

Jonathan W, Leff NF (2013) Bacterial communities associated with the surfaces of fresh fruits and vegetables. PLoS ONE 8(3):e59310

Joshi K, Mahendran R, Alagusundaram K et al (2013) Novel disinfectants for fresh produce. Trends Food Sci Technol 34(1):54–61

Ju SY, Ko JJ, Yoon HS et al (2017) Does electrolyzed water have different sanitizing effects than sodium hypochlorite on different vegetable types? Brit Food J 119(2):342–356

Jung Y, Gao J, Jang H et al (2017a) Sanitizer efficacy in preventing cross-contamination during retail preparation of whole and fresh-cut cantaloupe. Food Control 75:228–235

Jung Y, Jang H, Guo M et al (2017b) Sanitizer efficacy in preventing cross-contamination of heads of lettuce during retail crisping. Food Microbiol 64:179–185

Jungklang J, Saengnil K, Uthaibutra J (2017) Effects of water-deficit stress and paclobutrazol on growth, relative water content, electrolyte leakage, proline content and some antioxidant changes in *Curcuma alismatifolia* Gagnep. cv. Chiang Mai Pink. Saudi J Biol Sci 24(7):1505–1512

Keskinen LA, Burke A, Annous BA (2009) Efficacy of chlorine, acidic electrolyzed water and aqueous chlorine dioxide solutions to decontaminate *Escherichia coli* O157:H7 from lettuce leaves. Int J Food Microbiol 132(2–3):134–140

Khalid S, Malik AU, Khan AS et al (2017) Tree age and fruit size in relation to postharvest respiration and quality changes in 'Kinnow' mandarin fruit under ambient storage. Sci Hortic 220:183–192

Khan I, Tango CN, Miskeen S et al (2017) Hurdle technology: a novel approach for enhanced food quality and safety—a review. Food Control 73:1426–1444

Khayankarn S, Uthaibutra J, Setha S et al (2013) Using electrolyzed oxidizing water combined with an ultrasonic wave on the postharvest diseases control of pineapple fruit cv. 'Phu Lae'. Crop Prot 54:43–47

Kim C, Hung YC (2012) Inactivation of *E. coli* O157:H7 on blueberries by electrolyzed water, ultraviolet light, and ozone. J Food Sci 77(4):M206–M211

Kim MJ, Hung YC (2014) Effect of alkaline electrolyzed water as an inhibitor of enzymatic browning in red delicious apples. J Food Biochem 38(6):542–550

Kimura M, Chounan O, Takahashi R et al (2002) Effects of fermented milk containing γ-aminobutyric acid on normal adult subjects. Jpn J Food Chem 9(1):1–6

Klintham P, Tongchitpakdee S, Chinsirikul W et al (2017) Combination of microbubbles with oxidizing sanitizers to eliminate *Escherichia coli* and *Salmonella typhimurium* on Thai leafy vegetables. Food Control 77:260–269

Koide S, Takeda J, Shi J et al (2009) Disinfection efficacy of slightly acidic electrolyzed water on fresh cut cabbage. Food Control 20(3):294–297

Koide S, Shitanda D, Note M et al (2011) Effects of mildly heated, slightly acidic electrolyzed water on the disinfection and physicochemical properties of sliced carrot. Food Control 22(3–4):452–456

Koseki S, Yoshida K, Kamitani Y et al (2004) Effect of mild heat pre-treatment with alkaline electrolyzed water on the efficacy of acidic electrolyzed water against *Escherichia coli* O157:H7 and *Salmonella* on Lettuce. Food Microbiol 21(5):559–566

Laureano J, Giacosa S, Río Segade S et al (2016) Effects of continuous exposure to ozone gas and electrolyzed water on the skin hardness of table and wine grape varieties. J Texture Stud 47(1):40–48

Lee NY, Kim NH, Jang IS et al (2014) Decontamination efficacy of neutral electrolyzed water to eliminate indigenous flora on a large-scale of cabbage and carrot both in the laboratory and on a real processing line. Food Res Int 64:234–240

Lehto M, Kuisma R, Kymäläinen HR (2017) Neutral electrolyzed water (NEW), chlorine dioxide, organic acid based product, and ultraviolet-C for inactivation of microbes in fresh-cut vegetable washing waters. J Food Process Pres e13354

Li X, Hao J, Liu X et al (2015) Effect of the treatment by slightly acidic electrolyzed water on the accumulation of gamma-aminobutyric acid in germinated brown millet. Food Chem 186:249–255

Li H, Ren Y, Hao J et al (2017) Dual effects of acidic electrolyzed water treatments on the microbial reduction and control of enzymatic browning for fresh-cut lotus root. J Food Saf 37(3):e12333

Li L, Hao J, Song S et al (2018) Effect of slightly acidic electrolyzed water on bioactive compounds and morphology of broccoli sprouts. Food Res Int 105:102–109

Liao LB, Chen WM, Xiao XM (2007) The generation and inactivation mechanism of oxidation–reduction potential of electrolyzed oxidizing water. J Food Eng 78(4):1326–1332

Lin Q, Liang R, Williams PA et al (2018) Factors affecting the bioaccessibility of β-carotene in lipid-based microcapsules: digestive conditions, the composition, structure and physical state of microcapsules. Food Hydrocoll 77:187–203

Liu A, Niyongira R (2017) Chinese consumers food purchasing behaviors and awareness of food safety. Food Control 79:185–191

Liu R, Yu ZL (2017) Application of electrolyzed water on reducing the microbial populations on commercial mung bean sprouts. J Food Sci Technol 54(4):995–1001

Liu R, He X, Shi J et al (2013) The effect of electrolyzed water on decontamination, germination and γ-aminobutyric acid accumulation of brown rice. Food Control 33(1):1–5

Liu R, Zhang D, He X et al (2014) The relationship between antioxidant enzymes activity and mungbean sprouts growth during the germination of mungbean seeds treated by electrolyzed water. Plant Growth Regul 74(1):83–91

Luo K, Oh DH (2016) Inactivation kinetics of *Listeria monocytogenes* and *Salmonella enterica* serovar Typhimurium on fresh-cut bell pepper treated with slightly acidic electrolyzed water combined with ultrasound and mild heat. Food Microbiol 53:165–171

Luo K, Kim SY, Wang J et al (2016) A combined hurdle approach of slightly acidic electrolyzed water simultaneous with ultrasound to inactivate *Bacillus cereus* on potato. Food Sci Technol 73:615–621

Ma L, Zhang M, Bhandari B et al (2017) Recent developments in novel shelf life extension technologies of fresh-cut fruits and vegetables. Trends Food Sci Technol 64:23–38

Machado I, Meireles A, Fulgêncio R et al (2016) Disinfection with neutral electrolyzed oxidizing water to reduce microbial load and to prevent biofilm regrowth in the processing of fresh-cut vegetables. Food Bioprod Process 98:333–340

Mahajan PV, Caleb OJ, Gil MI et al (2017) Quality and safety of fresh horticultural commodities: recent advances and future perspectives. Food Packag Shelf 14(Part A):2–11

Malarkodi C, Rajeshkumar S, Annadurai G (2017) Detection of environmentally hazardous pesticide in fruit and vegetable samples using gold nanoparticles. Food Control 80:11–18

Mansur AR, Oh DH (2015) Combined effects of thermosonication and slightly acidic electrolyzed water on the microbial quality and shelf life extension of fresh-cut kale during refrigeration storage. Food Microbiol 51:154–162

Martínez-Hernández GB, Navarro-Rico J, Gómez PA et al (2015) Combined sustainable sanitising treatments to reduce *Escherichia coli* and *Salmonella enteritidis* growth on fresh-cut kailan-hybrid broccoli. Food Control 47:312–317

Martínez-Hernández GB, Artés-Hernández F, Gómez PA et al (2017) Postharvest treatments to control physiological and pathological disorders in lemon fruit. Food Packag Shelf 14:34–39

Masis N, Johnson SL, McCaffrey J et al (2017) Fruit and vegetable preferences and identification by kindergarteners through 2nd-graders with or without the US Department of Agriculture Fresh Fruit and Vegetable Program. J Nutr Educ Behav 49(9):752–758

Mditshwa A, Magwaza LS, Tesfay SZ et al (2017) Postharvest factors affecting vitamin C content of citrus fruits: a review. Sci Hortic 218:95–104

Meireles A, Giaouris E, Simões M (2016) Alternative disinfection methods to chlorine for use in the fresh-cut industry. Food Res Int 82:71–85

Meireles A, Ferreira C, Melo L et al (2017) Comparative stability and efficacy of selected chlorine-based biocides against *Escherichia coli* in planktonic and biofilm states. Food Res Int 102:511–518

Millan-Sango D, Sammut E, Van I et al (2017) Decontamination of alfalfa and mung bean sprouts by ultrasound and aqueous chlorine dioxide. Food Sci Technol 78:90–96

Mock JT, Chaudhari K, Sidhu A et al (2017) The influence of vitamins E and C and exercise on brain aging. Exp Gerontol 94:69–72

Moggia C, Graell J, Lara I et al (2016) Fruit characteristics and cuticle triterpenes as related to postharvest quality of highbush blueberries. Sci Hortic 211:449–457

Monnin A, Lee J, Pascall MA (2012) Efficacy of neutral electrolyzed water for sanitization of cutting boards used in the preparation of foods. J Food Eng 110(4):541–546

Nam TG, Lee SM, Park JH et al (2015) Flavonoid analysis of buckwheat sprouts. Food Chem 170:97–101

Navarro-Rico J, Artés-Hernández F, Gómez PA et al (2014) Neutral and acidic electrolysed water kept microbial quality and health promoting compounds of fresh-cut broccoli throughout shelf life. Innov Food Sci Emerg Technol 21:74–81

Ngnitcho PFK, Khan I, Tango CN et al (2017) Inactivation of bacterial pathogens on lettuce, sprouts, and spinach using hurdle technology. Innov Food Sci Emerg Tech 43:68–76

Ni L, Zheng W, Zhang Q et al (2016) Application of slightly acidic electrolyzed water for decontamination of stainless steel surfaces in animal transport vehicles. Prev Vet Med 133:42–51

Nieoczym D, Socała K, Raszewski G et al (2014) Effect of quercetin and rutin in some acute seizure models in mice. Prog Neuro-Psychoph 54(Supplement C):50–58

Okull DO, Laborde LF (2004) Activity of electrolyzed oxidizing water against *Penicilium expansum* in suspension and on wounded apples. J Food Sci 69(1):FMS23–FMS27

Ovissipour M, Al-Qadiri HM, Sablani SS et al (2015) Efficacy of acidic and alkaline electrolyzed water for inactivating *Escherichia coli* O104:H4, *Listeria monocytogenes*, *Campylobacter jejuni*, *Aeromonas hydrophila*, and *Vibrio parahaemolyticus* in cell suspensions. Food Control 53:117–123

Pangloli P (2009) Reduction of *Escherichia coli* O157:H7 on produce by use of electrolyzed water under simulated food service operation conditions. J Food Protect 72(9):1854–1861

Pangloli P, Hung YC (2011) Efficacy of slightly acidic electrolyzed water in killing or reducing *Escherichia coli* O157:H7 on iceberg lettuce and tomatoes under simulated food service operation conditions. J Food Sci 76(6):M361–M366

Pangloli P, Hung YC (2013) Reducing microbiological safety risk on blueberries through innovative washing technologies. Food Control 32(2):621–625

Park C, Hung Y, Doyle M et al (2001) Pathogen reduction and quality of lettuce treated with electrolyzed oxidizing and acidified chlorinated water. J Food Sci 66(9):1368–1372

Park H, Hung YC, Chung D (2004) Effects of chlorine and pH on efficacy of electrolyzed water for inactivating *Escherichia coli* O157:H7 and *Listeria monocytogenes*. Int J Food Microbiol 91(1):13–18

Park E, Alexander E, Taylor G et al (2008a) Effect of electrolyzed water for reduction of foodborne pathogens on lettuce and spinach. J Food Sci 73(6):M268–M272

Park E, Alexander E, Taylor G et al (2008b) Fate of foodborne pathogens on green onions and tomatoes by electrolysed water. Lett Appl Microbiol 46(5):519–525

Park E, Alexander E, Taylor G et al (2009) The decontaminative effects of acidic electrolyzed water for *Escherichia coli* O157:H7, *Salmonella typhimurium*, and *Listeria monocytogenes* on green onions and tomatoes with differing organic demands. Food Microbiol 26(4):386–390

Park K, Lim J, Jung H et al (2017) Disinfection efficacy of slightly acidic electrolyzed water (SlAEW) against some fresh vegetables. Korean J Food Preserv 24(2):312–319

Phillips K, Council-Troche M, McGinty R (2016) Stability of vitamin C in fruit and vegetable homogenates stored at different temperatures. J Food Compos Anal 45:147–162

Phua L, Neo S, Khoo G (2014) Comparison of the efficacy of various sanitizers and hot water treatment in inactivating inoculated foodborne pathogens and natural microflora on mung bean sprouts. Food Control 42:270–276

Pinto L, Ippolito A, Baruzzi F (2015) Control of spoiler *Pseudomonas* spp. on fresh cut vegetables by neutral electrolyzed water. Food Microbiol 50:102–108

Pinto L, Baruzzi F, Ippolito A (2016) Recent advances to control spoilage microorganisms in washing water of fruits and vegetables: the use of electrolyzed water. Acta Hort 1144:379–384

Pongrac P, Potisek M, Fraś A et al (2016) Composition of mineral elements and bioactive compounds in tartary buckwheat and wheat sprouts as affected by natural mineral-rich water. J Cereal Sci 69:9–16

Posada-Izquierdo G, Perez-Rodriguez F, Lopez-Galvez F et al (2014) Modeling growth of *Escherichia coli* O157:H7 in fresh-cut lettuce treated with neutral electrolyzed water and under modified atmosphere packaging. Int J Food Microbiol 177:1–8

Prasad K, Jacob S, Siddiqui M (2018) Chapter 2—fruit maturity, harvesting, and quality standards. In: Siddiqui MW (ed) Preharvest modulation of postharvest fruit and vegetable quality. Academic Press, pp 41–69

Puligundla P, Kim JW, Mok C (2018) Broccoli sprout washing with electrolyzed water: effects on microbiological and physicochemical characteristics. Food Sci technology 92:600–606

Qin P, Wei A, Zhao D et al (2017) Low concentration of sodium bicarbonate improves the bioactive compound levels and antioxidant and alpha-glucosidase inhibitory activities of tartary buckwheat sprouts. Food Chem 224:124–130

Rady A, Ekramirad N, Adedeji A et al (2017) Hyperspectral imaging for detection of codling moth infestation in GoldRush apples. Postharvest Biol Technol 129:37–44

Rahman S, Ding T, Oh DH (2010a) Effectiveness of low concentration electrolyzed water to inactivate foodborne pathogens under different environmental conditions. Int J Food Microbiol 139(3):147–153

Rahman S, Jin Y, Oh DH (2010b) Combined effects of alkaline electrolyzed water and citric acid with mild heat to control microorganisms on cabbage. J Food Sci 75(2):M111–M115

Rahman S, Ding T, Oh DH (2010c) Inactivation effect of newly developed low concentration electrolyzed water and other sanitizers against microorganisms on spinach. Food Control 21(10):1383–1387

Rahman S, Jin Y, Oh DH (2011) Combination treatment of alkaline electrolyzed water and citric acid with mild heat to ensure microbial safety, shelf-life and sensory quality of shredded carrots. Food Microbiol 28(3):484–491

Rahman S, Park J, Wang J et al (2012) Stability of low concentration electrolyzed water and its sanitization potential against foodborne pathogens. J Food Eng 113(4):548–553

Rahman S, Khan I, Oh DH (2016) Electrolyzed water as a novel sanitizer in the food industry: current trends and future perspectives. Compr Rev Food Sci F 15(3):471–490

Ramos B, Miller F, Brandão T et al (2013) Fresh fruits and vegetables—an overview on applied methodologies to improve its quality and safety. Innov Food Sci Emerg Technol 20:1–15

Ramsay S, Shriver L, Taylor C (2017) Variety of fruit and vegetables is related to preschoolers' overall diet quality. Prev Med Rep 5(Supplement C):112–117

Ren SC, Sun JT (2014) Changes in phenolic content, phenylalanine ammonia-lyase (PAL) activity, and antioxidant capacity of two buckwheat sprouts in relation to germination. J Funct Foods 7:298–304

Rico D, Martín-Diana A, Barry-Ryan C et al (2008) Use of neutral electrolysed water (EW) for quality maintenance and shelf-life extension of minimally processed lettuce. Innov Food Sci Emerg Technol 9(1):37–48

Robinson G, Thorn R, Reynolds D (2012) The effect of long-term storage on the physiochemical and bactericidal properties of electrochemically activated solutions. Int J Mol Sci 14(1):457–469

Rodriguez-Garcia O, Gonzalez-Romero V, Fernandez-Escartin E (2011) Reduction of *Salmonella enterica*, *Escherichia coli* O157:H7, and *Listeria monocytogenes* with electrolyzed oxidizing water on inoculated has avocados (*Persea americana* var. Hass). J Food Prot 74(9):1552–1557

Rolny N, Costa L, Carrion C et al (2011) Is the electrolyte leakage assay an unequivocal test of membrane deterioration during leaf senescence? Plant Physiol Biochem 49(10):1220–1227

Romanazzi G, Sanzani S, Tian S et al (2016a) Induced resistance to control postharvest decay of fruit and vegetables. Postharvest Biol Technol 122:82–94

Romanazzi G, Smilanick J, Feliziani E (2016b) Integrated management of postharvest gray mold on fruit crops. Postharvest Biol Technol 113:69–76

Rui L, Jianxiong H, Haijie L et al (2011) Application of electrolyzed functional water on producing mung bean sprouts. Food Control 22(8):1311–1315

Santo D, Graça A, Nunes C et al (2016) Survival and growth of *Cronobacter sakazakii* on fresh-cut fruit and the effect of UV-C illumination and electrolyzed water in the reduction of its population. Int J Food Microbiol 231:10–15

Santo D, Graça A, Nunes C et al (2018) *Escherichia coli* and *Cronobacter sakazakii* in 'Tommy Atkins' minimally processed mangos: survival, growth and effect of UV-C and electrolyzed water. Food Microbiol 70:49–54

Sardella D, Gatt R, Valdramidis V (2017) Physiological effects and mode of action of ZnO nanoparticles against postharvest fungal contaminants. Food Res Int 101:274–279

Scharff R, Besser J, Sharp D et al (2016) An economic evaluation of PulseNet: a network for Foodborne disease surveillance. Am J Prev Med 50(5 Suppl 1):S66–S73

Shahidi F, Ambigaipalan P (2015) Phenolics and polyphenolics in foods, beverages and spices: antioxidant activity and health effects—a review. J Funct Foods 18:820–897

Shigenobu K, Seiichiro I, Kazuhiko I (2004) Efficacy of acidic electrolyzed water for microbial decontamination of cucumbers and strawberries. J Food Prot 67(6):1247–1251

Silva B, Cadavez V, Teixeira J et al (2017) Meta-analysis of the incidence of foodborne pathogens in vegetables and fruits from retail establishments in Europe. Curr Opin Food Sci 18:21–28

Solomon S, Sugar P (2009) Efficacy of electrolyzed water to minimize postharvest sucrose losses in sugarcane. Sugar Technol 11(2):228–230

Souza L, Faroni L, Heleno F et al (2018) Ozone treatment for pesticide removal from carrots: optimization by response surface methodology. Food Chem 243:435–441

Stopforth JD, Kottapalli TMB, Samadpour M (2008) Effect of acidified sodium chlorite, chlorine, and acidic electrolyzed water on *Escherichia coli* O157:H7, *Salmonella*, and *Listeria monocytogenes* inoculated onto leafy greens. J Food Prot 71(3):625–628

Sun J, Zhang S, Chen J et al (2012) Efficacy of acidic and basic electrolyzed water in eradicating *Staphylococcus aureus* biofilm. Can J Microbiol 58(4):448–454

Sun J, Wang M, Liu H et al (2018) Acidic electrolysed water delays browning by destroying conformation of polyphenoloxidase. J Sci Food Agric 98(1):147–153

Tango C, Mansur A, Oh D (2015) Fumaric acid and slightly acidic electrolyzed water inactivate gram positive and gram negative Foodborne pathogens. Microorganisms 3(1):34–46

Tango C, Khan I, Ngnitcho K et al (2017) Slightly acidic electrolyzed water combined with chemical and physical treatments to decontaminate bacteria on fresh fruits. Food Microbiol 67:97–105

Teng Y, Sheu M, Hsieh Y et al (2016) β-carotene reverses multidrug resistant cancer cells by selectively modulating human P-glycoprotein function. Phytomed 23(3):316–323

Tirawat D, Phongpaichit S, Benjakul S et al (2016) Microbial load reduction of sweet basil using acidic electrolyzed water and lactic acid in combination with mild heat. Food Control 64:29–36

Tkhawkho L, Jackson K, Nitzan O et al (2017) Destruction of Clostridium difficile spores colitis using acidic electrolyzed water. Am J Infect Control 45(1):1053

Toivonen P, Brummell D (2008) Biochemical bases of appearance and texture changes in fresh-cut fruit and vegetables. Postharvest Biol Technol 48(1):1–14
Tomás-Callejas A, Martínez-Hernández G, Artés F et al (2011) Neutral and acidic electrolyzed water as emergent sanitizers for fresh-cut mizuna baby leaves. Postharvest Biol Technol 59(3):298–306
Torlak E (2014) Inactivation of *Alicyclobacillus acidoterrestris* spores in aqueous suspension and on apples by neutral electrolyzed water. Int J Food Microbiol 185:69–72
Tsurunaga Y, Takahashi T, Katsube T et al (2013) Effects of UV-B irradiation on the levels of anthocyanin, rutin and radical scavenging activity of buckwheat sprouts. Food Chem 141(1):552–556
Tuan P, Thwe A, Kim J et al (2013) Molecular characterisation and the light-dark regulation of carotenoid biosynthesis in sprouts of tartary buckwheat (*Fagopyrum tataricum* Gaertn.). Food Chem 141(4):3803–3812
Udompijitkul P, Daeschel M, Zhao Y (2007) Antimicrobial effect of electrolyzed oxidizing water against *Escherichia coli* O157:H7 and *Listeria monocytogenes* on fresh strawberries (Fragaria x ananassa). J Food Sci 72(9):M397–M406
Vandekinderen I, Van C, Meulenaer B et al (2009a) Moderate and high doses of sodium hypochlorite, neutral electrolyzed oxidizing water, peroxyacetic acid, and gaseous chlorine dioxide did not affect the nutritional and sensory qualities of fresh-cut Iceberg lettuce (*Lactuca sativa* Var. capitata L.) after washing. J Agric Food Chem 57(10):4195–4203
Vandekinderen I, Van C, Devlieghere F et al (2009b) Evaluation of the use of decontamination agents during fresh-cut leek processing and quantification of their effect on its total quality by means of a multidisciplinary approach. Innov Food Sci Emerg Technol 10(3):363–373
Vandekinderen I, Van J, Devlieghere F et al (2009c) Effect of decontamination on the microbial load, the sensory quality and the nutrient retention of ready-to-eat white cabbage. Eur Food Res Technol 229(3):443–455
Vasquez-Lopez A, Villarreal-Barajas T, Rodriguez-Ortiz G (2016) Effectiveness of neutral electrolyzed water on incidence of fungal rot on tomato fruits (*Solanum lycopersicum* L.). J Food Prot 79(10):1802–1806
Vojkovska H, Myskova P, Gelbicova T et al (2017) Occurrence and characterization of food-borne pathogens isolated from fruit, vegetables and sprouts retailed in the Czech Republic. Food Microbiol 63:147–152
Wang H, Feng H, Luo Y (2004) Microbial reduction and storage quality of fresh-cut cilantro washed with acidic electrolyzed water and aqueous ozone. Food Res Int 37(10):949–956
Wang H, Feng H, Luo Y (2007) Control of browning and microbial growth on fresh-cut apples by sequential treatment of sanitizers and calcium ascorbate. J Food Sci 72(1):M001–M007
Whangchai K, Saengnil K, Singkamanee C et al (2010) Effect of electrolyzed oxidizing water and continuous ozone exposure on the control of *Penicillium digitatum* on tangerine cv. 'Sai Nam Pung' during storage. Crop Protect 29(4):386–389
Xanthopoulos G, Templalexis C, Aleiferis N (2017) The contribution of transpiration and respiration in water loss of perishable agricultural products: the case of pears. Biosys Eng 158:76–85
Xie Z, Xia S, Le G (2014) Gamma-aminobutyric acid improves oxidative stress and function of the thyroid in high-fat diet fed mice. J Funct Foods 8:76–86
Xiong K, Li X, Guo S et al (2014) The antifungal mechanism of electrolyzed oxidizing water against *Aspergillus flavus*. Food Sci Biotech 23(2):661–669
Xuan X, Wang M, Ahn J et al (2016) Storage stability of slightly acidic electrolyzed water and circulating electrolyzed water and their property changes after application. J Food Sci 81(3):E610–E617
Xuan X, Ding T, Li J et al (2017) Estimation of growth parameters of *Listeria monocytogenes* after sublethal heat and slightly acidic electrolyzed water (SAEW) treatment. Food Control 71:17–25
Yamamoto T, Nakano T, Yamaguchi M et al (2012) Disinfective process of strongly acidic electrolyzed product of sodium chloride solution against Mycobacteria. Med Mol Morphol 45(4):199–205
Yoon J, Lee S (2017) Review: comparison of the effectiveness of decontaminating strategies for fresh fruits and vegetables and related limitations. Crit Rev Food Sci Nutr 1–20

Zhang C, Lu Z, Li Y et al (2011) Reduction of *Escherichia coli* O157:H7 and *Salmonella enteritidis* on mung bean seeds and sprouts by slightly acidic electrolyzed water. Food Control 22(5):792–796
Zhang M, Zeiss M, Geng S (2015) Agricultural pesticide use and food safety: California's model. J Integ Agr 14(11):2340–2357
Zhang C, Cao W, Hung Y et al (2016a) Application of electrolyzed oxidizing water in production of radish sprouts to reduce natural microbiota. Food Control 67:177–182
Zhang C, Cao W, Hung Y et al (2016b) Disinfection effect of slightly acidic electrolyzed water on celery and cilantro. Food Control 69:147–152
Zhi H, Liu Q, Dong Y et al (2017) Effect of calcium dissolved in slightly acidic electrolyzed water on antioxidant system, calcium distribution, and cell wall metabolism of peach in relation to fruit browning. J Hortic Sci Biote 92(6):621–629
Zhou R, Zhang G, Hu Y et al (2012) Reductions in flesh discolouration and internal morphological changes in Nanhui peaches (*Prunus persica* (L.) Batsch, cv. Nanhui) by electrolysed water and 1-methylcyclopropene treatment during refrigerated storage. Food Chem 135(3):985–992
Zhou X, Guo H, Li Z et al (2015) Experimental study on the disinfection efficiencies of a continuous-flow ultrasound/ultraviolet baffled reactor. Ultrason Sonochem 27:81–86
Zhu W, Zhu B, Li Y et al (2016) Acidic electrolyzed water efficiently improves the flavour of persimmon (*Diospyros kaki* L. cv. Mopan) wine. Food Chem 197(Pt A):141–149
Zou H, Wang L (2017) The disinfection effect of a novel continuous-flow water sterilizing system coupling dual-frequency ultrasound with sodium hypochlorite in pilot scale. Ultrason Sonochem 36:246–252

Chapter 5
Application of Electrolyzed Water in Red Meat and Poultry Processing

Yanhong Bai, Liyuan Niu and Qisen Xiang

5.1 Introduction

Foods with high nutritional value and of animal origin, such as red meat and poultry products, are primary sources of superior protein for humans. Meat products provide a relatively rich source of well-absorbed iron and are concentrated sources of B vitamins, including vitamin B_{12}, which is absent from plant foods. Worldwide, 40% of all the meat consumed is pork, followed by poultry and beef at approximately 30 and 25%, respectively (Rahman et al. 2016). Microbial contamination in red meat and poultry is a crucial issue related to meat safety and quality. Carcasses during slaughtering, chilling and cutting processes and meat products during processing, storage and handling are subjected to contamination from various sources. Meat contaminated with *Listeria monocytogenes*, *Escherichia coli* O157:H7, *Salmonella* spp., *Campylobacter coli* and *Staphylococcus aureus* have been reported as the casual agents of food-borne diseases (Mies et al. 2004; Koohmaraie et al. 2005). People infected with these bacteria suffer from gastroenteritis or even haemorrhagic colitis (Carlson et al. 2008). Therefore, the prevention of microbial spoilage of meat and outbreaks of food-borne diseases are very important.

Y. Bai (✉) · L. Niu · Q. Xiang
College of Food and Biological Engineering, Zhengzhou University of Light Industry, Zhengzhou 450001, China
e-mail: baiyh212@163.com

Y. Bai · L. Niu · Q. Xiang
Henan Key Laboratory of Cold Chain Food Quality and Safety Control, Zhengzhou 450001, China

Y. Bai · L. Niu · Q. Xiang
Henan Collaborative Innovation Center of Food Production and Safety, Zhengzhou 450001, China

T. Ding et al. (eds.), *Electrolyzed Water in Food: Fundamentals and Applications*,
https://doi.org/10.1007/978-981-13-3807-6_5

A variety of processing or storage techniques have been developed for the purpose of improving microbial safety and extending shelf life (Hwang and Beuchat 1995; Goksoy et al. 2000; González-Fandos and Dominguez 2007; Kang et al. 2007). Chemical decontaminants, such as organic acids and chlorine compounds, are commonly used in the red meat and poultry industries (Chantarapanont et al. 2004; Stopforth et al. 2004; Anang et al. 2007; Alonso-Hernando et al. 2013). However, some of these methods possess many drawbacks, such as the remaining chemical residues, high cost, or limited effectiveness, and chemical and physical changes to the meat products (Al-Haq et al. 2005; Rahman et al. 2013).

Currently, EW has awakened high interest as a novel disinfectant in the food industry to reduce or eliminate microbial contaminants on food products, processed food surfaces, and food-contact surfaces. EW is produced by electrolysis of a solution of deionized water and salt (NaCl or $KCl–MgCl_2$) in an electrolysis chamber, within which a diaphragm membrane separates the anode and cathode. Depending on the pH values, EW is classified into AEW, SAEW, NEW, or AlEW (Tango et al. 2014; Rahman et al. 2016). Previous studies have demonstrated that various EW possesses antimicrobial efficacy on food matrices (Ding et al. 2010) and equipment (Venkitanarayanan et al. 1999a, b). EW is generally recognized as safe (GRAS) and is created with an environmentally friendly technology. The simplified production and small investment costs make it an especially promising antimicrobial in food processing (Kim et al. 2000; Bosilevac et al. 2005). The purpose of this chapter is to provide an overview of the effects of EW treatment on red meat and poultry products in terms of antimicrobial activity and changes in the physicochemical and sensory qualities of the food.

5.2 Application of EW in the Processing and Storage of Fresh Pork

Global pork consumption accounts for approximately 40% of all meat consumed, followed by poultry (30%) and beef (25%) (FAO 2006). In spite of the increasing consumption of pork products, the microbial safety of pork during processing and storage remains an issue. Pork products have different species of microbes, from various sources, which cause outbreaks of food-borne diseases and contribute to losses to human health and the economy. For years, treatments with EW have been applied to the decontamination of fresh pork (Brychcy et al. 2015a, b; Fabrizio and Cutter 2004; Rahman et al. 2013; Rigdon et al. 2017; Wang et al. 2012; Mansur et al. 2015), and these have proven effective against the *E. coli*, *L. monocytogenes*, *Salmonella typhimurium*, *C. coli* and *S. aureus* associated with pork products.

5.2.1 Microbial Decontamination Efficacy of EW Treatment of Fresh Pork

As shown in Table 5.1, Fabrizio and Cutter (2004) demonstrated that spraying with AEW (pH 2.6, ORP 1150 mV, ACC 50 mg/L) for 15 s can reduce *C. coli* (1.8 $\log_{10}$ CFU/cm^2) that was inoculated onto fresh pork surfaces, but longer contact times may be needed for the reduction of *coliforms*, *E. coli*, *L. monocytogenes* and *S. typhimurium*. Under a similar treatment condition (pH 2.5, ORP 1130 mV, ACC 50.2 mg/L), Rahman et al. (2013) conducted a decontamination study with AEW for 5 min, and both *E. coli* O157:H7 and *L. monocytogenes* in fresh pork were significantly reduced by 1.7 $\log_{10}$ CFU/g. AEW is an effective microbial decontaminant for red meat, poultry carcasses, eggs, fruits, fresh cut vegetables and food processing equipment (Huang et al. 2008; Hati et al. 2012). However, the potential application of AEW is limited because of its low pH (2.2–2.7). The electrolyzed water with high acidity could be corrosive to the metal equipment, and the free chlorine (Cl_2) in this product could cause cytotoxicity and genotoxicity (Rahman et al. 2010a).

SAEW is a newly developed sanitizer that has proven promising and safe in several previous studies. SAEW treatment showed strong inactivation effects on various pathogens inoculated on the oyster mushroom, lettuce and spinach (Ding et al. 2011; Rahman et al. 2010a, b). Rahman et al. (2013) reported that SAEW (pH 6.8, ORP 700–720 mV, ACC 10 mg/L) inactivated *E. coli* O157:H7 and *L. monocytogenes* in fresh pork as effectively as the AEW treatment (Table 5.1). Wang et al. (2012) successfully modelled the growth of *L. monocytogenes* on pork with SAEW treatment as a function of storage temperatures (4–30 °C). The modelling results made it clear that the specific growth rate (SGR) of this microbe increased and the lag time (LT) decreased with an elevation in temperature.

Because treatment with AEW has some disadvantages, Brychcy et al. (2015a, b) proposed the use of low-concentrated acidic electrolyzed water (low concentration of sodium chloride and low current of electrolysis). The EW was generated by electrolyzing low-concentration sodium chlorine (0.001%, 0.01% or 0.1% NaCl) for 0, 5 or 10 min. Brychcy et al. (2015b) analysed the effect of EW against yeast and mould and the effect of psychrotrophs on pork muscle. The greatest reductions in total viable number (TVC) (3.3 $\log_{10}$ reduction) for yeast and mould (2.7 $\log_{10}$ reduction) and psychrotrophs (3.1 $\log_{10}$ reduction) were observed after the pork muscle was sprayed with 0.1% NaCl electrolyzed for 10 min (pH 2.2, ORP 1159 mV, ACC 16.6 mg/L). These results revealed that the usage of more concentrated salt solution and longer electrolysis time in the generation of EW may exhibit more efficient decontamination.

Recently, Han et al. (2018) evaluated the antimicrobial efficacy of NEW on pathogens on inoculated pork skin samples and pork chops. After treatment of the skin and pork for 2 min, significant reductions in the populations of *E. coli* O157:H7 (2.1 $\log_{10}$ CFU/cm^2), *Salmonella enteritidis* (2.2 $\log_{10}$ CFU/cm^2) and *Yersinia enterocolitica* (2.7 $\log_{10}$ CFU/cm^2) were obtained. However, for the inoculated pork chops, the reduction was much smaller than for the skin samples, and the extended

Table 5.1 Antimicrobial efficacy of various types of electrolyzed water treatments against pathogens on fresh pork samples

Product	Microorganism	Treatment condition	Exposure time (min)	Temp. (°C)	Reduction (log_{10} CFU)	EW property			Reference
						pH	ORP (mV)	ACC (mg/L)	
Fresh Pork meat	*Campylobacter coli*	AEW	0.25	23	1.8 cm^2	2.6	1150	50	Fabrizio and Cutter (2004)
	Coliforms	AEW	0.25	23	1.2 cm^2	2.6	1150	50	Fabrizio and Cutter (2004)
	Escherichia coli	AEW	0.25	23	1.1 cm^2	2.6	1150	50	Fabrizio and Cutter (2004)
	E. coli O157:H7	SAEW	5	23	1.7 g	6.8	700	10	Rahman et al. (2013)
		AEW	5	23	1.9 g	2.5	1130	50.2	Rahman et al. (2013)
	Listeria monocytogenes	AEW	0.25	23	1.2 cm^2	2.6	1150	50	Fabrizio and Cutter (2004)
		SAEW	2	23	1.7 g	6.8	700	10	Wang et al. (2012)
		SAEW	5	23	1.8 g	6.8	700–720	10	Rahman et al. (2013)
		AEW	5	23	1.9 g	2.5	1130	50	Rahman et al. (2013)

(continued)

Table 5.1 (continued)

Product	Microorganism	Treatment condition	Exposure time (min)	Temp. (°C)	Reduction (log_{10} CFU)	EW property			Reference
						pH	ORP (mV)	ACC (mg/L)	
	Psychrotrophs	SAEW (0.001% NaCl, electrolysis for 10 min)	2	23	1.5 cm^2	3.6	957	0.8	Brychcy et al. (2015b)
		AEW (0.01% NaCl, electrolysis for 10 min)	2	23	1.6 cm^2	2.9	1052	8.5	Brychcy et al. (2015b)
		AEW (0.1% NaCl, electrolysis for 10 min)	2	23	3.1 cm^2	2.2	1159	16.6	Brychcy et al. (2015b)
	Salmonella typhimurium	AEW	0.25	23	1.7 cm^2	2.6	1150	50	Fabrizio and Cutter (2004)
	Total viable counts	AEW	0.25	23	1.2 cm^2	2.6	1150	50	Fabrizio and Cutter (2004)
		SAEW (0.001% NaCl, electrolysis for 10 min)	2	23	1.7 cm^2	3.6	957	0.8	Brychcy et al. (2015b)
		AEW (0.01% NaCl, electrolysis for 10 min)	2	23	1.6 cm^2	2.9	1052	8.5	Brychcy et al. (2015b)

(continued)

Table 5.1 (continued)

Product	Microorganism	Treatment condition	Exposure time (min)	Temp. (°C)	Reduction (log_{10} CFU)	EW property			Reference
						pH	ORP (mV)	ACC (mg/L)	
		AEW (0.1% NaCl, electrolysis for 10 min)	2	23	3.3 cm^2	2.2	1159	16.6	Brychcy et al. (2015b)
	Yeast and mould	SAEW (0.001% NaCl, electrolysis for 10 min)	2	23	1.6 cm^2	3.6	957	0.8	Brychcy et al. (2015b)
		AEW (0.01% NaCl, electrolysis for 10 min)	2	23	1.7 cm^2	2.9	1052	8.5	Brychcy et al. (2015b)
		AEW (0.1% NaCl, electrolysis for 10 min)	2	23	2.7 cm^2	2.2	1159	16.6	Brychcy et al. (2015b)
Pork skin	*E. coli* O157:H7	10 mL NEW	2	NA	2.1 cm^2	7.6	818	74	Han et al. (2018)

(continued)

Table 5.1 (continued)

Product	Microorganism	Treatment condition	Exposure time (min)	Temp. (°C)	Reduction (log_{10} CFU)	EW property			Reference
						pH	ORP (mV)	ACC (mg/L)	
	Salmonella enteritidis	10 mL NEW	2	NA	2.2 cm^2	7.6	818	74	Han et al. (2018)
	Yersinia enterocolitica	10 mL NEW	2	NA	1.7 cm^2	7.6	818	74	Han et al. (2018)
Pork Chops	*E. coli* O157:H7	10 mL NEW	2	NA	0.3 cm^2	7.6	818	74	Han et al. (2018)
	S. enteritidis	10 mL NEW	2	NA	0.2 cm^2	7.6	818	74	Han et al. (2018)
	Yersinia enterocolitica	10 mL NEW	2	NA	ND	7.6	818	74	Han et al. (2018)

ORP Oxidation reduction potential; *ACC* Available chlorine concentration; *AEW* Acidic electrolyzed water; *SAEW* Slightly acidic electrolyzed water; *NEW* Neutral electrolyzed water; *ND* Not detectable treatment time (10 min) did not exert significant impact on the antimicrobial efficacy. The differences in protein content and structure between pork chops and pork skin samples were considered to be the major factors affecting the efficacy of NEW. Unfortunately, the formation of viable but nonculturable (VBNC) pathogens was observed after the NEW treatment. Pathogens in the VBNC state may revive under appropriate conditions and pose a risk to public health (Aurass et al. 2011). The flow cytometry results showed that <50% diluted NEW water led to the formation of VBNC cells, and higher levels of VBNC cells were observed in the *Yersinia* cultures.

5.2.2 *Hurdle Enhancement of EW Treatment with Other Interventions on Fresh Pork*

More and more studies have focused on the application of novel combinations of natural antimicrobials and other food preservatives to improve food quality and microbial stability. The microbial decontamination efficacy of EW and other measures against pathogens on fresh pork have also been examined (Table 5.2). Due to their ability to increase flavour, prolong shelf life, and improve the microbial safety of meat products, organic acid salts such as calcium lactate (CaL), acetic acid (AA), lactic acid (LA) and fumaric acid (FA) have been applied as a spray or dip to decontaminate fresh meat (Huang et al. 2008; Lawrence et al. 2003; Naveena et al. 2006; Selgas et al. 2009).

Rahman et al. (2013) treated inoculated boneless pork loins with SAEW, AEW, 3% CaL, SAEW + CaL, 3% LA, aqueous ozone (AO), and sodium hypochlorite solution (NaOCl) for 5 min at room temperature. Among these treatments, a combination treatment showed the greatest reduction of *E. coli* O157:H7 and *L. monocytogenes* (3.0–3.2 $\log_{10}$ CFU/g) in fresh pork and increased the shelf life of samples at refrigerated temperatures compared with the EW treatment, AO or organic acid salt alone. As one of the strongest weak organic acids, FA has remarkable bactericidal effects (Comes and Beelman 2002), and when present at a low pH, its undissociated form shows bactericidal activity against pathogens in meat (Bell et al. 1986). The combination of SAEW and 0.5% FA also showed greater bactericidal effects (>2.5 $\log_{10}$ reduction) on the pathogens *E. coli* O157:H7, *L. monocytogenes*, *S. aureus* and *S. typhimurium* on fresh pork after the 3 min treatment at 40 °C than observed with the individual treatments (Mansur et al. 2015). EW treatment in combination with other decontamination measures can be a suitable hurdle technology for reducing microorganisms in fresh pork and providing excellent opportunities for food preservation.

Brychcy et al. (2016) attempted to cover fresh pork samples with carrageenan and gelatine sols containing AEW solutions to protect against yeast and mould and psychrotrophs. However, the initial antimicrobial effects were not observed immediately after treatment. With an extended storage period, treatment with AEW (0.1% NaCl electrolysis for 10 min)-containing hydrosols was observed to cause significant decreases in the total number of microorganisms (2.1 $\log_{10}$ CFU/g), yeast and mould

Table 5.2 Microbial decontamination efficacy of electrolyzed water and other interventions against pathogens on fresh pork

Product	Microorganism	Treatment condition	Exposure time (min)	Temp. (°C)	Reduction (log_{10} CFU) (g)	EW property			Reference
						pH	ORP (mV)	ACC (mg/L)	
Fresh Pork meat	*E. coli* O157:H7	SAEW	5	23	1.7	6.8	700	10	Rahman et al. (2013)
		3% CaL	5	23	1.3	6.5	290	ND	Rahman et al. (2013)
		SAEW + 3% CaL	5	23	3.0	6.8	245	1	Rahman et al. (2013)
		AO	5	23	0.4	6.6	1245	5.2 mg O_3/L of water	Rahman et al. (2013)
		3% LA	5	23	1.0	2.4	613	ND	Rahman et al. (2013)
		NaOCl	5	23	1.4	10.6	630	100	Rahman et al. (2013)
		SAEW	3	40	1.2–1.6	6.3	826	30	Mansur et al. (2015)
		0.5% FA	3	40	1.2–1.4	2.3	570	ND	Mansur et al. (2015)
		SAEW + 0.5% FA	3	40	2.6	3.0	1091	15	Mansur et al. (2015)
	Listeria monocytogenes	SAEW	5	23	1.7	6.8	700–720	10	Rahman et al. (2013)

(continued)

Table 5.2 (continued)

Product	Microorganism	Treatment condition	Exposure time (min)	Temp. (°C)	Reduction (log_{10} CFU) (g)	EW property			Reference
						pH	ORP (mV)	ACC (mg/L)	
		3% CaL	5	23	1.3	6.5	290	ND	Rahman et al. (2013)
		SAEW + 3% CaL	5	23	3.2	6.8	245	1	Rahman et al. (2013)
		AO	5	23	0.4	6.6	1245	5.2 mg O_3/L of water	Rahman et al. (2013)
		3% LA	5	23	1.0	2.4	613	ND	Rahman et al. (2013)
		NaOCl	5	23	1.5	10.6	630	100	Rahman et al. (2013)
		SAEW	3	40	1.2–1.6	6.3	826	30	Mansur et al. (2015)
		0.5% FA	3	40	1.2–1.4	2.3	570	ND	Mansur et al. (2015)
		SAEW + 0.5% FA	3	40	2.7	3.0	1091	15	Mansur et al. (2015)
	Salmonella typhimurium	SAEW	3	40	1.2–1.6	6.3	826	30	Mansur et al. (2015)
		0.5% FA	3	40	1.2–1.4	2.3	570	ND	Mansur et al. (2015)
		SAEW + 0.5% FA	3	40	3.0	3.0	1091	15	Mansur et al. (2015)

(continued)

Table 5.2 (continued)

Product	Microorganism	Treatment condition	Exposure time (min)	Temp. (°C)	Reduction ($\log_{10}$ CFU) (g)	EW property			Reference
						pH	ORP (mV)	ACC (mg/L)	
	Staphylococcus aureus	SAEW	3	40	1.2–1.6	6.3	826	30	Mansur et al. (2015)
		0.5% FA	3	40	1.2–1.4	2.3	570	ND	Mansur et al. (2015)
		SAEW – 0.5% FA	3	40	2.4	3.0	1091	15	Mansur et al. (2015)
	Total viable counts	3% CaL	5	23	1.0	6.5	290	ND	Rahman et al. (2013)
		SAEW + 3% CaL	5	23	2.2	6.8	245	1	Rahman et al. (2013)
	Yeast and mould	SAEW + 3% CaL	5	23	1.6	6.8	245	1	Rahman et al. (2013)

FA Fumaric acid; *CaL* Calcium lactate

(1.8 log_{10} CFU/g), and psychrotrophs (1.4 log_{10} CFU/g) recovered from meat after 7 days of refrigerated storage. The results showed that the carrageenan and gelatine hydrosols containing AEW solutions could effectively extend the shelf life of fresh pork meat due to the continuous antimicrobial effects.

5.2.3 *Effect of EW Treatment on the Quality and Microbial Stability of Fresh Pork*

5.2.3.1 Effect of EW Treatment on Physiochemical Properties of Fresh Pork

Lipid oxidation is related to food quality degradation, which is a major cause of the oxidative deterioration of meat products. A thiobarbituric acid-reactive substances (TBARS) value was used to evaluate the degree of lipid oxidation (Botsoglou et al. 2002; Gatellier et al. 2007), and the values in meat products exceeding 2.0 mg/kg may induce undesirable off-odours and flavours (Byun et al. 2003). No significant differences were observed in the initial TBARS values between the controls and the EW-treated samples (Brychcy et al. 2016; Rahman et al. 2013). However, TBARS values increased as the storage time increased, and the increases in meat products treated with EW were slower than those of untreated samples. Rahman et al. (2013) observed significant changes in the TBARS values in the refrigerated, untreated, fresh pork after 6 days of storage. During the 12-day storage, the smallest changes were observed in the SAEW + CaL treatment, followed by treatment with SAEW alone. These results demonstrate that lipid oxidation could be inhibited by the EW treatment, thereby extending the shelf life.

Proteins, especially peptides produced from the hydrolysed reaction of proteins, have excellent potential as antioxidants in food because they can inhibit lipid oxidation through multiple mechanisms, including scavenging free radicals, inactivation of reactive oxygen species, chelation of pro-oxidative transition metals, reduction of hydroperoxides, and alteration of the physical properties of food systems (Elias et al. 2008; Serpen et al. 2012). Proteins and peptides are rich in meat products. Therefore, Brychy et al. (2016) measured the antioxidant activity of pork meat using a radical scavenging assay (DPPH) and reducing ability assay (FRAP). They confirmed that the electrolysis of 0.1% NaCl solution significantly improved the antioxidant activity of meat samples compared to treatment with hydrosols containing non-electrolysed salt solutions. In comparison with the DPPH scavenging capacities, the reductive antioxidant power of the meat samples was lower. This may result from the poor ability of meat antioxidants to reduce ferric ion to its ferrous form.

The initial pH values of fresh pork samples after the EW treatment did not alter much and were slightly acidic (Rahman et al. 2013; Mansur et al. 2015). Rahman et al. (2013) reported that the pH of fresh pork with AEW, SAEW or SAEW + CaL treatment slowly increased during refrigerated storage, with an increase during the

12-day storage of only 0.17 pH units. Mansur et al. (2015) observed pH changes in all pork samples after treatment with SAEW or SAEW + 0.5% FA at 40 °C. Similarly, the pH values increased gradually over time and then became slightly acidic or close to neutral. The pH values could increase along with an increase in the total bacterial counts (Rousset and Renerre 1991; Rahman et al. 2013). The bacteria break down the protein compounds of meat products and then produce amines, which could be the cause of the increase in pH during storage (Gill 1983). Holmer et al. (2009) noted that higher pH and longer ageing periods will cause increased microbial proliferation and decreased shelf life. These studies suggest that an increase in pH and the subsequent deterioration in the quality of the meat can be improved by the EW treatment.

Few studies have focused on moisture content changes after EW treatment. Sodium chloride (NaOH) has been reported to successfully increase the water-holding capacity of a meat system when used in conjunction with salts or phosphates. Therefore, Rigdon et al. (2017) evaluated the effects of AlEW and AlEW incorporated with 2.5% potassium-lactate on the water-holding capacity and Warner-Bratzler shear force (WBSF) to determine if AlEW solutions could be used as a replacement for traditional enhancement solutions. However, the research showed that the use of AlEW or AlEW + potassium lactate did not improve the tenderness or juiciness of the pork chops, with some samples losing even more moisture than the control (no enhancement) during the storage, thawing or cooking.

Moreover, Brychcy et al. (2015b) detected haem pigment changes in fresh pork after spraying the samples for 120 s with 0.001%, 0.01% or 0.1% NaCl solution electrolyzed for 5 or 10 min. The results showed that electrolysis of more concentrated NaCl solutions (0.01% or 0.1%) caused a decrease in deoxymyoglobin (DeoMb) and metmyoglobin (MetMb), and an increase in the oxymyoglobin (OxyMb) content. The gelatine hydrosols containing electrolyzed 0.1% NaCl caused a significant increase in the OxyMb and a significant decrease in the DeoMb concentrations in pork relative to samples treated by sols containing NaCl solution without electrolysis (Brychcy et al. 2016). However, after 7 days in storage, samples treated with gelatine AEW-containing sols showed decreased OxyMb contents relative to samples treated with sols containing NaCl solution without electrolysis.

5.2.3.2 Effect of EW Treatment on Sensory Attributes of Fresh Pork

The Hunter $L^*a^*b^*$ values were measured to determine whether EW treatment significantly oxidized the pigments on the pork meat surface (Brychcy et al. 2015b, 2016; Fabrizio and Cutter 2005; Han et al. 2018; Shimamura et al. 2016). In the study by Brychcy et al. (2015b), the L^* (brightness) value of fresh pork was not significantly affected by the electrolysis time of the NaCl solution, but this value declined with a decrease in the NaCl concentration. A decrease in the L^* value of the samples sprayed with EW solutions can be associated with the oxidative properties of the AEW. The parameter a^* (redness/greenness) and b^* (yellowness/blueness) values from the EW-treated samples were significantly different from those of the untreated samples. In the case of pork samples treated with carrageenan or gelatine sols containing AEW solutions, the electrolysis time influenced the L^* value during the 7-day storage (Brychcy et al. 2016). The L^* value significantly improved with an extension of the electrolysis time. Changes in the $L^*a^*b^*$ colour parameters favourably affected sensory evaluations of the meat quality, especially for samples treated with the AEW and gelatine-containing hydrosols.

In some studies (Brychcy et al. 2015b, 2016; Fabrizio and Cutter 2005; Rahman et al. 2013; Rigdon et al. 2017; Mansur et al. 2015), the sensory quality of the meat products during the shelf life was evaluated by panellists to better understand the consumer perspective and acceptance of EW-treated meat for colour, freshness, texture, decay and off-flavour (odour). Although the EW treatment slightly influenced the colour parameters of the pork, the sensory studies proved that these colour changes were not significant for consumers and even resulted in higher acceptability of the meat products. The residual NaCl content of EW (although very low) gives meat samples fresh colour and keeps the meat odour as well. No significant difference was observed in the odour, texture and overall appearance of these samples immediately after EW treatment. However, sensory scores of the treated fresh pork decreased more slowly than those of the control as the storage time increased. Rigdon et al. (2017) evaluated the sensory characteristics of fresh pork treated with the AlEW, which shows antimicrobial activity. No differences were detected for off-flavours with the AlEW-treated pork chops and non-enhanced controls. Moreover, the AlEW-treated pork chops showed no improvement in tenderness or juiciness. The use of the AlEW treatment even showed slightly adverse effects on the sensory properties of the pork products.

5.2.3.3 Effect of EW Treatment on Microbial Stability of Fresh Pork During Storage

Generally, the end of shelf life is dependent on the microbial proliferation during storage; therefore, investigation of microbial stability of EW-treated products is also quite important for food safety. In accordance with the International Commission on Microbiological Specifications for Foods (ICMSF 1986), a TVC value $\geq 7 \log_{10}$ CFU/g is considered as the upper microbiological limit for fresh meat of good quality. The end of the shelf life occurs when the count of a group of microorganisms reaches an unacceptable level or when the sensory panel rejects the samples. Studies have demonstrated that SAEW or AEW treatments effectively improved microbial stability and extended the shelf life of fresh pork at different temperatures with marginal changes to the sensory quality (Brychcy et al. 2016; Rahman et al. 2013).

5.3 Application of EW Treatment in Decontamination of Fresh Beef

Beef, a source of high-quality protein and nutrients, is the third most widely consumed meat in the world, accounting for approximately one-quarter of meat production worldwide. During slaughtering and processing of beef products, the meat is easily subjected to contaminations from various sources. A few studies (Al-Holy and Rasco 2015; Ding et al. 2010; Tango et al. 2014; Veasey and Muriana 2016) have examined the microbial decontamination efficacy of the EW treatment on fresh beef (Table 5.3). Changes in physiochemical qualities and microbial stability of fresh beef after the EW treatment during storage have also been investigated.

5.3.1 Microbial Decontamination Efficacy of EW Treatment on Fresh Beef

Al-Holy and Rasco (2015) reported the effectiveness of AEW (pH 2.3, ACC 38 mg/L) against *E. coli* O157:H7, *L. monocytogenes* and *S. typhimurium* inoculated on beef surfaces. The *L. monocytogenes* was found more resistant (1.3 $\log_{10}$ reduction) than *E. coli* O157:H7 and *S. typhimurium* to the AEW treatment even though the AEW contact time was prolonged, whereas the SAEW treatment (pH 6.3, ORP 820–934 mV, ACC 25 mg/L) exhibited significantly bactericidal effects against *S. aureus*, *L. monocytogenes* and *E. coli* O157:H7 on fresh beef (Tango et al. 2014). The resistance of *L. monocytogenes* to EW treatment seems to be dependent on the pH values. Furthermore, Tango et al. (2014) observed reductions in bacterial counts by 0.63–2.52 $\log_{10}$ CFU/g with different times, but increases in the contact time, compared to increases in the temperature, were not effective for all treatments.

The efficacy of the EW treatment on beef cheek meat and beef liver has also been reported (Table 5.4). Kalchayanand et al. (2008) investigated the effectiveness of AEW alone (13 s) or AlEW (13 s) + AEW (13 s) against *E. coli* O157:H7 inoculated on beef cheek meat. However, the counts of *E. coli* O157:H7 were reduced to less than 0.5 $\log_{10}$ CFU/g. Neither treatment was effective as a decontamination strategy for microbial reduction. Shimamura et al. (2016) sterilized *S. enteritidis*, *E. coli* and *S. aureus* inoculated on beef liver with AlEW and AEW at 4 and 25 °C for 3 min. After treatment, the numbers of bacteria were significantly reduced by more than 3.0 $\log_{10}$ CFU/g.

Notably, less effective bactericidal effects on foods of animal origin have been observed than were observed with these suspensions. This fact may be due to factors such as organic matter (Hricova et al. 2008). The presence of organic matter reduced the ACCs and ORPs rapidly, and the ACCs and the antimicrobial activity of the AEW were reduced in proportion to the amount of organic residue present on the food surfaces (Ayebah et al. 2006; Oomori et al. 2005). Ayebah et al. (2006) suggested

Table 5.3 Antimicrobial efficacy of the various types of electrolyzed water treatments against pathogens on fresh beef

Microorganism	Treatment condition	Exposure time (min)	Temp. (°C)	Reduction (log_{10} CFU) (g)	EW property			Reference
					pH	ORP (mV)	ACC (mg/L)	
E. coli O157:H7	AEW	3	23	1.6	2.3–2.7	1110–1200	50	Ding et al. (2010)
	SAEW	3	23	1.7	6.2	500–520	5	Ding et al. (2010)
	AEW	10	22	>1.5	2.3	NA	38	Al-Holy and Rasco (2015)
	EW	0.5	23	<1	NA	NA	50	Veasey and Muriana (2016)
	SAEW	1	40	1.1	6.3	820–934	25	Tango et al. (2014)
	SAEW	3	40	1.7	6.3	820–934	25	Tango et al. (2014)
	SAEW	5	40	1.8	6.3	820–934	25	Tango et al. (2014)
L. monocytogenes	AEW	10	22	1.3	2.3	NA	38	Al-Holy and Rasco (2015)
	SAEW	1	40	1.1	6.3	820–934	25	Tango et al. (2014)
	SAEW	3	40	1.9	6.3	820–934	25	Tango et al. (2014)
	SAEW	5	40	1.9	6.3	820–934	25	Tango et al. (2014)

(continued)

Table 5.3 (continued)

Microorganism	Treatment condition	Exposure time (min)	Temp. (°C)	Reduction ($\log_{10}$ CFU) (g)	EW property			Reference
					pH	ORP (mV)	ACC (mg/L)	
S. typhimurium	AEW	10	22	1.4	2.3	NA	38	Al-Holy and Rasco (2015)
	SAEW	1	40	0.8	6.3	820–934	25	Tango et al. (2014)
	SAEW	3	40	1.6	6.3	820–934	25	Tango et al. (2014)
	SAEW	5	40	1.7	6.3	820–934	25	Tango et al. (2014)
S. aureus	SAEW	1	40	0.7	6.3	820–934	25	Tango et al. (2014)
	SAEW	3	40	1.3	6.3	820–934	25	Tango et al. (2014)
	SAEW	5	40	1.3	6.3	820–934	25	Tango et al. (2014)
Total mesophiles	AEW	10	22	2.0	2.3	NA	38	Al-Holy and Rasco (2015)
TVC	SAEW	1	40	1.5	6.3	820–934	25	Tango et al. (2014)
	SAEW	3	40	2.3	6.3	820–934	25	Tango et al. (2014)
	SAEW	5	40	2.2	6.3	820–934	25	Tango et al. (2014)

Table 5.4 Antimicrobial efficacy of electrolyzed water treatments against pathogens on beef cheek and beef liver

Product	Microorganism	Treatment condition	Exposure time	Temp. (°C)	Reduction (log_{10} CFU)	EW properties			Reference
						pH	ORP (mV)	ACC (mg/L)	
Beef cheek meat	*E. coli* O157:H7	AEW	26 s	25	0.3 cm^2	2.8	1190	60	Kalchayanand et al. (2008)
		AlEW + AEW	AlEW, 13 s AEW, 13 s	25	0.2 cm^2	AEW 2.8	AEW 1190	AEW 60	Kalchayanand et al. (2008)
Beef liver (10 g)	*E. coli*	AlEW + AEW	AlEW, 3 min AEW, 3 min	4	3.6 g	AlEW 11.5 AEW 2.5	AEW 975	AEW 30	Shimamura et al. (2016)
		AlEW + AEW	AlEW, 3 min AEW, 3 min	25	3.7 g	AlEW 11.5 AEW 2.5	AEW 960	AEW 14	Shimamura et al. (2016)
	S. enteritidis	AlEW + AEW	AlEW, 3 min AEW, 3 min	4	3.5 g	AlEW 11.5 AEW 2.5	AEW 975	AEW 30	Shimamura et al. (2016)
		AlEW + AEW	AlEW, 3 min AEW, 3 min	25	3.2 g	AlEW 11.5 AEW 2.5	AEW 960	AEW 14	Shimamura et al. (2016)
	S. aureus	AlEW + AEW	AlEW, 3 min AEW, 3 min	4	3.6 g	AlEW 11.5 AEW 2.5	AEW 975	AEW 30	Shimamura et al. (2016)
		AlEW + AEW	AlEW, 3 min AEW, 3 min	25	3.1 g	AlEW 11.5 AEW 2.5	AEW 960	AEW 14	Shimamura et al. (2016)

AlEW Alkaline electrolyzed water

treating food samples with AlEW prior to AEW. AlEW helps remove food residues and make the adherent bacteria more susceptible to AEW.

5.3.2 *Hurdle Enhancement of EW Treatment with Other Interventions on Fresh Beef*

A variety of hurdle technologies, physical, chemical, physiochemical methods, etc., are currently under development for the decontamination of raw meat products. In general, these methods are mild because they guarantee natural appearance while eliminating pathogens. Leistner (2000) proposed that the combination of different hurdle technologies may enhance antimicrobial efficacy. Huang et al. (2008) reported that the stability and colour of fresh meat are often improved by dipping the meat into solutions containing organic acid. The FA treatment effectively reduced counts of *L. monocytogenes*, *E. coli* O157:H7 and *S. typhimurium* on beef during storage (Podolak et al. 1996). Therefore, Tango et al. (2014) studied the hurdle-processing effect of SAEW and FA on several pathogens on fresh beef and compared the efficacy of the SAEW with the AEW and FA (Table 5.5). The synergic effect of the SAEW combined with FA at 40 °C for 3 min exhibited a reduction of 1.10 $\log_{10}$ CFU/g more than the sum of reductions caused by the single treatments. The SAEW treatment followed by the FA treatment significantly improved the efficacy of the SAEW for eliminating pathogens on red meat including fresh pork and beef. The pH and undissociated form of FA may help disrupt the outer membrane of the bacteria to allow the HOCl to penetrate the cytoplasmic membrane.

5.3.3 *Effect of EW Treatment on the Quality and Microbial Stability of Fresh Beef*

Most of the published reports on fresh beef decontamination focus on the antimicrobial activity immediately after the EW treatment. Relatively few studies (Botta et al. 2018; Tango et al. 2014) have monitored the effects on the physiochemical, organoleptic, nutritional properties and microbial stability of meat following the EW treatment.

5.3.3.1 Effect of EW Treatment on Physiochemical Properties of Fresh Beef

Tango et al. (2014) evaluated the changes in colour, odour and texture of the EW-treated fresh beef compared to the control by the sensory index (SI) and pH. The storage temperature of 4 °C resulted in a slower decline in sensory acceptance than

Table 5.5 Microbial decontamination efficacy of electrolyzed water and other interventions against pathogens on fresh beef

Microorganism	Treatment condition	Exposure time (min)	Temp. (°C)	Reduction (log_{10} CFU) (g)	EW property			Reference
					pH	ORP (mV)	ACC (mg/L)	
E. coli O157:H7	SAEW	3	25	1.1	6.3	820–934	25	Tango et al. (2014)
	0.5% FA	3	25	1.1	2.3	ND	ND	Tango et al. (2014)
	SAEW + 0.5% FA	3	25	1.5	6.3	820–934	25	Tango et al. (2014)
	SAEW	3	40	1.8	6.3	820–934	25	Tango et al. (2014)
	0.5% FA	3	40	1.7	2.3	ND	ND	Tango et al. (2014)
	SAEW + 0.5% FA	3	40	3.0	6.3	820–934	25	Tango et al. (2014)
	SAEW	3	60	2.1	6.3	820–934	25	Tango et al. (2014)
	0.5% FA	3	60	2.0	2.3	ND	ND	Tango et al. (2014)
	SAEW + 0.5% FA	3	60	3.2	6.3	820–934	25	Tango et al. (2014)
L. monocytogenes	SAEW	3	25	1.1	6.3	820–934	25	Tango et al. (2014)
	0.5% FA	3	25	0.8	2.3	ND	ND	Tango et al. (2014)

(continued)

Table 5.5 (continued)

Microorganism	Treatment condition	Exposure time (min)	Temp. (°C)	Reduction (log_{10} CFU) (g)	EW property			Reference
					pH	ORP (mV)	ACC (mg/L)	
	SAEW + 0.5% FA	3	25	1.7	6.3	820–934	25	Tango et al. (2014)
	SAEW	3	40	1.9	6.3	820–934	25	Tango et al. (2014)
	0.5% FA	3	40	1.8	2.3	ND	ND	Tango et al. (2014)
	SAEW + 0.5% FA	3	40	2.8	6.3	820–934	25	Tango et al. (2014)
	SAEW	3	60	2.2	6.3	820–934	25	Tango et al. (2014)
	0.5% FA	3	60	2.1	2.3	ND	ND	Tango et al. (2014)
	SAEW + 0.5% FA	3	60	3.2	6.3	820–934	25	Tango et al. (2014)
S. typhimurium	SAEW	3	25	0.7	6.3	820–934	25	Tango et al. (2014)
	0.5% FA	3	25	0.6	2.3	ND	ND	Tango et al. (2014)
	SAEW + 0.5% FA	3	25	1.5	6.3	820–934	25	Tango et al. (2014)
	SAEW	3	40	1.6	6.3	820–934	25	Tango et al. (2014)
	0.5% FA	3	40	1.5	2.3	ND	ND	Tango et al. (2014)

(continued)

Table 5.5 (continued)

Microorganism	Treatment condition	Exposure time (min)	Temp. (°C)	Reduction (log_{10} CFU) (g)	EW property			Reference
					pH	ORP (mV)	ACC (mg/L)	
	SAEW + 0.5% FA	3	40	2.7	6.3	820–934	25	Tango et al. (2014)
	SAEW	3	60	1.8	6.3	820–934	25	Tango et al. (2014)
	0.5% FA	3	60	1.7	2.3	ND	ND	Tango et al. (2014)
	SAEW + 0.5% FA	3	60	3.0	6.3	820–934	25	Tango et al. (2014)
S. aureus	SAEW	3	25	0.8	6.3	820–934	25	Tango et al. (2014)
	0.5% FA	3	25	0.7	2.3	ND	ND	Tango et al. (2014)
	SAEW + 0.5% FA	3	25	1.2	6.3	820–934	25	Tango et al. (2014)
	SAEW	3	40	1.3	6.3	820–934	25	Tango et al. (2014)
	0.5% FA	3	40	1.3	2.3	ND	ND	Tango et al. (2014)
	SAEW + 0.5% FA	3	40	2.7	6.3	820–934	25	Tango et al. (2014)
	SAEW	3	60	1.6	6.3	820–934	25	Tango et al. (2014)

(continued)

Table 5.5 (continued)

Microorganism	Treatment condition	Exposure time (min)	Temp. (°C)	Reduction (log_{10} CFU) (g)	EW property			Reference
					pH	ORP (mV)	ACC (mg/L)	
	0.5% FA	3	60	1.6	2.3	ND	ND	Tango et al. (2014)
	SAEW + 0.5% FA	3	60	2.9	6.3	820–934	25	Tango et al. (2014)
TVC	SAEW	3	25	1.9	6.3	820–934	25	Tango et al. (2014)
	0.5% FA	3	25	1.7	2.3	ND	ND	Tango et al. (2014)
	SAEW + 0.5% FA	3	25	2.8	6.3	820–934	25	Tango et al. (2014)
	SAEW	3	40	2.3	6.3	820–934	25	Tango et al. (2014)
	0.5% FA	3	40	2.2	2.3	ND	ND	Tango et al. (2014)
	SAEW + 0.5% FA	3	40	3.7	6.3	820–934	25	Tango et al. (2014)
	SAEW	3	60	2.7	6.3	820–934	25	Tango et al. (2014)
	0.5% FA	3	60	2.6	2.3	ND	ND	Tango et al. (2014)
	SAEW + 0.5% FA	3	60	3.7	6.3	820–934	25	Tango et al. (2014)

10 °C. Based on the evaluated sensory quality, the acceptability limit was set at a value of 2.5 (Kreyenschmidt et al. 2010). The difference between the SI of treated beef was longer than untreated beef ($p < 0.05$). However, no significant difference was observed between the combined treatment with SAEW and AEW at 4 °C. The difference in SI could be caused by the presence of different species of TVC with different metabolic activities and sensory quality deterioration depending on the type of fresh products (Jacxsens et al. 2003). The pH values of meat samples ranged from 4.33 to 5.58 after the EW treatment and did not differ remarkably between the treatments during 8 days of storage at 10 °C. However, on the 12th day of the experiment at 4 °C, a significant difference was observed among the SAEW + FA, AEW and SAEW treatments.

5.3.3.2 Effect of EW Treatment on Microbial Stability of Fresh Beef During Storage

The synergic effect of SAEW and 0.5% FA on the microbial stability on fresh beef during storage was remarkably effective (Tango et al. 2014). The TVC of fresh beef meat treated with SAEW + 5% FA reached approximately 7.0 $\log_{10}$ CFU/g on 9 and 7 days at 4 and 10 °C, respectively. Under the same conditions of storage, the TVC was less under the SAEW or AEW treatments, which reached the shelf life limit over a period of 6 and 5 days, respectively, whereas the untreated beef samples reached 7.0 $\log_{10}$ CFU/g at almost 4 and 2 days, respectively. Thus, a shelf life extension of 5 days was obtained with the combined treatment, and this could be due to the bactericidal effect of the FA, which can passively diffuse through the bacteria cell wall in its undissociated forms. Furthermore, Tango et al. (2014) calculated the end of the shelf life in terms of sensory acceptance (in days) using regression analysis. The correlations between the TVC growth and the SI throughout the shelf life experiment at 4 and 10 °C storage temperatures were significant.

In another study, to extend the shelf life of beefsteaks, Botta et al. (2018) conducted experiments with aqueous ozone (AO) (6.00 ± 0.25 mg/L) and EW (ACC 100 mg/L) for 90 s at 4 °C before vacuum packaging (VP). However, the results showed limited effectiveness in reducing the initial microbial counts of the products. Moreover, treatments with AO and EW were incapable of substantially modifying the microbiota composition, dynamics and the related volatility during 15 days of VP storage under refrigerated conditions (4 °C).

5.4 Application of EW Treatment in Decontamination of Poultry Products

5.4.1 Application of EW Treatment in Decontamination of Poultry

Poultry carcasses can be potentially contaminated with food-borne pathogens as a result of exposure to faeces during and after slaughter. *Campylobacter*, which is often implicated as a risk factor for food-borne diseases in humans, is especially important. Outbreaks of *S. enteritidis* have been associated with contaminated meat, particularly with poultry (Friedman et al. 2000). EW as a novel antimicrobial has been reported to be effective against *Campylobacter jejuni*, *E. coli*, *L. monocytogenes*, *Salmonella* spp, *S. aureus* and mesophiles (Al-Holy and Rasco 2015; Duan et al. 2017; Fabrizio et al. 2002; Kim et al. 2005; Northcutt et al. 2007; Rahman et al. 2012; Shimamura et al. 2016) that are associated with poultry carcasses and chicken wings (Table 5.6).

5.4.1.1 Microbial Decontamination Efficacy of EW Treatment on Poultry

To explore the use of the EW treatment against pathogens associated with the poultry, broiler carcasses were submerged (4 °C, 45 min) in various antimicrobial interventions, spray-washed (25 °C, 15 s) by these antimicrobials or subjected to multiple interventions (AlEW spray + immersed in AEW) (Fabrizio et al. 2002). Spray-washing treatments with any compounds did not reduce *S. typhimurium* at day 0, whereas submersion in the AEW and combination treatment (AlEW + AEW) reduced *S. typhimurium* by approximately 0.8 and 2.1 $\log_{10}$/mL of rinsate, respectively. Chicken wings artificially contaminated with *C. jejuni* were also treated with AEW for 10 min, which resulted in a reduction in the bacterial load by 3.0 $\log_{10}$ CFU/g (Park et al. 2002). Similarly, the bactericidal effect (2.3 $\log_{10}$ CFU/g reduction) of AEW on *C. jejuni* in chicken carcasses by the immersion treatment was reported by Kim et al. (2005), and the combination of the immersion and spraying treatment did not improve the bactericidal effects of the EW solutions. Northcutt et al. (2007) investigated the effects of spray washing broiler carcasses with AEW for 5, 10, or 15 s. The results showed that AEW washing spray significantly reduced the levels of bacteria recovered from carcasses. Increasing the washing time from 5 to 10 s lowered the levels of total aerobic bacteria, *E. coli*, *Campylobacter*, and *Salmonella*, but no further microbiological reductions were observed when the washing time was extended from 10 to 15 s. Rahman et al. (2012) evaluated the effectiveness of SAEW on *L. monocytogenes* and *S. typhimurium* inoculated on fresh chicken breast meat. After 10 min of treatment with SAEW containing 10 ppm of active chlorine at 23 °C, reductions of 2.32 and 1.9 $\log_{10}$ CFU/g were recorded for *L. monocytogenes* and *S. typhimurium*, respectively. However, Al-Holy and Rasco (2015) reported that

Table 5.6 Antimicrobial efficacy of various types of electrolyzed water treatments against pathogens on poultry

Product	Microorganism	Treatment condition	Exposure time	Temp. (°C)	Reduction (log_{10} CFU)	EW property			Reference
						pH	ORP (mV)	ACC (mg/L)	
Chicken carcasses	Aerobic bacteria	AEW	45 min	4	1.3 mL of rinsate	2.6	1150	50	Fabrizio et al. (2002)
		25 mL AlEW + AEW	NA	4	2.4 mL of rinsate	AlEW, 11.6 AEW, 2.6	AlEW, −795 AEW, 1150	AEW, 50	Fabrizio et al. (2002)
	Coliforms	AEW	45 min	4	1.1 mL of rinsate	2.6	1150	50	Fabrizio et al. (2002)
		25 mL AlEW + AEW	NA	4	1.6 mL of rinsate	AlEW, 11.6 AEW, 2.6	AlEW, −795 AEW, 1150	AEW, 50	Fabrizio et al. (2002)
		AEW	15 s	4	0.7–1.1 cm^2	2.46	1126	58	Duan et al. (2017)
		SAEW	15 s	NA	0.6–1.1 cm^2	5.98	865	30	Duan et al. (2017)
	Campylobacter jejuni	AEW	40 min	NA	2.3 g	2.5	1140	47.2	Kim et al. (2005)
		6 L AEW	10 s	NA	2.1 mL	2.4	1180	NA	Northcutt et al. (2007)
	E. coli	AEW	45 min	4	1.1 mL of rinsate	2.6	1150	50	Fabrizio et al. (2002)

(continued)

Table 5.6 (continued)

Product	Microorganism	Treatment condition	Exposure time	Temp. (°C)	Reduction (log_{10} CFU)	EW property			Reference
						pH	ORP (mV)	ACC (mg/L)	
		25 mL AlEW + AEW	NA	4	1.5 mL of rinsate	AlEW, 11.6 AEW, 2.6	AlEW, – 795 AEW, 1150	AEW, 50	Fabrizio et al. (2002)
		AlEW + AEW	AlEW, 3 min AEW, 3 min	25	3.6 g	AlEW 11.5 AEW 2.5	960	14	Shimamura et al. (2016)
		6 L AEW	10 s	NA	1.8 mL	2.4	1180	NA	Northcutt et al. (2007)
	E. coli O157:H7	AEW	10 min	22	1.5–1.6 g	2.3	NA	38	Al-Holy and Rasco (2015)
	L. monocytogenes	SAEW	10 min	25	2.3 g	6.2–6.5	760–770	10	Rahman et al. (2012)
		AEW	10 min	25	2.3 g	2.5	1100–1120	50	Rahman et al. (2012)
		AEW	10 min	22	1.1 g	2.3	NA	38	Al-Holy and Rasco (2015)

(continued)

Table 5.6 (continued)

Product	Microorganism	Treatment condition	Exposure time	Temp. (°C)	Reduction (log_{10} CFU)	EW property			Reference
						pH	ORP (mV)	ACC (mg/L)	
	S. typhimurium	AEW	45 min	4	0.8 mL of rinsate	2.6	1150	50	Fabrizio et al. (2002)
		25 mL AlEW + AEW	NA	4	2.1 mL of rinsate	AlEW, 11.6 AEW, 2.6	AlEW, −795 AEW, 1150	AEW, 50	Fabrizio et al. (2002)
		SAEW	10 min	25	1.9 g	6.2–6.5	760–770	10	Rahman et al. (2012)
		AEW	10 min	25	1.9 g	2.5	1100–1120	50	Rahman et al. (2012)
		AEW	10 min	22	1.5 g	2.3	NA	38	Al-Holy and Rasco (2015)
		6 L AEW	10 s	NA	0.9 mL	2.4	1180	NA	Northcutt et al. (2007)
	S. enteritidis	AlEW + AEW	AlEW, 3 min AEW, 3 min	25	3.3 g	AlEW 11.5 AEW 2.5	960	14	Shimamura et al. (2016)

(continued)

Table 5.6 (continued)

Product	Microorganism	Treatment condition	Exposure time	Temp. (°C)	Reduction (log_{10} CFU)	EW property			Reference
						pH	ORP (mV)	ACC (mg/L)	
	Salmonella spp.	6 L AEW	10 s	NA	2.7 mL	2.4	1180	NA	Northcutt et al. (2007)
	S. aureus	AlEW + AEW	AlEW, 3 min AEW, 3 min	25	3.9 g	AlEW 11.5 AEW 2.5	960	14	Shimamura et al. (2016)
	Total viable counts	SAEW	10 min	25	1.4 g	6.2–6.5	760–770	10	Rahman et al. (2012)
		AEW	10 min	25	1.4 g	2.5	1100–1120	50	Rahman et al. (2012)
		AEW	15 s	NA	0.6–0.8 cm^2	2.5	1126	58	Duan et al. (2017)
		SAEW	15 s	NA	0.5–0.8 cm^2	6.0	865	30	Duan et al. (2017)
	Total mesophiles	AEW	10 min	22	2 g	2.3	NA	38	Al-Holy and Rasco (2015)
Chicken wings	*C. jejuni*	AEW	10 min	23	3.0 g	2.6	1082	51.6	Park et al. (2002)

bactericidal effect of the AEW on *L. monocytogenes* (1.1 $\log_{10}$ reduction) was less than the effect on *E. coli* O157:H7 (1.5–1.6 $\log_{10}$ reduction) and *S. typhimurium* (1.5 $\log_{10}$ reduction) on chicken carcass samples.

Shimamura et al. (2016) sterilized pathogens inoculated on chicken breasts with AlEW and AEW at 4 and 25 °C for 3 min. After treatment, the numbers of *S. enteritidis*, *E. coli* and *S. aureus* were significantly reduced by more than 3.0 $\log_{10}$ CFU/g. The effects of the various disinfectants on the initial microbial load on the chilled chicken carcasses were evaluated and compared. Sprays of 2% LA, AEW and SAEW were the most effective treatments, showing reductions of 0.5–0.8 $\log_{10}$ CFU/cm^2 and 0.5–1.0 $\log_{10}$ CFU/cm^2 in the TVC and total coliforms, respectively (Duan et al. 2017).

5.4.1.2 Effect of EW Treatment on the Quality and Microbial Stability of Poultry

Effect of EW Treatment on Physiochemical Properties of Poultry

Spoilage organisms are associated with undesirable biochemical changes in meat products, such as protein degradation, biogenic amines, volatile organic compounds and so on (Nychas et al. 2013). The total volatile basic nitrogen (TVB-N) is commonly used as an important indicator of meat product freshness. According to the Chinese Standard Fresh and Frozen Poultry Product (GB 16869-2005) guidelines, the values of TVB-N should be less than 15.0 mg N/100 g of fresh chicken meat. Duan et al. (2017) evaluated the TVB-N values of chicken carcass samples after AEW or SAEW treatments had been evaluated during refrigerated storage. The results showed that the TVB-N values immediately after EW treatment with 15-s sprays did not alter significantly. However, during the refrigerated storage, the TVB-N values of the controls increased rapidly and exceeded the limit at day 6. After chicken samples were treated with AEW or SAEW, the increases in the TVB-N values for the chicken samples occurred much more slowly and had not reached the limit at the end of the 8-day storage. The EW treatment effectively maintained a safe level of TVB-N during refrigerated storage, especially when samples were treated with SAEW.

Several studies have also measured the TBARS values of poultry products (Rahman et al. 2012; Shimamura et al. 2016; Duan et al. 2017). The trend observed for the TBARS was similar in chicken carcass samples to that observed in red meat. Rahman et al. (2012) reported that TBARS values in chicken breast for the SAEW or AEW treatment varied between 0.8 and 2.0 malonaldehyde (MDA)/kg meat. Loss of oxidation stability in the control and EW treated samples occurred after 6 and 10 days of storage, respectively, at 5 °C, with corresponding increases in the TBARS to $\geq$2.0 mg (MDA)/kg. The combined AlEW and AEW treatment also exhibited similar results in chicken breast meat (Shimamura et al. 2016). This result demonstrates that lipid oxidation could be inhibited by the EW treatment, thereby extending the shelf life.

The pH changes in fresh or chilled chicken meat samples treated with EW during the refrigerated storage were similar to changes observed in red meat (Rahman et al. 2012; Shimamura et al. 2016; Duan et al. 2017). The pH values in the chicken meat treated by EW increased within a narrow range during 7 days of refrigerated storage. Shimamura et al. (2016) measured the pH values of chicken breast dipped in AlEW solution (pH 11.5), followed by dipping in AEW solution for 3 min, and observed that pH values with or without the combined treatment of AlEW and AEW changed from 5.3 to 5.6 and 5.4 to 5.7. Generally, the pH value of the meat products increased with storage time, protein destruction and amine production (Lu and Wu 2012). These results suggest that the increase in pH and deterioration of meat quality can be improved by a combination of AlEW and AEW.

Effect of EW Treatment on Nutritional Attributes of Poultry

Quality changes in meat products are known to affect the concentrations of free amino acids (Watanabe et al. 2004). The taste and nutrition of meat are associated with some free amino acids (Nishimura and Kato 1988). Shimamura et al. (2016) examined the changes in free amino acids and dipeptides of chicken breast meats after the combination treatment with AlEW and AEW at day 0. The results did not show a significant difference between the control and treated samples.

Effect of EW Treatment on Sensory Attributes of Poultry

The surface colour of chicken breasts treated with a combination of AlEW and AEW was evaluated by Shimamura et al. (2016). Colour variables did not show significant difference from those in the untreated samples, although after the AlEW and AEW treatment, the meat was brighter than the untreated samples, with an average L^* value of the untreated samples being 51.5, whereas the brightness of the treated samples was approximately 55.4. The colour inside the chicken breast was not changed by combination treatment of AlEW and AEW.

Poultry products are highly perishable foods, and the spoilage time usually varies between 4 and 10 days under refrigerated conditions (Marenzi 1986). The sensory quality of chicken meat during the shelf life, such as colour, odour, texture and overall appearance, was evaluated by panellists to better understand the consumer perspective and acceptance of EW-treated meat. The EW treatment showed slight influence on the colour parameters of the chicken breasts, but the sensory studies showed that colour changes were not significant for consumers (Rahman et al. 2012; Shimamura et al. 2016). Rahman et al. (2012) observed that chicken breasts with SAEW or AEW treatment obtained better sensory scores than the control after 7 days of storage at 5 °C. Both –OH and HOCl in the EW solutions have strong antimicrobial activity and antioxidant effects that help to maintain oxidation stability. Moreover, the residual NaCl content of EW (very little) preserves the colour and odour of meat samples for freshness.

Effect of EW Treatment on Microbial Stability of Poultry During Storage

Rahman et al. (2012) reported that SAEW treatment for 10 min at room temperature could extend the shelf life of chicken breast meat from 1 to 4 days depending on the storage temperature. Lower storage temperature (4 °C) yielded lower SGRs and LTs of the pathogens. During the 10-d storage at 5 °C, chicken samples treated with SAEW or AEW never reached the limit of 7.0 $\log_{10}$ CFU/g. Treated samples never reached the unacceptable limit during the 5 days of storage at 15 °C, whereas the upper limit count was exceeded for TVC on day 4 or 5 in the control meat samples. Similarly, chilled chicken carcasses treated with AEW or SAEW for 15 s at 4 °C showed a 2-day extension of shelf life compared with the control, which exceeded the TVC limit of 7.0 $\log_{10}$ CFU/cm^2 at day 6 (Duan et al. 2017).

5.4.2 *Application of EW Treatment in Decontamination of Shell Eggs*

The egg surface may be contaminated with pathogens, primarily from infected hen's faeces but secondarily by the contaminated environments of egg production facilities (Golden et al. 1988; Humphrey 1994). EW has been reported to be very effective in removing most food-borne microbial loads on shell egg surfaces (Table 5.7) while maintaining the egg quality and extending the shelf life (Achiwa and Nishio 2003; Bialka et al. 2004; Cao et al. 2009; Fasenko et al. 2009; Ma et al. 2009; Ni et al. 2014; Park et al. 2005; Russell 2003).

5.4.2.1 Microbial Decontamination Efficacy of EW Treatment on Shell Eggs

Electrostatic spraying of shell eggs with AEW (hourly for 24 h) reduced *E. coli*, *S. aureus* and *S. typhimurium* by 3–6 orders of magnitude (Russell 2003). The effectiveness of SAEW at different temperatures (4, 20 and 45 °C) for decontamination of the surface of shell eggs was evaluated by Cao et al. (2009). A reduction of 6.5 $\log_{10}$ CFU/g of *S. enteritidis* on shell eggs was achieved by SAEW containing 15 mg/L ACC for 3 min, but for the untreated samples, a reduction of only 0.9–1.2 $\log_{10}$ CFU/g was achieved. Ni et al. (2014) reported that after shell eggs were sprayed with AlEW (10 s), followed by SAEW (10 s) at ACCs of 60, 80 and 100 mg/L, the populations of *S. enteritidis*, *E. coli* O157:H7, *S. aureus* and natural microflora on the eggshells were reduced by 2.4–2.8, 2.2–3.5, 2.2–2.8 and 1.4–3.4 $\log_{10}$ CFU/g, indicating that this combination treatment was more effective than AlEW or SAEW alone. An increasing reduction of *L. monocytogenes* on shell eggs was observed as the ACC of AEW was increased from 16 to 77 mg/L, and the treatment time was

Table 5.7 Antimicrobial efficacy of various types of electrolyzed water treatments against pathogens on shell eggs

Microorganism	EW Type	Application	Exposure time (min)	Temp. (°C)	Reduction (log_{10} CFU)	EW property			Reference
						pH	ORP (mV)	ACC (mg/L)	
Coliforms	SAEW	Spray	10 s	23	1.4 g	5.7	926	100	Ni et al. (2014)
	AlEW + SAEW	Spray	AlEW, 10 s SAEW, 10 s	23	1.2 g	AlEW, 10.5 SAEW, 5.7	AlEW, −755 SAEW, 926	AlEW, 28 SAEW, 100	Ni et al. (2014)
E. coli	AEW	Spray	15 s (hourly)	NA	4–6 egg	2.1	1150	8	Russell (2003)
	AEW	Immersion	3 min	45	3.6 egg	2.7	1150	70–80	Bialka et al. (2004)
	AlEW	Immersion	3 min	45	≥2.3 egg	11.4	−795	NA	Bialka et al. (2004)
	AlEW + AEW	AlEW, immersion AEW, spray	AlEW, 3 min AEW, 1 min	AlEW, 45 AEW, 50	≥3.0 egg	AlEW, 11.4 AEW, 2.7	AlEW, −795 AEW, 1150	AEW, 70–80	Bialka et al. (2004)
E. coli O157:H7	SAEW	Spray	10 s	23	2.7 g	5.7	926	100	Ni et al. (2014)
	AlEW + SAEW	Spray	AlEW, 10 s SAEW, 10 s	23	3.5 g	AlEW, 10.5 SAEW, 5.7	AlEW, −755 SAEW, 926	AlEW, 28 SAEW, 100	Ni et al. (2014)

(continued)

Table 5.7 (continued)

Microorganism	EW Type	Application	Exposure time (min)	Temp. (°C)	Reduction (log_{10} CFU)	EW property			Reference
						pH	ORP (mV)	ACC (mg/L)	
L. monocytogenes	AEW	Spray	15 s (hourly)	NA	1–4 egg	2.1	1150	8	Russell (2003)
	AEW	Immersion	5 min	23	4.0 egg	2.5	1121	77	Park et al. (2005)
	AlEW + AEW	Immersion	AlEW, 1 min AEW, 1 min	23	4.4 egg	AlEW, 11.2 AEW, 2.5	AlEW, – 940 AEW, 1121	AEW, 77	Park et al. (2005)
Natural microflora	SAEW	Spray	10 s	23	3.0 g	5.67	926	100	Ni et al. (2014)
	AlEW + SAEW	Spray	AlEW, 10 s SAEW, 10 s	23	3.4 g	AlEW, 10.5 SAEW, 5.7	AlEW, – 755 SAEW, 926	AlEW, 28 SAEW, 100	Ni et al. (2014)
S. aureus	AEW	Spray	15 s (hourly)	NA	3–6 egg	2.1	1150	8	Russell (2003)
	SAEW	Spray	10 s	23	2.8 g	5.7	926	100	Ni et al. (2014)
	AlEW + SAEW	Spray	AlEW, 10 s SAEW, 10 s	23	2.8 g	AlEW, 10.5 SAEW, 5.7	AlEW, – 755 SAEW, 926	AlEW, 28 SAEW, 100	Ni et al. (2014)
Staphylococci	SAEW	Spray	10 s	23	1.2 g	5.7	926	100	Ni et al. (2014)

(continued)

Table 5.7 (continued)

Microorganism	EW Type	Application	Exposure time (min)	Temp. (°C)	Reduction (log_{10} CFU)	EW property			Reference
						pH	ORP (mV)	ACC (mg/L)	
	AlEW + SAEW	Spray	AlEW, 10 s SAEW, 10 s	23	2.1 g	AlEW, 10.5 SAEW, 5.7	AlEW, – 755 SAEW, 926	AlEW, 28 SAEW, 100	Ni et al. (2014)
S. enteritidis	AEW	Immersion	3 min	20	3.8 g	NA	NA	12	Cao et al. (2009)
	AEW	Immersion	3 min	20	6.5 g	2.5	1106.6	15	Cao et al. (2009)
	SAEW	Immersion	3 min	20	4.2 g	NA	NA	12	Cao et al. (2009)
	SAEW	Immersion	3 min	20	6.5 g	6.12	206.1	15	Cao et al. (2009)
	AEW	Immersion	3 min	20	4.6 g	NA	NA	10	Ma et al. (2009)
	AEW	Spray	3 min	20	3.8 g	NA	NA	10	Ma et al. (2009)
	AlEW + AEW	Immersion	AlEW, 3 min AEW, 3 min	20	4.8 g	AlEW, 11.8 AEW, NA	AlEW, – 872.5 AEW, NA	10	Ma et al. (2009)
	SAEW	Spray	10 s	23	2.4 g	5.7	926	100	Ni et al. (2014)

(continued)

Table 5.7 (continued)

Microorganism	EW Type	Application	Exposure time (min)	Temp. (°C)	Reduction (log_{10} CFU)	EW property			Reference
						pH	ORP (mV)	ACC (mg/L)	
	AlEW + SAEW	Spray	AlEW, 10 s SAEW, 10 s	23	2.8 g	AlEW, 10.5 SAEW, 5.7	AlEW, −755 SAEW, 926	AlEW, 28 SAEW, 100	Ni et al. (2014)
	AEW	Immersion	5 min	23	3.5 egg	2.5	1121	77	Park et al. (2005)
	AlEW + AEW	Immersion	AlEW, 1 min AEW, 1 min	23	3.7 egg	AlEW, 11.2 AEW, 2.5	AlEW, −940 AEW, 1121	AEW, 77	Park et al. (2005)
	AEW	Immersion	3 min	45	≥2.1 egg	2.7	1150	70–80	Bialka et al. (2004)
	AlEW	Immersion	3 min	45	1.7 egg	11.4	−795	NA	Bialka et al. (2004)
Salmonella pullorum	AEW	Immersion	3 min	20	5.0 g	NA	NA	10	Ma et al. (2009)
	AEW	Spray	3 min	20	3.8 g	NA	NA	10	Ma et al. (2009)
	AlEW + AEW	Immersion	AlEW, 3 min AEW, 3 min	20	5.1 g	AlEW, 11.8 AEW, NA	AlEW, −872.5 AEW, NA	10	Ma et al. (2009)
S. typhimurium	AEW	Spray	15 s (hourly)	NA	4–6 egg	2.1	1150	8	Russell (2003)

(continued)

Table 5.7 (continued)

Microorganism	EW Type	Application	Exposure time (min)	Temp. (°C)	Reduction (log_{10} CFU)	EW property			Reference
						pH	ORP (mV)	ACC (mg/L)	
Total aerobic bacteria	AEW	Spray	30 min	24	0.5 cm^2	5.5	1140	NA	Fasenko et al. (2009)
	AEW	Spray	60 min	24	1.5 cm^2	5.5	1140	NA	Fasenko et al. (2009)
	AEW	Spray	90 min	24	1.0 cm^2	5.5	1140	NA	Fasenko et al. (2009)
	AlEW + AEW	Spray	AlEW, 3 s AEW, 10 s	AlEW, 50 AEW, 23	3.1 egg	AlEW, 11.3–11.6 AEW, 2.5–2.7	NA	20–30	Achiwa and Nishio (2003)
	SAEW	Spray	10 s	23	3.0 g	5.7	926	100	Ni et al. (2014)
	AlEW + SAEW	Spray	AlEW, 10 s SAEW, 10 s	23	3.4 g	AlEW, 10.5 SAEW, 5.7	AlEW, −755 SAEW, 926	AlEW, 28 SAEW, 100	Ni et al. (2014)
Yeast and mould	SAEW	Spray	10 s	23	2.5 g	5.7	926	100	Ni et al. (2014)
	AlEW + SAEW	Spray	AlEW, 10 s SAEW, 10 s	23	4.3 g	AlEW, 10.5 SAEW, 5.7	AlEW, −755 SAEW, 926	AlEW, 28 SAEW, 100	Ni et al. (2014)

increased from 1 to 5 min, leading to a maximal reduction of 4.0 $\log_{10}$ CFU/egg compared to the shell egg without treatment (Park et al. 2005).

5.4.2.2 Effect of EW Treatment on the Quality and Microbial Stability of Shell Eggs

Bialka et al. (2004) evaluated efficacy of EW for the microbial safety of shell eggs at 45 °C. Log_{10} reductions of $\geq$2.1 and $\geq$2.3 for *S. enteritidis* and *E. coli* K12, respectively, were achieved after AEW or AlEW followed by AEW treatments. Furthermore, they measured the quality of eggs after the EW treatment for albumen height, presence of cuticle, and eggshell strength. The results demonstrate that the treatment had no negative effects on the egg quality except for cuticle destruction, and it affected the cuticle destruction as similarly as commercial detergent-sanitizer did. A study on the effects of AEW on hatching eggs was conducted by Fasenko et al. (2009). The AEW with a pH value of 5.5 and 1140 mV of ORP for more than 60 min of treatment significantly reduced the total aerobic bacteria on the eggshell by >1.0 $\log_{10}$ CFU/cm^2, with no effect on the cuticle structure, normal embryonic development, hatchability, and chick quality. In addition, broiler mortality during the first 2 weeks of the production period was significantly reduced in the chicks hatched from eggs sprayed with AEW in comparison to the controls.

5.5 Application of EW for Decontamination of Ready-to-Eat Meat Products

As a result of outbreaks and recalls associated with *L. monocytogenes*, the U.S. Department of Agriculture Food Safety Inspection Service (USDA-FSIS) has issued a zero-tolerance policy for this pathogen in ready-to-eat (RTE) meats. The recent USDA regulation, entitled "Compliance Guideline: Controlling *L. monocytogenes* in Post-Lethality Exposed Ready-to-Eat Meat and Poultry Products," has incentivized and enhanced the application of post-lethality treatments and/or antimicrobials or surface treatments to promote safety and reduce risk from pathogenic bacteria during the processing of RTE meat products (USDA-FSIS 2014). Interventions including steam or heat treatment could be applied to reduce microbial contamination on RTE meat surfaces following packaging in meat and poultry processing industries (Cygnarowicz-Provost et al. 1994; Muriana et al. 2002). Other compounds, such as sodium or potassium lactate and liquid smoke, have also been investigated (Ryser and Marth 1999). AEW treatment has proven effective in reducing *L. monocytogenes* in cell culture or on cutting boards (Venkitanarayanan et al. 1999a, b). A few studies have also investigated the application potential of EW treatment in the decontamination of RTE meat products (Fabrizio and Cutter 2005; Veasey and Muriana 2016).

5.5.1 Microbial Decontamination Efficacy of EW Treatment on Ready-to-Eat Meat Products

Veasey and Muriana (2016) examined antimicrobial activity of EW with different pH values (4, 5 and 6) against *E. coli* O157:H7 on fresh beef and *L. monocytogenes* on ready-to-eat meat (Frankfurters) for 30 s. Little or no reduction relative to that observed in the control was observed after treatment. Similar antimicrobial effects of AEW or AlEW on *L. monocytogenes* inoculated on RTE meats including frankfurters and ham were reported by Fabrizio and Cutter (2005). The *L. monocytogenes* was reduced by nearly 2 $\log_{10}$ CFU/g immediately following dipping treatment with AEW at 25 °C for 15 min. When RTE meat samples were sprayed with AEW, AlEW, 2% AA or 10% trisodium phosphate, no statistically significant reductions in aerobic mesophilic plate counts were observed on either day 0 or 7. The combination spray of AlEW/AEW also resulted in a slight reduction of *L. monocytogenes* on RTE meat products. Even with a prolonged contact time, the number of *L. monocytogenes* on meat surfaces did not meet the regulatory requirements for control of *L. monocytogenes* on RTE meats.

5.5.2 Effect of EW Treatment on the Quality and Microbial Stability of Ready-to-Eat Meat Products

When compared to untreated samples, no significant difference was noted for spraying with AEW or AlEW treatment for L^*, a^* and b^* values in surfaces of ready-to-eat meat (frankfurters and ham) (Fabrizio and Cutter 2005). With the combination AEW and AlEW treatment, the a^* and b^* values, but not the L^* value, were significantly higher than for the untreated samples. After 7 days of storage, differences were not observed in the L^*, a^* and b^* values between the treated and untreated frankfurters. The Hunter $L^*a^*b^*$ values indicate that EW treatment does not have an adverse "bleaching" effect on the surface of ready-to-eat meats.

With respect to the effect of EW treatment on microbial stability, Fabrizio and Cutter (2005) observed that the reduction of *L. monocytogenes* was maintained for up to 7 days at 4 °C after the dipping treatment with the AEW. The treated frankfurters have lower counts of *L. monocytogenes* compared to the controls, although this reduction had declined by 14 days.

5.6 Conclusions

The use of EW to decontaminate foods of animal origin, such as fresh red meat, ready-to-eat meat, poultry and shell eggs, was effective in reducing pathogenic microorganisms such as *C. jejuni*, *E. coli* O157:H7, *Salmonella* spp., *S. aureus*, and

L. monocytogenes. The application of multiple EW types or EW combined with other decontaminants has shown to be more efficient in terms of microbial reduction, shelf life extension, and food quality maintenance. EW could serve as a promising alternative during the processing of meat and poultry to decrease food-borne pathogens risk. There was no complete elimination of the pathogens in published studies on red meat and chicken meat treated with EW; the fewer bactericidal effects of EW may be due to the presence of organic matter that could generate monochloramines or blood residue in the immersion solutions. Consequently, a pretreatment of meat products with clean water and frequent change of EW immersion solutions could be necessary. Although EW water reduced somewhat the bacterial loads on animal origin foods, strict manufacturing and slaughter hygiene practices always must be considered as important parts of an integral food safety system.

References

Achiwa N, Nishio T (2003) The use of electrolyzed water for sanitation control of eggshells and GP center. Food Sci Technol Res 9(1):100–103

Al-Haq MI, Sugiyama J, Isobe S (2005) Applications of electrolyzed water in agriculture and food industries. Food Sci Technol Res 11(2):135–150

Al-Holy MA, Rasco BA (2015) The bactericidal activity of acidic electrolyzed oxidizing water against *Escherichia coli,* O157:H7, *Salmonella typhimurium*, and *Listeria monocytogenes* on raw fish, chicken and beef surfaces. Food Control 54:317–321

Alonso-Hernando A, Guevara-Franco JA, Alonso-Calleja C et al (2013) Effect of the temperature of the dipping solution on the antimicrobial effectiveness of various chemical decontaminants against pathogenic and spoilage bacteria on poultry. J Food Prot 76(5):833–842

Anang DM, Rusul G, Bakar J et al (2007) Effects of lactic acid and lauricidin on the survival of *Listeria monocytogenes, Salmonella enteritidis and Escherichia coli* O157:H7 in chicken breast stored at 4 °C. Food Control 18(8):961–969

Aurass P, Prager R, Flieger A (2011) EHEC/EAEC O104:H4 strain linked with the 2011 German outbreak of haemolytic uremic syndrome enters into the viable but non-culturable state in response to various stresses and resuscitates upon stress relief. Environ Microbiol 13(12):3139–3148

Ayebah B, Hung YC, Kim C et al (2006) Efficacy of electrolyzed water in the inactivation of planktonic and biofilm listeria monocytogenes in the presence of organic matter. J Food Prot 69(9):2143–2150

Bell MF, Marshall RT, Anderson ME (1986) Microbiological and sensory tests of beef treated with acetic and formic acids. J Food Prot 49(3):207–210

Bialka KL, Demirci A, Knabel SJ et al (2004) Efficacy of electrolyzed oxidizing water for the microbial safety and quality of eggs. J Poult Sci 83(12):2071–2078

Bosilevac JM, Shackelford SD, Brichta DM et al (2005) Efficacy of ozonated and electrolyzed oxidative waters to decontaminate hides of cattle before slaughter. J Food Prot 68(7):1393–1398

Botsoglou NA, Christaki E, Fletouris DJ et al (2002) The effect of dietary oregano essential oil on lipid oxidation in raw and cooked chicken during refrigerated storage. Meat Sci 62(2):259–265

Botta C, Ferrocino I, Cavallero MC et al (2018) Potentially active spoilage bacteria community during the storage of vacuum packaged beefsteaks treated with aqueous ozone and electrolyzed water. Int J Food Microbiol 266:337–345

Brychcy E, Jarmoluk A, Marycz K (2015a) Impact of low-concentrated acidic electrolysed water obtained by membrane electrolysis on the decontamination of meat microbiota. B Vet I Pulawy 59(3):369–376

Brychcy E, Król Ż, Kulig D et al (2016) The effect of carrageenan and gelatine hydrosols incorporated with acidic electrolysed water on surface microbiota and quality changes on pork meat. Int J Food Sci Tech 51(7):1618–1629

Brychcy E, Malik M, Drożdżewski P et al (2015b) Low-concentrated acidic electrolysed water treatment of pork: inactivation of surface microbiota and changes in product quality. Int J Food Sci Tech 50(11):2340–2350

Byun JS, Min JS, Kim IS et al (2003) Comparison of indicators of microbial quality of meat during aerobic cold storage. J Food Prot 66(9):1733–1737

Cao W, Zhu ZW, Shi ZX et al (2009) Efficiency of slightly acidic electrolyzed water for inactivation of *Salmonella enteritidis* and its contaminated shell eggs. Int J Food Microbiol 130(2):88–93

Carlson BA, Ruby J, Smith GC et al (2008) Comparison of antimicrobial efficacy of multiple beef hide decontamination strategies to reduce levels of *Escherichia coli* O157:H7 and *Salmonella*. J Food Prot 71(11):2223–2227

Chantarapanont W, Berrang ME, Frank JF (2004) Direct microscopic observation of viability of *Campylobacter jejuni* on chicken skin treated with selected chemical sanitizing agents. J Food Prot 46(3):1146–1152

Comes JE, Beelman RB (2002) Addition of fumaric acid and sodium benzoate as an alternative method to achieve a 5-log reduction of *Escherichia coli* O157:H7 populations in apple cider. J Food Prot 65(3):476–483

Cygnarowicz-Provost M, Whiting RC, Craig JC (1994) Steam surface pasteurization of beef frankfurters. J Food Sci 59(1):1–5

Ding T, Rahman SME, Oh DH (2011) Inhibitory effects of low concentration electrolyzed water and other sanitizers against foodborne pathogens on oyster mushroom. Food Control 22(2):318–322

Ding T, Rahman SME, Purev U et al (2010) Modelling of *Escherichia coli* O157:H7 growth at various storage temperatures on beef treated with electrolyzed oxidizing water. J Food Eng 97(4):497–503

Duan D, Wang H, Xue S et al (2017) Application of disinfectant sprays after chilling to reduce the initial microbial load and extend the shelf-life of chilled chicken carcasses. Food Control 75:70–77

Elias RJ, Kellerby SS, Decker EA (2008) Antioxidant activity of proteins and peptides. Crit Rev Food Sci 48(5):430–441

Fabrizio KA, Cutter CN (2004) Comparison of electrolyzed oxidizing water with other antimicrobial interventions to reduce pathogens on fresh pork. Meat Sci 68(3):463–468

Fabrizio KA, Cutter CN (2005) Application of electrolyzed oxidizing water to reduce *Listeria monocytogenes* on ready-to-eat meats. Meat Sci 71(2):327–333

Fabrizio KA, Sharma RR, Demirci A et al (2002) Comparison of electrolyzed oxidizing water with various antimicrobial interventions to reduce *Salmonella* species on poultry. J Poult Sci 81(10):1598–1605

FAO (2006) Databases: food balance sheets. Available via http://www.fao.org/faostat/en/#country. Accessed 16 December 2018

Fasenko GM, O'Dea Christopher EE, Mcmullen LM (2009) Spraying hatching eggs with electrolyzed oxidizing water reduces eggshell microbial load without compromising broiler production parameters. J Poult Sci 88(5):1121–1127

Friedman CR, Neimann J, Wegener HC et al (2000) Epidemiology of *Campylobacter jejuni* infections in the United States and other industrialized nations. In: Nachamkin I, Blaser MJ (eds) Campylobacter, 2nd edn. ASM, Washington, DC

Gatellier P, Gomez S, Gigaud V et al (2007) Use of a fluorescence front face technique for measurement of lipid oxidation during refrigerated storage of chicken meat. Meat Sci 76(3):543–547

Gill CO (1983) Meat spoilage and evaluation of the potential storage life of fresh meat. J Food Prot 46(5):444–452

Göksoy EO, James C, Corry JEL (2000) The effect of short-time microwave exposures on inoculated pathogens on chicken and the shelf-life of uninoculated chicken meat. J Food Eng 45(3):153–160

Golden DA, Beuchat LR, Brackett RE (1988) Evaluation of selective direct plating media for their suitability to recover un-injured, heat-injured, and freeze-injured *Listeria monocytogenes* from foods. Appl Environ Microbiol 54(6):1451–1456

González-Fandos E, Dominguez JL (2007) Effect of potassium sorbate washing on the growth of *Listeria monocytogenes* on fresh poultry. Food Control 18(7):842–846

Han D, Hung YC, Wang L (2018) Evaluation of the antimicrobial efficacy of neutral electrolyzed water on pork products and the formation of viable but nonculturable (VBNC) pathogens. Food Microbiol 73:227–236

Hati S, Mandal S, Minz P et al (2012) Electrolyzed oxidized water (EOW): non-thermal approach for decontamination of food borne microorganisms in food industry. Food Nutr Sci 3(6):760–768

Holmer SF, McKeith RO, Boler DD et al (2009) The effect of pH on shelf-life of pork during aging and simulated retail display. Meat Sci 82(1):86–93

Hricova D, Stephan R, Zweifel C (2008) Electrolyzed water and its application in the food industry. J Food Prot 71(9):1934–1947

Huang Y, Hung Y, Hsu S et al (2008) Application of electrolyzed water in the food industry. Food Control 19(4):329–345

Humphrey TJ (1994) Contamination of egg shell and contents with *Salmonella enteritidis*: a review. Int J Food Microbiol 21(1–2):31–40

Hwang CA, Beuchat LR (1995) Efficacy of selected chemicals for killing pathogenic and spoilage microorganisms on chicken skin. J Food Prot 58(1):19–23

ICMSF (1986) Microorganisms in Foods 2: sampling for microbiological analysis: principles and scientific applications. University of Toronto, Toronto

Jacxsens L, Devlieghere F, Ragaert P et al (2003) Relation between microbiological quality, metabolite production and sensory quality of equilibrium modified atmosphere packaged fresh-cut produce. Int J Food Microbiol 83(3):263–280

Kalchayanand N, Arthur TM, Bosilevac JM et al (2008) Evaluation of various antimicrobial interventions for the reduction of *Escherichia coli* O157:H7 on bovine heads during processing. J Food Prot 71(3):621–624

Kang SC, Kim MJ, Choi UK (2007) Shelf-life extension of fresh-cut iceberg lettuce (*Lactuca satival*) by different antimicrobial films. J Microbiol Biotechnol 17(8):1284–1290

Kim C, Hung YC, Brackett RE (2000) Efficacy of electrolyzed oxidizing (EO) and chemically modified water on different types of foodborne pathogens. Int J Food Microbiol 61(2):199–207

Kim C, Hung YC, Russell SM (2005) Efficacy of electrolyzed water in the prevention and removal of fecal material attachment and its microbicidal effectiveness during simulated industrial poultry processing. Poultry Sci 84(11):1778–1784

Koohmaraie M, Arthur TM, Bosilevac JM et al (2005) Post-harvest interventions to reduce/eliminate pathogens in beef. Meat Sci 71(1):79–91

Kreyenschmidt J, Hübner A, Beierle E et al (2010) Determination of the shelf life of sliced cooked ham based on the growth of lactic acid bacteria in different steps of the chain. J Appl Microbiol 108(2):510–520

Lawrence TE, Dikeman ME, Hunt MC et al (2003) Effects of calcium salts on beef longissimus quality. Meat Sci 64(3):299–308

Leistner L (2000) Basic aspects of food preservation by hurdle technology. Int J Food Microbiol 55(1):181–186

Lu Y, Wu C (2012) Reductions of *Salmonella enterica* on chicken breast by thymol, acetic acid, sodium dodecyl sulfate or hydrogen peroxide combinations as compared to chlorine wash. Int J Food Microbiol 152(1):31–34

Mansur AR, Tango CN, Kim GH et al (2015) Combined effects of slightly acidic electrolyzed water and fumaric acid on the reduction of foodborne pathogens and shelf life extension of fresh pork. Food Control 47:277–284

Ma QJ, Li BM, Wang CY et al (2009) Efficiency of electrolyzed oxidizing water for inactivation of *salmonella* spp. and inoculated shell eggs. Int J Food Eng 5(3):64–67

Marenzi C (1986) Proper meat storage prevents spoilage. Poultry-Misset 6:12–15

Mies PD, Covington BR, Harris KB et al (2004) Decontamination of cattle hides prior to slaughter using washes with and without antimicrobial agents. J Food Prot 67(3):579–582

Muriana PM, Quimby W, Davidson CA et al (2002) Post-package pasteurization of ready-to-eat deli meats by submersion heating for reduction of *Listeria monocytogenes*. J Food Prot 65(6):963–969

Naveena BM, Sen AR, Muthukumar M et al (2006) The effect of lactates on the quality of microwave-cooked chicken patties during storage. J Food Sci 71(9):603–608

Ni L, Cao W, Zheng WC et al (2014) Efficacy of slightly acidic electrolyzed water for reduction of foodborne pathogens and natural microflora on shell eggs. Food Sci Technol Res 20(1):93–100

Nishimura T, Kato H (1988) Taste of free amino acids and peptides. Food Rev Int 4(2):175–194

Northcutt J, Smith D, Ingram KD et al (2007) Recovery of bacteria from broiler carcasses after spray washing with acidified electrolyzed water or sodium hypochlorite solutions. Poultry Sci 86(10):2239–2244

Nychas GJE, Marshall DL, Sofos JN et al (2013) Meat, poultry and seafood. In: Doyle MP, Buchanan RL (eds) Food microbiology: fundamentals and frontiers, 4th edn. ASM, Washington, DC

Oomori T, Oka T, Inuta T et al (2005) The efficiency of disinfection of acidic electrolyzed water in the presence of organic materials. Anal Sci 16(4):365–369

Park H, Hung YC, Brackett RE (2002) Antimicrobial effect of electrolyzed water for inactivating *Campylobacter jejuni* during poultry washing. Int J Food Microbiol 72(1–2):77–83

Park CM, Hung YC, Lin CS et al (2005) Efficacy of electrolyzed water in inactivating *Salmonella enteritidis* and *Listeria monocytogenes* on shell eggs. J Food Prot 68(5):986–990

Podolak R, Zayas J, Kastner C et al (1996) Reduction of bacterial populations on vacuum-packaged ground beef patties with fumaric and lactic acids. J Food Prot 59(10):1037–1040

Rahman SME, Ding T, Oh DH (2010a) Effectiveness of low concentration electrolyzed water to inactivate foodborne pathogens under different environmental conditions. Int J Food Microbiol 139(3):147–153

Rahman SME, Ding T, Oh DH (2010b) Inactivation effect of newly developed low concentration electrolyzed water and other sanitizers against microorganisms on spinach. Food Control 21(10):1383–1387

Rahman SME, Khan I, Oh DH (2016) Electrolyzed water as a novel sanitizer in the food industry: current trends and future perspectives. Compr Rev Food Sci F 15(3):471–490

Rahman SME, Park J, Song KB et al (2012) Effects of slightly acidic low concentration electrolyzed water on microbiological, physicochemical, and sensory quality of fresh chicken breast meat. J Food Sci 77(1):M35–M41

Rahman SME, Wang J, Oh DH (2013) Synergistic effect of low concentration electrolyzed water and calcium lactate to ensure microbial safety, shelf life and sensory quality of fresh pork. Food Control 30(1):176–183

Rigdon M, Hung YC, Stelzleni AM (2017) Evaluation of alkaline electrolyzed water to replace traditional phosphate enhancement solutions: effects on water holding capacity, tenderness, and sensory characteristics. Meat Sci 123:211–218

Rousset S, Renerre M (1991) Effect of CO_2 or vacuum packaging on normal and high pH meat shelf-life. Int J Food Sci Tech 26(6):641–652

Russell SM (2003) The effect of electrolyzed oxidative water applied using electrostatic spraying on pathogenic and indicator bacteria on the surface of eggs. Poultry Sci 82(1):158–162

Ryser ET, Marth EH (eds) (1999) Listeria, listeriosis and food safety. Marcel Dekker, New York

Selgas MD, Salazar ML, García ML (2009) Usefulness of calcium lactate, citrate and gluconate for calcium enrichment of dry fermented sausages. Meat Sci 82(4):478–480

Serpen A, Gökmen V, Fogliano V (2012) Total antioxidant capacities of raw and cooked meats. Meat Sci 90(1):60–65

Shimamura Y, Shinke M, Hiraishi M et al (2016) The application of alkaline and acidic electrolyzed water in the sterilization of chicken breasts and beef liver. Food Sci Nutr 4(3):431–440

Stopforth JD, Yoon Y, Belk KE et al (2004) Effect of simulated spray chilling with chemical solutions on acid-habituated and non-acid-habituated *Escherichia coli* O157:H7 cells attached to beef carcass tissue. J Food Prot 67(10):2099–2106

Tango CN, Mansur AR, Kim GH et al (2014) Synergetic effect of combined fumaric acid and slightly acidic electrolysed water on the inactivation of food-borne pathogens and extending the shelf life of fresh beef. J Appl Microbiol 117(6):1709–1720

USDA-FSIS (2014) FSIS Compliance guideline: controlling Listeria monocytogenes in Post-Lethality Exposed Ready-to-Eat Meat and Poultry Products. Available via https://www.fsis.usda.gov/wps/portal/fsis/topics/regulatory-compliance/compliance-guides-index/bacteria-guidance. Accessed 16 December 2018

Veasey S, Muriana PM (2016) Evaluation of electrolytically-generated hypochlorous acid ('Electrolyzed Water') for sanitation of meat and meat-contact surfaces. Foods 5(2):42

Venkitanarayanan KS, Ezeike GO, Hung YC et al (1999a) Inactivation of *Escherichia coli* O157:H7 and *Listeria monocytogenes* on plastic kitchen cutting boards by electrolyzed oxidizing water. J Food Prot 62(8):857–860

Venkitanarayanan KS, Ezeike GO, Hung YC et al (1999b) Efficacy of electrolyzed oxidizing water for inactivating *Escherichia coli* O157:H7, *Salmonella enteritidis*, and *Listeria monocytogenes*. Appl Environ Microb 65(9):4276–4279

Wang J, Rahman SME, Park MS et al (2012) Modeling the response of *Listeria monocytogenes* at various storage temperatures in pork with/without electrolyzed water treatment. Food Sci Biotechnol 21(6):1549–1555

Watanabe A, Ueda Y, Higuchi M (2004) Effects of slaughter age on the levels of free amino acids and dipeptides in fattening cattle. Anim Sci J 75(4):361–367

Chapter 6
Application of Electrolyzed Water on Aquatic Product

Yong Zhao, Zhaohuan Zhang, Pradeep K. Malakar, Siqi Wang and Li Zhao

6.1 Introduction

Aquatic products which include fish, shrimps, crabs and shellfish are high-quality sources of dietary protein and these proteins are also able to fulfill additional amino acid requirements for humans. Aquatic products accounted for about 17% of global protein intake from non-plant-derived foods and for 6.7% of total protein intake in 2013. Global aquatic product capture and aquaculture production are projected to increase annually and according to FAOSTAT (2016), the total output of global capture fisheries and aquaculture production was 93.4 million tons and 73.8 million tons in 2014.

Aquatic products are likely to be contaminated by pathogenic or spoilage bacteria during the process of production, transportation and sales. These bacteria reduce seafood quality, shorten shelf life and also pose a potential risk to human health. Substantial capture and aquaculture production are based in developing countries and ensuring product integrity is a challenge when the food safety control system in these countries is not well established. Therefore, development of efficient and internationally accepted sterilization and preservation techniques in seafood products would be of great value to the seafood industry and for improving public health (Lenz et al. 2015; Sevenich et al. 2015).

Y. Zhao (✉) · Z. Zhang · P. K. Malakar · S. Wang · L. Zhao
College of Food Science and Technology, Shanghai Ocean University, 999 Hu Cheng Huan Road, 201306 Shanghai, China
e-mail: yzhao@shou.edu.cn

Y. Zhao · Z. Zhang · S. Wang · L. Zhao
Laboratory of Quality and Safety Risk Assessment for Aquatic Products on Storage and Preservation, Ministry of Agriculture, Hu Cheng Huan Road, 201306 Shanghai, China

Y. Zhao · Z. Zhang · S. Wang · L. Zhao
Shanghai Engineering Research Center of Aquatic-Product Processing and Preservation, Hu Cheng Huan Road, 201306 Shanghai, China

T. Ding et al. (eds.), *Electrolyzed Water in Food: Fundamentals and Applications*,
https://doi.org/10.1007/978-981-13-3807-6_6

A variety of intervention methods have been established to control pathogenic or spoilage bacteria in seafood. These methods include ultra-high pressure, ozone sterilization, microwave sterilization, irradiation sterilization technology and ohmic heating technology. Electrolyzed water (EW), a novel sanitizer, has been widely used in the seafood industries of many countries, including the USA, China, Japan, etc. EW has been proven to improve the quality of aquatic products by eliminating pathogenic or spoilage bacteria in seafood. Recent research indicates that this novel technology can effectively remove bacterial biofilm produced by these bacteria (Han et al. 2017). In addition, the environmental foot print of the EW sanitizer is minor when compared to conventional sanitizers (Xie et al. 2012a, b; Lin et al. 2013a, b; Wang et al. 2015a, b).

Electrolyzed water ice (EW ice), which is produced by freezing EW, has been shown to be an emerging method for markedly improving the safety and quality of seafood. Kim et al. (2006) showed that acidic electrolytic water ice (AEW ice) markedly retarded the growth of *aerobic* and *psychrotrophic* bacteria, formation of total volatile basic nitrogen (TVB-N) and thiobarbituric acid (TBA) in Pacific saury (Kim et al. 2006). The shelf life of Pacific saury could be prolonged by 4–5 days compared with tap water (TW) ice storage. Phuvasate and Su (2010) suggested that AEW ice can also be used as bactericidal treatment for reducing histamine-producing bacteria on fish skin (Phuvasate and Su 2010). Xie et al. (2012b), Lin et al. (2013a, b), Wang et al. (2015a, b) demonstrated that AEW ice could be used for preserving the freshness of shrimp by investigating the microbiological, physical and chemical changes during storage. The results indicated that AEW ice is effective in inhibiting the proliferation of bacteria, changes of pH and colour difference and the formation of TVB-N (Xie et al. 2012b; Lin et al. 2013a, b; Wang et al. 2015a, b).

This chapter summarizes the current state of the art for the use of EW in aquatic products for reducing spoilage microbes and pathogens and analyzes its utility for removing biofilm produced by seafood-derived bacteria. In addition, we introduce the application of EW ice in sterilization and preservation of aquatic products. We hope this chapter provides comprehensive and useful data for promoting the use of EW and EW ice in the seafood industry.

6.2 Applications of EW for Inactivation of Bacteria Contaminating Aquatic Products

6.2.1 Natural Contamination of Aquatic Products and EW

According to incomplete statistics provided by China fisheries yearbook 2015, the annual incidence of aquaculture diseases in China was more than 50%, and the loss rate was about 20%, in which aquatic products hygiene problems caused by biological factors are the main problems. Biological pollution of aquatic products included bacterial pollution, fungal pollution, viral pollution, parasitic pollution, etc. (China fisheries yearbook 2002). The most prominent source pollution was pathogenic bacteria which, when ingested over the infective dose, causes human disease or deaths. In 2014, there were 1480 recorded foodborne disease outbreaks in China resulting in 7420 cases of disease. Seafood-derived pathogenic bacteria accounted for 8.4% of the cases (Wang et al. 2015a, b).

The common pathogens in aquatic products can be divided into two categories: The first category is natural contamination from bacteria which are ubiquitous in the environment, e.g. *Clostridium botulinum*, *Listeria monocytogenes*, *Vibrio* spp, etc. The second category is cross-contamination resulting from handling, transportation and processing of aquatic products, such as *Salmonella* spp, *Escherichia coli*, *Shigella* spp, *C. perfringens* and *Staphylococcus* spp (Novotny et al. 2004).

V. parahaemolyticus is one of the most important pathogen causing seafood-borne gastroenteritis in Shanghai, and the symptoms are diarrhea, vomiting, headache, etc. Xie et al. studied the inactivation kinetics of *V. parahaemolyticus* using three strains of ATCC and two strains derived from seafood products treated by acidic electrolyzed water. Xie used different concentrations of NaCl solution to make acidic EW and compared the inactivation of *V. parahaemolyticus* between AEW and basic electrolyzed water (BEW), neutral electrolyzed water (NEW) and the various diluted AEW. Compared with control, NEW prepared by more than 0.05% concentration of sodium chloride and AEW prepared by different electrolyte of NaCl, KCl, $CaCl_2$ and $MgCl_2$ could reduce *V. parahaemolyticus* by 7.81–8.70 $\log_{10}$ CFU/mL in 2 min. AEW when diluted 10:1 also could reduce *V. parahaemolyticus* by 5.62 $\log_{10}$ CFU/mL. AEW diluted up to 50 or 100 multiples could only reduce *V. parahaemolyticus* by less than 1 $\log_{10}$ CFU/mL. Scanning electron microscopy (SEM) results showed that after *V. parahaemolyticus* was treated by AEW for 2 min, the outer-cell membrane was damaged and some cell contents were released (Fig. 6.1). Therefore, AEW can inactivate *V. parahaemolyticus* effectively and can be used to disinfect seafood product (Xie et al. 2012b). Besides *V. parahaemolyticus*, EW also had the strong antimicrobial activity for other seafood-derived bacteria, such as *L. monocytogenes*, *E.coli* and *Staphylococcus aureus* (Mikš-Krajnik et al. 2016; Zhang et al. 2017; Hao et al. 2016).

6.2.2 Applications of EW Against Biofilm of Seafood-Derived Bacteria

Biofilms are firm three-dimensional, complex, self-assembled structures that contain EPS (exopolysaccharides, proteins, extracellular DNA, etc.) which are attached to biotic or abiotic surfaces and support communities of microorganisms (Costerton et al. 1999; Hall-Stoodley et al. 2004). More than 80% of the bacterial infections in the human population are associated with biofilms and approximately 60% of foodborne outbreaks are caused by biofilms (Wolcott et al. 2008; Bridier et al. 2015).

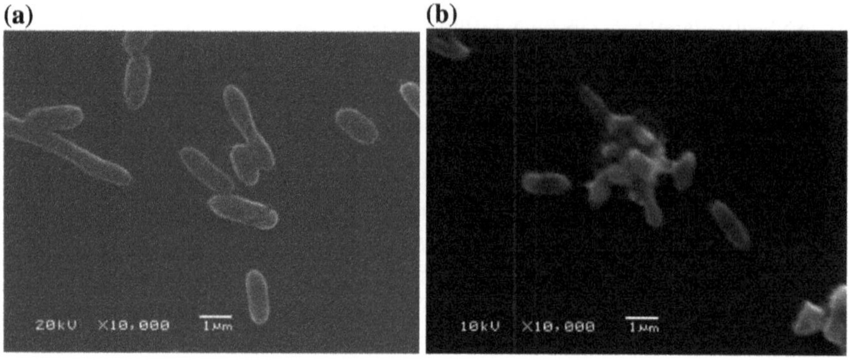

Fig. 6.1 Scanning electron micrographs of *V. parahaemolyticus* cell before and after AEW treatment, **a** cell before AEW treatment, **b** cell after AEW treatment

Biofilm-associated cells are more resistant to external stresses such as antibiotics and detergents when compared to planktonic cells and, therefore, are extremely difficult to eliminate (Hoiby et al. 2010; Costerton et al. 1987).

It is well-documented that biofilm development is one of the most complex physiological processes and five stages have been proposed (Stoodley et al. 2002). Stages one and two are generally identified by a loose or transient association with the surface, followed by robust adhesion. Stages three and four involve the aggregation of cells into microcolonies and subsequent growth and maturation. Biofilm structures can be flat or mushroom-shaped depending on the nutrient source, which seems to influence the interactions between localized clonal growth and the subsequent rearrangement of cells through type IV pilus-mediated gliding motility in response to the nutritional cues (Klausen et al. 2003). Stage five is characterized by a return to transient motility where biofilm cells are sloughed or shed. Importantly, in the dynamic process, EPS facilitates the trapping of nutrients and maintenance of the structural integrity of the biofilm while also providing a sanctuary for the encased bacterial cells (Bassin et al. 2012). Pathogens encased in the EPS-rich matrix, therefore, provide a source of contamination when the biofilm interacts with food materials (Carpentier and Cerf 1993). In addition, the structure of EPS reduces disinfectants access and possibly triggers bacterial tolerance to commonly used sanitizers (Flemming and Wingender 2010; Koo et al. 2013; Xiao et al. 2012).

A number of studies have demonstrated that EW can remove biofilms. Kim et al. (2001) investigated the resistance of *L. monocytogenes* biofilms on stainless steel surfaces to EW (pH of 2.60, ORP of 1160 mV and chlorine of 56 mg/L) and found that a 300-s treatment on a stainless steel surface, could reduce the *L. monocytogenes* population from 1.9×10^{10} CFU/82.5 cm^2 to below detection levels (5 CFU/coupon). Ayebah et al. (2005) inactivated *L. monocytogenes* biofilms on stainless steel surfaces with a combination of ER and EW. They found that ER water alone did not significantly reduce the size of the *L. monocytogenes* biofilms. Treatment with EW for only 30–120 s reduced the viable bacteria populations in biofilms by 4.3–5.2 log

CFU per coupon (2 by 5 cm), whereas the combined treatment of ER water followed by EW could produce an additional reduction by 0.3–1.2 log CFU per coupon. Sun et al. (2012) showed that AEW efficiently inactivated biofilm-imbedded *S. aureus*. In our laboratory (Han et al. 2017), we investigated the removal efficiency of food-borne pathogen biofilm using AEW in a model system and we achieved more than 67% reduction in the population of biofilm cells. In addition, we found that AEW triggered EPS disruption, through deformation of the carbohydrate C–O–C bond and deformation of the aromatic rings in the amino acids tyrosine and phenylalanine.

Number of authors have shown that high ACC causes changes in metabolic compounds within biofilm cells, causing cell death and removal (Han et al. 2017; Vázquez-Sánchez et al. 2014). However, the pH and ORP of AEW were not the main factors contributing to bactericidal ability. Sun et al. (2012) concluded that the available chlorine in AEW might be one of the main factors for the inactivation of *S. aureus* biofilms. In addition, AEW inactivated bacteria due to the oxidative ability of ACC against the cell membrane (Huang et al. 2008). In conclusion, AEW principally causes the degradation of bacterial protective barriers like EPS and increases membrane permeability. Other effects include the leakage of cellular inclusions and decrease of activity of some key enzymes such as dehydrogenase (Zeng et al. 2010). This degradative process possibly increases the detachment of bacterial cells from EPS matrix and makes these cells more vulnerable to sanitizer agents (Kumar and Anand 1998).

6.2.3 Applications of EW Against Bacterial Pathogens Cross Contaminating Aquatic Products

Aquatic products are also likely to be contaminated by pathogenic microorganisms in the supply chain from farming, capture, storage, processing, transportation and distribution. The common contaminants are *V. parahaemolyticus, L monocytogenes* and *Escherichia coli* (Xie et al. 2012a, b; Wang et al. 2014a, b, c; Wu et al. 2014; Al-Qadiri et al. 2016). In these supply chains, hygienic processing is an essential tool for reducing or eliminating the contamination of pathogenic bacteria and EW has been widely adopted (Xie et al. 2012a, b; Al-Qadiri et al. 2016) (Table 6.1).

Wang et al. (2014a) modelled the inactivation of *V. parahaemolyticus* on cooked shrimp by AEW. This study measured the log reductions of *V. parahaemolyticus* versus NaCl concentration (0.7–2.4 g/L), treatment time (3.6–10.4 min) and temperature (23–57 °C) using a central composite design and presented the results via a response surface model. Wang et al. (2014b) then studied the fate of *V. parahaemolyticus* on shrimp during storage at different temperatures (4–30 °C) after AEW treatment. Since Wang et al. (2014c) hypothesized that the rate of chlorine loss in AEW increased under bright conditions and they studied the effect of AEW ice, the frozen form of AEW, on quality of shrimp (*Litopenaeus vannamei*) which was kept in the dark. This detailed study confirmed that AEW reduced the numbers of *V. parahaemolyticus* on

Table 6.1 Inactivation of pathogens on aquatic products by electrolyzed water

Aquatic products	Pathogens	Treatment solution	Exposure time (min)	Reduction (log CFU)	E.W. water index			
					ACC (ppm)	pH	ORP (mV)	Temp. (°C)
Shrimp	VP	NEW10	15	4.16	10.27 ± 0.12	7.11 ± 0.01	841 ± 2	25 ± 2
		NEW50	15	2.07	50.67 ± 0.29	7.05 ± 0.03	863 ± 5	25 ± 2
Shrimp	VP	AEW	5	3.1	21	2.3	1127	50
Shrimp	LM	AEW	5	1.9	21	2.3	1127	50
Shellfish	VP	WAEW	60 or 120	0.7–1.3	10	3.55	950	22
		SAEW	60 or 120	1.0–1.5	20	3.1	1150	22
Shellfish	EC	WAEW	60 or 120	0.7–1.1	10	3.55	950	22
		SAEW	60 or 120	1.4–1.7	20	3.1	1150	22
Shellfish	LM	WAEW	60 or 120	0.6–0.9	10	3.55	950	22
		SAEW	60 or 120	1.0–1.6	20	3.1	1150	22
Shellfish	AH	WAEW	60 or 120	0.6–1.3	10	3.55	950	22
		SAEW	60 or 120	1.3–1.6	20	3.1	1150	22
Shellfish	CJ	WAEW	60 or 120	0.8–1.9	10	3.55	950	22
		SAEW	60 or 120	1.5–2.2	20	3.1	1150	22
Atlantic salmon	LM	EO	2	1.41	60	2.7	1150	20
		EO	6	1.55	60	2.7	1150	20
		EO	10	1.58	60	2.7	1150	20
Atlantic salmon	LM	EO	2	1.52	60	2.7	1150	30
		EO	6	1.82	60	2.7	1150	30
		EO	10	2.26	60	2.7	1150	30
Atlantic salmon	LM	EO	2	1.94	60	2.7	1150	40
		EO	6	2.36	60	2.7	1150	40
		EO	10	2.85	60	2.7	1150	40

VP represents *Vibrio parahaemolyticus*; LM represents *Listeria monocytogenes*; CJ represents *Campylobacter jejuni*; EC represents *Escherichia coli*; AH represents *Aeromonas hydrophila* shrimps significantly and would be useful for the improving quality and safety in aquatic products.

AEW can also inhibit and reduce the viability of microorganisms on other aquatic products, in addition to reducing the microorganisms on shrimp. Al-Qadiri et al. (2016) investigated the effects of AEW to saline solutions treatments (1–2% NaCl) and untreated controls on inducing bacterial inhibition and injury in live clams (*Venerupis philippinarum*) and mussels (*Mytilus edulis*). They used two types of AEW, where one type was labelled strong (SAEW) with an ACC of 20 mg/L, pH = 3.1, and an ORP of 1150 mV, another type was labelled weak (WAEW) at ACC of 10 mg/L, pH = 3.55 and ORP of 950 mV. They found that both AEW types (10–20 mg/L ACC) reduces bacterial contamination in live clam and mussel and SAEW showed the largest inhibitory activity and the extent of reduction for *Escherichia coli O104:H4*, *Listeria monocytogenes*, *Aeromonas hydrophila*, *Vibrio parahaemolyticus* and *Campylobacter jejuni*.

A variety of novel physical methods, such as high hydrostatic pressure (HHP), ultrasound (US) and ultraviolet light (UV) are being used for controlling pathogens in aquatic products. Inactivation of pathogens on aquatic products by EW has been studied in combination with these non-thermal sterilization methods by Park and Ha (2015), Du et al. (2016), Mikš-Krajnik et al. (2016).

Optimal operating parameters of electrolyzed water and ultra-high pressure (AEW-HHP) were obtained by Du et al. (2016) using response surface methodology. These optimal operating parameters are NaCl concentration for electrolysis 1.5 g/L, treatment pressure of 400 MPa and a treatment time 10 min. The combination, AEW-HHP, significantly reduced the numbers of *V. parahaemolyticus* and *L. monocytogenes* by 6.08 and 5.71 $\log_{10}$ CFU/g respectively. In comparison, HHP only inactivated the two pathogens by 4.74 and 4.31 $\log_{10}$ CFU/g. This study showed that AEW can be used synergistically to ensure the safety of aquatic products in a non-thermal processing environment.

The effect of SAEW combined with US on *Escherichia coli* and *Vibrio parahaemolyticus* on freshly sliced shad (*Konosirus punctatus*) was investigated by Park and Ha (2015). SAEW (15 or 30 ppm chlorine) and US (50 or 100 min) when used synergistically was able to reduce the population of the two pathogens by $1 - \log_{10}$. Individual treatments of SAEW and US under the same conditions were inconclusive. In another study, Mikš-Krajnik et al. (2016) found that AEW in combination with other inactivating methods, ultraviolet light (UV), ultrasound (US) (UV + US + AEW) showed significantly ($P \leq 0.05$) higher reduction in *L. monocytogenes* of 0.75 log CFU/g when compared to control (0.17 log CFU/g). These results indicate that UV+US+AEW was most effective at reducing the populations of *L. monocytogenes* on raw salmon fillets.

6.2.4 Applications of EW Against Spoilage Microbes on Aquatic Products

Aquatic products are highly perishable foods which are susceptible to bacterial spoilage and this spoilage accelerates quality loss and increases public health risks (Chen et al. 2016). Microbial spoilage of seafood products also decreases shelf lives and is a significant contributor to food waste. In fish products, bacteria such as *Pseudomonas* spp, *Acinetobacter*, *Moraxella*, *Flavobacterium*, *Shewanella*, *Alcaligenes*, *Vibrio* and *coliform* are the main spoilage bacteria (Ray and Bhunia 2004). The degree of spoilage is dependent on the total microbial flora, quality of the aquatic products, handling and packaging methods, and storage temperature. In fish packed in modified atmosphere, *Photobacterium* and *lactic acid* bacteria are reported to be the primary bacteria that are responsible for spoilage. In aerobically packed fish, the main spoilage bacteria come from the genera *Pseudomonas* and *Shewanella* (Dalgaard et al. 1993; Emborg et al. 2002; Gram and Huss 2000). For unpreserved fish, spoilage is a result of gram-negative, fermentative bacteria (such as *Vibrionacaeae*), whereas psychrotolerant gram-negative bacteria (such as *Pseudomonas* spp. and *Shewanella* spp.) tend to spoil chilled fish (Gram and Huss 2000). Furthermore, water-borne bacteria that belong to the genera *Flavobacterium*, *Chryseobacterium* and *Photobacterium* are the core community of seafood bacteria (Chaillou et al. 2015). The specific spoilage microorganisms known to be involved in seafood spoilage are listed in Table 6.2.

The use of EW for improving the shelf life of aquatic products is fairly recent. Studies by Phuvasate and Su found that soaking Atlantic salmon and yellowfin tuna skin in EW (100 ppm chlorine for 120 min) could reduce *Enterobacter aerogenes* and *Morganella morganii* populations by 1.3 and 2.2 log CFU/cm^2, respectively. These are spoilage bacteria that can produce significant amounts of histamine (Phuvasate and Su 2010). McCarthy and Burkhardt-III (2012) carried out research on the efficacy of EW (50 ppm chlorine) against *M. morganii* biofilms on *mahi–mahi* surfaces and *L. monocytogenes* biofilms on salmon surfaces. Their results showed that EW reduced the number of *M. morganii* and *L. monocytogenes* from 4.02 to 2.48 log_{10} CFU/g and from 4.47 to 2.33 log_{10} CFU/g, respectively. Khazandi et al. (2017) used EW sanitizer for increasing the shelf life of Southern Australian King George Whiting (KGW) and Tasmanian Atlantic Salmon (TAS) fillets stored at 4 °C and found that this sanitizer was effective again the dominant spoilage microbiota (*Pseudomonas* spp. and *Shewanella* spp) found in these products. EW significantly reduced total bacterial load and specific spoilage organisms on KGW and TAS fillets during storage and extended the shelf life of the fillets by 2 and 4 days.

In seafood processing, maintaining the cold-chain is a prerequisite for ensuring product quality and freshness. EW ice has been proposed as a progressive tool for preserving quality and suppressing microbial growth in aquatic products during processing. It not only has the advantage of tap water ice but also has the potential to be bactericidal (Lin et al. 2013a, b). Feliciano et al. (2010) investigated the efficacy of EW ice for inhibiting the microbial growth on whole tilapia, tilapia fillet and water during storage. The result showed that EW ice has the ability to reduce the numbers

Table 6.2 List of specific spoilage microorganisms involved in different seafood products spoilage

Product	Spoilage activity	Specific spoilage bacteria	Spoilage compounds
Fresh chilled products	High	*Shewanella putrifaciens* *Pseudomonas putrifaciens* *Pseudomonas fluorescens* *Fluorescent pseudomonads*	TMA H_2S CH_3SH $(CH_3)_2S$ HX
Fresh fish	High	*Vibrionacaeae* *Aeromonas*	Ketones, aldehydes Esters Non-H2S Sulphides
Fresh chilled fish	Moderate	*Photobacterium phosphoreum* *Moraxella* *Acinetobacter* *Alcaligene*	TMA HX
Chill vacuum-packaged cold-smoked	Low	*Lactobacillus* *Aerobacter* *Flavobacterium* *Micrococcus* *Bacillus* *Staphylococcus* *Mesophilic*	NH_3 Acetic Butyric acid Propionic acid

Source Adopted and modified from Ghaly et al. (2010)
TMA Trimethylamine, H_2S Hydrogen sulphide, CH3SH Methyl mercaptan, $(CH_3)_2S$ Dimethylsulfide, HX Hypoxanthine, NH_3 Ammonia

of *E. coli K12*, *Listeria innocua* and *Pseudomonas putida* on whole tilapia, tilapia fillets and the water during 72 h storage. Kim et al. (2006) used EW ice to preserve the quality of Pacific saury and their results showed that EW ice could markedly decrease the growth of aerobic and psychrotrophic bacteria. In this study, the shelf life of Pacific saury was increased by 4–5 days when compared with tap water (TW) ice storage.

Most peer-reviewed studies explore EW efficacy against pathogenic and aerobic bacteria and research on the efficacy of EW against specific spoilage organisms (SSO), particularly psychrotrophic and lactic acid-producing bacteria, remains to be investigated.

6.3 Applications of EW and EW Ice for Preservation of Aquatic Products

6.3.1 The Definition of EW Ice

Electrolyzed water ice (EW ice) is an emerging sterilization and preservation technique and is produced by freezing EW. EW ice is being applied in production, transportation and storage of aquatic products, and is seen to be beneficial for improving the safety and quality of seafood (Li et al. 2013). EW ice not only has the characteristics of the preservation and sterilization potential of EW but also maintains the advantages of the traditional low-temperature preservation. Figure 6.2 showed the different specifications of electrolyzed water ice and Fig. 6.3 showed the application of electrolyzed water ice for preservation of aquatic products.

Figure 6.4 provides a timeline for the application of electrolyzed water ice in aquatic products. In 2006, Kim et al. applied EW ice for preserving the freshness of Pacific saury. Results showed that EW ice significantly inhibited the growth of both aerobic and psychrotrophic bacteria in the tissues of Pacific saury and reduced the formation of TVBN and TBA. In 2013, Lin et al. used EW ice for preserving the quality of *Litopenaeus vannamei*. Results showed that EW ice limits pH and TVBN changes of shrimp and was also capable of inhibiting growth of bacteria during storage. In addition, EW ice was clearly more efficient in maintaining the initial attachments between muscle fibres in shrimp. In 2010, Feliciano et al. used EW ice for reducing inoculated populations of *Escherichia coli* K12, *Listeria innocua* and *Pseudomonas putida* in Tilapia fillets and results showed that all bacterial population was reduced during the study. Phuvasate and Su (2010) investigated efficacy of EW ice treatments in reducing histamine-producing bacteria on fish skin. A treatment of EW ice (100 ppm chlorine) for 24 h was capable of reducing *E. aerogenes* and *M. morganii* on tuna skin by 2.4 and 3.5 log CFU/cm^2, respectively. As previously reported, Wang et al. (2014a, b, c) investigated the effect of acidic EW ice on quality

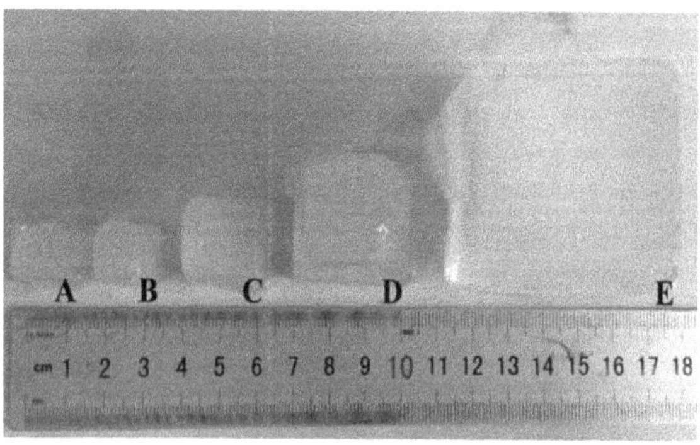

Fig. 6.2 Different specifications of electrolyzed water ice

Fig. 6.3 The application of electrolyzed water ice for preservation of aquatic products

of shrimp stored in the dark and showed that EW ice significantly inhibited changes of pH, formation of TVBN and proliferation of total bacteria counts in shrimp.

In the last few years, researchers have begun to focus on the underlying inactivation mechanism of AEW ice on bacteria. Results show that AEW ice while limiting the changes of pH, shrinkage of muscle fibres and bacteria in shrimp also displays inhibitory activity towards cathepsin B and polyphenol oxidase. In 2017, Sun et al. reports showed that EW ice could prevent browning in *Litopenaeus vannamei* by

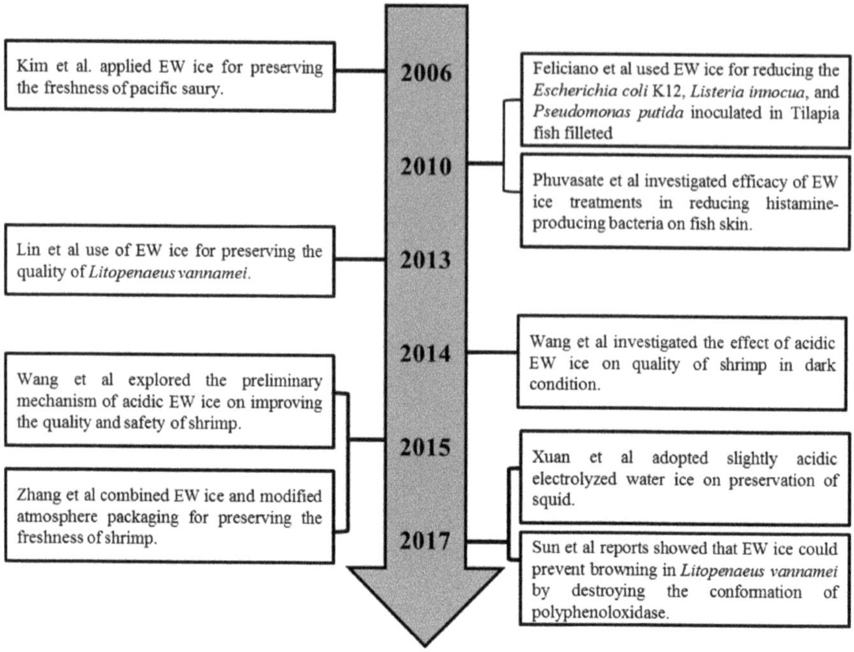

Fig. 6.4 The history of the application of electrolyzed water ice in aquatic products

reducing or stopping the formation of polyphenol oxidase. Xuan et al. (2017) adopted slightly acidic electrolyzed water ice on preservation of squid. Results showed the total bacterial counts were significantly reduced by SAEW-ice. It also delayed the appearance of browning and softening. Furthermore, SAEW-ice treatment also inhibited the increase of POV and maintained relatively low TBA and TVBN contents.

The hurdle effect was also explored in the application of EW ice. In 2015, Zhang et al. combined EW ice and modified atmosphere packaging for preserving the freshness of shrimp. Results showed that WAEW ice-glazing combined with MAP was highly effective in maintaining lower TVBN, TMA and TBARS values in frozen shrimp. They also displayed higher inhibitor effects on total aerobes and *Staphylococcus aureus*.

6.3.2 *Applications of EW and EW Ice on Shrimp*

Shrimp is one of the most valuable fishery products in the South and Southeastern parts of Asia and accounts for 90% of the global aquaculture shrimp production. In addition, shrimp is also the leading seafood consumed worldwide because it has high levels of omega-3 fatty acids and low levels of mercury. However, rapid spoilage of shrimp due to microbiological, chemical and physical changes leads to a short shelf life and poses a challenge for getting this product to market. EW ice, a novel preservation technique, has been widely used in during the storage of shrimp (Li et al. 2013). In this section, we discuss the effect of EW ice on shrimp preservation through changes of microbiology (total viable count and bacterial diversity), chemical property (pH, TVB-N, enzymatic browning via polyphenol oxidase) and protein degradation (myofibrillar proteins).

Bacteria which are present in harvested shrimp (shell, muscle, hepatopancreas, etc.) including pathogens and spoilage organisms, will rapidly reproduce and cause deterioration of quality during processing and transportation. Soaking in AEW for 5 min at 20 °C, decreased the bacterial population on shrimp surfaces by 1 log CFU/g (Xie et al. 2012b). Figure 6.5 maps and contrasts the diversity of bacterial flora in shrimp after treatment with AEW and TW ice using PCR-DGGE and treatment with AEW ice clearly reduced bacterial diversity when measured using the Shannon index and average similarity coefficient (Wang et al. 2015a). In general, AEW and AEW ice exhibit a strong bacteriostatic effect by reducing the total viable count and decreasing bacterial diversity.

The post-mortem chemical properties of shrimp such as pH and production of total volatile basic nitrogen TVB-N are the most important criteria for evaluating the quality of the shrimp. The pH of shrimp treated with AEW ice was minimally increased when compared with untreated shrimp and showed no significant difference from TW ice treatment (Zhao et al. 2016). Increase in the pH value is mainly due to decomposition of the shrimp meat protein as the basic, alkaline nitrogen small molecules (such as ammonia compounds, trimethylamine, etc.) are released (Zhou et al. 2011).

Total volatile basic nitrogen (TVBN) of meat and fish samples refers to the total amount of nitrogen that can be distilled with water vapour and this nitrogen exists mainly as ammonia and amine (trimethylamine and dimethylamine). The concentration of TVBN is a measure of the quality of shrimp and according to Lin et al. (2013a, b), during storage, the TVBN value of AEW-ice treated shrimp

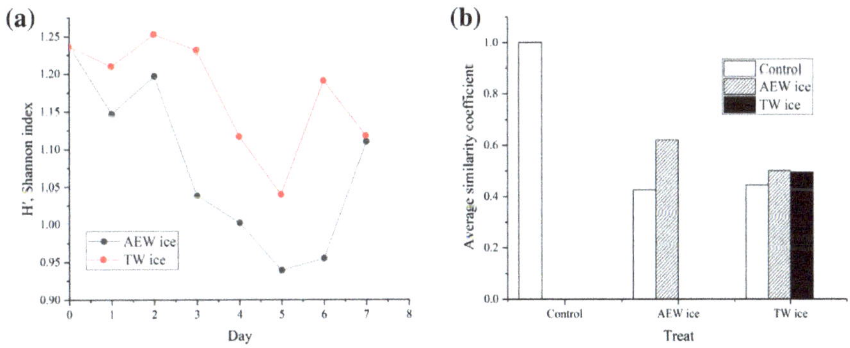

Fig. 6.5 The Shannon index (**a**) and average similarity coefficient (**b**) of microbial communities on shrimp under AEW ice and TW ice storage

increased from 10.21 to 19.84 mg/100 g, while the TVBN value TW-ice group increased from 10.21 to 22.89 mg/100 g. After storage of 3 days, the TVBN value of AEW-ice stored shrimps was significantly (P <0.05) lower than TW-ice. In particular, on the sixth day, the TVBN value of TW-ice stored shrimps was 1.2 times that of AEW-ice. Therefore, AEW-ice treatment is more effective in controlling the formation of TVBN in shrimp meat.

Generally, myofibrillar proteins are the key proteins related to seafood spoilage and these proteins include myosin heavy chains, actin and tropomyosin. Research has demonstrated that degradation of the above proteins can lead to loss of muscle texture and signals the onset of spoilage (Delbarre-ladrat et al. 2006). Electrolyzed water is also proven to decrease the rate of degradation of myofibrillar proteins in shrimps. As shown by the shrimp histological sections, storage in AEW ice maintains the attachments between muscle fibres in shrimp. The intermuscular space only increases significantly after 6 days of storage on AEW ice and TW ice. However, the AEW-ice treated shrimp muscle fibre gap (9.6 μm) was significantly (P<0.05) less than that of TW-ice treatment (15.6 m). In addition, compared with TW-ice treatment, AEW-ice treated shrimp meat fibre had better integrity. Based on the above analysis, AEW-ice can delay the change of muscle fibre gap and maintain the integrity of muscle fibre (Lin et al. 2013a, b).

Furthermore, browning is one of the major concerns that frequently occur in shrimps, which will reduce the edibility and cause considerable financial loss. The major biochemical reaction in browning formation is induced by polyphenoloxidase (PPO), which is a copper-containing metalloprotein, and easily catalyzed phenols to colourless quinones. As shown in Fig. 6.6, AEW ice exhibited a good ability in delaying browning in shrimp. Further analysis revealed that AEW ice can significantly inhibit the activity of PPO, which provided strong evidence to elucidate the mechanism of AEW ice for delaying the shrimp browning.

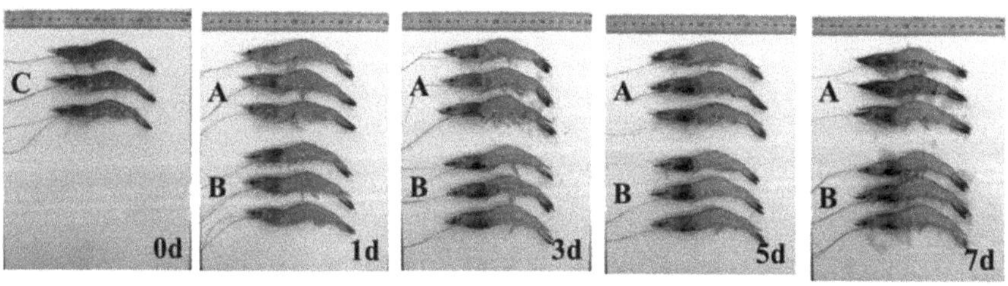

Fig. 6.6 Digital photograph of shrimp stored in AEW ice (**A**) and TW ice (**B**) for 7 days, (**C**) represents untreated control at 0 day

6.3.3 Applications of EW and EW Ice on Fish

Fish are popular in contemporary life for its nutrition and delicacy. As a new disinfectant for seafood, especially fish, EW has shown promising results to prevent outbreaks of foodborne diseases and the activity of spoilage bacteria in fish (Gram and Huss 2000; Xie et al. 2012b). EW ice can effectively inhibit the growth of microbes on fish at low temperature. Bacteria could breed when the fish were exposed to ambient temperature and cause the corruption. In order to guarantee the better quality of the aquatic products as well as its shelf life, fresh aquatic products should be suggested to wash by EW prior to freezing with EW ice. Slightly acidic electrolyzed water (SAEW) is already advised as a novel non-thermal disinfecting agent (Athayde et al. 2017). For effectively inhibiting the growth of microbes and related biochemical reactions, the bactericidal activity of EW ice has been reported (Al-Holy and Rasco 2015). Recently, SAEW is creatively made into ice for the quality maintenance and preservation of seafood on account of its rapid, harmless and broad-spectrum disinfection characteristics (Zhang et al. 2015). Comparing with tap water (TW) ice, weakly acidic electrolyzed water (WAEW) ice has great efficiency in inhibiting the growth of aerobic and psychrotrophic bacteria on Pacific saury and extending the shelf life (4–5 days) (Kim et al. 2006). Phuvasate and Su (2010) reported that EW ice reduced the numbers of *Enterobacter aerogenes* and *Morganella morganii* (both of which were histamine-producing bacteria) 2.4 and 3.5 log CFU/cm^2, respectively, on the surface of tuna for 24 h. EW and EW ice should be suggested as a measure of post-harvest treatment on food contact surfaces and on the skin to reduce histamine-producing bacteria. Feliciano et al. (2010), Xuan et al. (2017) indicated that SAEW-ice treatment delayed the appearance of browning and softening and inhibited the increase of peroxide value (POV) and maintained relatively low thiobarbituric acid (TBA) and total volatile basic nitrogen (TVBN) contents. EW ice has the potential to inhibit the deterioration of seafood and guarantee the food safety during storage, which could be a new approach worthy of further investigation.

6.3.4 Applications of EW and EW Ice on Shellfish

Molluscan shellfish have held an important place in the diets throughout the world. Shellfish dishes are a feature of almost all the cuisines of the world, providing an important source of protein in many cuisines around the world, especially in the countries with coastal areas (Dugo et al. 2017). However, shellfish are particularly essential for clean handling after catch because they are perishable and some of them are eaten raw in the world. For the supply of shellfish to consumers in fresh condition, clean handling after catch from the sea is essential. It is important to meet such requirements by keeping shellfish under a certain low temperature and clean conditions after catching. EW could be regarded as a good choice for handling shellfish combined with low temperature. Also EW ice is used for keeping shellfish during the transportation of shellfish (Al-Qadiri et al. 2016; Kasai et al. 2011). The application of EW ice is superior to keeping shellfish below 10 °C during transportation and handling after catch (Liu et al. 2014).

The clam is one of the most important aquaculture shellfish species and has an important socioeconomic aspect because it is easily grown and harvested and has a good nutritional value. The blue mussel is a euryhaline and eurythermal shellfish cultivated in intensive aquaculture and is usually consumed raw, smoked or canned. AEW could be used to reduce bacterial contamination in live clam and mussel, which may help control possible unhygienic practices during production and processing of shellfish without apparent changes in the quality of the shellfish (Fan et al. 2015). The Japanese scallop is one of the most important cultured shellfish, of which filter feeder is easy to gather harmful microorganisms. Hence, procedures of purification and sterilization are necessitated before it is ready for the market. Slightly acidic brine electrolysis is applied for its qualities of minor irritation, non-residue and bactericidal broad-spectrum (Li 2016).

6.3.5 *Summary and Prospect*

Aquatic products are high-value foods all over the world, but the pathogenic microorganism and short shelf life will limit the development of aquaculture industry. Sterilization and preservation of aquatic products posed a big challenge for food researches from every country; especially the food security systems in these countries are not well established. Previous studies demonstrated that EW and EW ice maintains the quality of seafood by inhibiting the proliferation of bacteria, changes in pH, formation of total volatile basic nitrogen (TVB-N) and activities of endogenous enzymes. However, these findings only proved that EW and EW ice were a promising technique for improving the quality of shrimps, while the in-depth sterilization and preservation mechanism is not yet fully understood. Nowadays, the rapid development of omics technologies provides strong tools to further reveal the sterilization and preservation mechanism of EW and EW ice. Knowledge about these potential mechanisms of EW and EW ice on the changes of proteins and bacteria will facilitate improving these techniques, which will contribute to minimize risk and prolong the shelf life of seafood products foods.

References

Al-Holy MA, Rasco BA (2015) The bactericidal activity of acidic electrolyzed oxidizing water against *Escherichia coli*, O157: H7, *Salmonella*, *Typhimurium*, and *Listeria monocytogenes*, on raw fish, chicken and beef surfaces. Food Control 54:317–321

Al-Qadiri HM, Al-Holy MA, Shiroodi SG et al (2016) Effect of acidic electrolyzed water-induced bacterial inhibition and injury in live clam (*venerupis philippinarum*) and mussel (*mytilus edulis*). Int J Food Microbiol 231:48–53
Athayde DR, Flores D, Da JS et al (2017) Application of electrolyzed water for improving pork meat quality. Food Res Int 100:757–763
Ayebah B, Hung YC, Frank JF (2005) Enhancing the bactericidal effect of electrolyzed water on *Listeria monocytogenes* biofilms formed on stainless steel. J Food Prot 68:1375–1380
Bassin JP, Kleerebezem R, Rosado AS et al (2012) Effect of different operational conditions on biofilm development, nitrification, and nitrifying microbial population in moving-bed biofilm reactors. Environ Sci Technol 46(3):1546–1555
Bridier A, Sanchez-Vizuete P, Guilbaud M et al (2015) Biofilm-associated persistence of food-borne pathogens. Food Microbio 45:167–178
Carpentier B, Cerf O (1993) Biofilms and their consequences, with particular reference to hygiene in the food industry. J Appl Microbiol 75(6):499–511
Chaillou S, Chaulot-Talmon A, Caekebeke H et al (2015) Origin and ecological selection of core and food-specific bacterial communities associated with meat and seafood spoilage. ISME J 9(5):1105–1118
Chen J, Xu B, Deng S et al (2016) Effect of combined pretreatment with slightly acidic electrolyzed water and botanic biopreservative on quality and shelf life of Bombay duck (*harpadon nehereus*). J Food Qual 39(2):116–125
Costerton JW, Cheng KJ, Geesey GG et al (1987) Bacterial biofilms in nature and disease. Ann Rev Microbiol 41(1):435–464
Costerton JW, Stewart PS, Greenberg EP (1999) Bacterial biofilms: a common cause of persistent infections. Science 284(5418):1318–1322
Dalgaard P, Gram L, Huss HH (1993) Spoilage and shelf-life of cod fillets packed in vacuum or modified atmospheres. Int J Food Microbiol 19(4):283–294
Delbarre-ladrat C, Chéret R, Taylor R et al (2006) Trends in postmortem aging in fish: understanding of proteolysis and disorganization of the myofibrillar structure. Crit Rev Food Sci Nutr 46(5):409–421
Du SP, Zhang ZH, Xiao LL et al (2016) Acidic electrolyzed water as a novel transmitting medium for high hydrostatic pressure reduction of bacterial loads on shelled fresh shrimp. Front Microbiol 7:305
Dugo G, Mondello L, Costa R et al (2017) Potential use of proteomics in shellfish aquaculture: from assessment of environmental toxicity to evaluation of seafood quality and safety. Curr Org Chem 21(5):402–425
Emborg J, Laursen BG, Rathjen T et al (2002) Microbial spoilage and formation of biogenic amines in fresh and thawed modified atmosphere-packed salmon (salmo salar) at 2 °C. J Appl Microbiol 92(4):790–799
Fan W, Run-Ze FU, Shen J et al (2015) Changes of *Patinopecten yessoensis* microbial ecology and nutrition components during the process of depuration. Sci Technol Food Ind 36(19):78–83
FAOSTAT (2016) World fisheries and aquaculture status in 2016: contributing to the full realization of food and nutritional safety, Rome, 200
Feliciano L, Lee J, Lopes JA et al (2010) Efficacy of sanitized ice in reducing bacterial load on fish fillet and in the water collected from the melted ice. J Food Sci 75(4):M231–M238
Fisheries Bureau of the Ministry of Agriculture (2002) China Fisheries Yearbook. China Agriculture Press
Flemming HC, Wingender J (2010) The biofilm matrix. Nat Rev Microbiol 8(9):623–633
Ghaly AE, Dave D, Budge S et al (2010) Fish spoilage mechanisms and preservation techniques: review. Am J Appl Sci 7(7):859–877
Gram L, Dalgaard P (2002) Fish spoilage bacteria–problems and solutions. Curr Opin Biotechnol 13(3):262–266
Gram L, Huss HH (2000) Fresh and processed fish and shellfish

Han Q, Song XY, Zhang ZH et al (2017) Removal of foodborne pathogen biofilms by acidic electrolyzed water. Front Microbiol 8:988

Hall-Stoodley L, Costerton JW, Stoodley P (2004) Bacterial biofilms: from the natural environment to infectious diseases. Nat Rev Microbiol 2(2):95–108

Hao J, Wu T, Li H et al (2016) Differences of bactericidal efficacy on *Escherichia coli*, *Staphylococcus aureus*, and *Bacillus subtilis*, of slightly and strongly acidic electrolyzed water. Food Bioprocess Technol 10:1–10

Hoiby N, Bjarnsholt T, Givskov M et al (2010) Antibiotic resistance of bacterial biofilms. Int J Antimicrob AG 35:322e332

Huang YR, Hung YC, Hsu SY et al (2008) Application of electrolyzed water in the food industry. Food Control 19:329–345

Kasai H, Kawana K, Labaiden M et al (2011) Elimination of *Escherichia coli* from oysters using electrolyzed seawater. Aquaculture 319(3):315–318

Khazandi M, Deo P, Ferro S et al (2017) Efficacy evaluation of a new water sanitizer for increasing the shelf life of Southern Australian King George whiting and Tasmanian Atlantic Salmon fillets. Food Microbiol 68:51–60

Kim C, Hung YC, Brackett RE et al (2001) Inactivation of *Listeria monocytogenes*, biofilms by electrolyzed oxidizing water. J Food Process Preserv 25(2):91–100

Kim WT, Lim YS, Shin IS et al (2006) Use of electrolyzed water ice for preserving freshness of pacific saury (*cololabis saira*). J Food Prot 69(9):2199–2204

Klausen M, Heydorn A, Ragas P, Lambertsen L et al (2003) Biofilm formation by *Pseudomonas aeruginosa* wild type, flagella and type IV pili mutants. Mol Microbiol 48:1511–1524

Koo H, Falsetta ML, Klein MI (2013) The exopolysaccharide matrix a virulence determinant of cariogenic biofilm. J Dent Res 92(12):1065–1073

Kumar CG, Anand SK (1998) Significance of microbial biofilms in food industry: a review. Int J Food Microbiol 42(1):9–27

Lenz CA, Kai R, Knorr D et al (2015) High pressure thermal inactivation of clostridium botulinum type e endospores—kinetic modeling and mechanistic insights. Front Microbiol 6:652

Li G (2016) Sterilization process of slightly acidic electrolyzed water in the purification of Japanese *Scallop*. Guangdong Ocean University

Li JB, Lin T, Liao C et al (2013) Effect of physicochemical properties of electrolyzed water ice on inactivation of *Vibrio parahaemolyticus* during storage. J Food Sci Biotechnol 32(11):1169–1175

Lin T, Wang JJ, Li JB et al (2013a) Use of acidic electrolyzed water ice for preserving the quality of shrimp. J Agric Food Chem 61(36):8695–8702

Lin T, Wang JJ, Pan YJ et al (2013b) Comparison of the bacterial activity of acidic electrolyzed water against foodborne pathogenic bacteria in pure culture and foods. Food Sci 34(15):69–74

Liu SY, Liu JR, Ma YS (2014) Analysis of intestinal microflora in all stages of the live delivery of *Scallop*. Fish Sci 33(10):000626–630

Mccarthy S (2012) Efficacy of electrolyzed oxidizing water against *Listeria monocytogenes* and *Morganella morganii* on conveyor belt and raw fish surfaces. Food Control 24(1–2):214–219

Mikš-Krajnik M, Feng LX, Bang WS et al (2016) Inactivation of *Listeria monocytogenes*, and natural microbiota on raw salmon fillets using acidic electrolyzed water, ultraviolet light or/and ultrasounds. Food Control 74:54–60

Novotny L, Dvorska L, Lorencova A et al (2004) Fish: a potential source of bacterial pathogens for human beings: a review. Veterinární Medicína 49(9):343–358

Park SY, Ha SD (2015) Reduction of *Escherichia coli*, and *Vibrio parahaemolyticus*, counts on freshly sliced shad (*Konosirus punctatus*) by combined treatment of slightly acidic electrolyzed water and ultrasound using response surface methodology. Food Bioprocess Technol 8(8):1762–1770

Phuvasate S, Su YC (2010) Effects of electrolyzed oxidizing water and ice treatments on reducing histamine-producing bacteria on fish skin and food contact surface. Food Control 21(3):286–291

Ray B, Bhunia R (2004) Fundamental food microbiology. CRC Press, Boca Raton

Sevenich R, Bark F, Kleinstueck E, Crews C et al (2015) The impact of high pressure thermal sterilization on the microbiological stability and formation of food processing contaminants in selected fish systems and baby food puree at pilot scale. Food Control 50:539–547

Stoodley P, Sauer K, Costerton Davies DG et al (2002) Biofilms as complex differentiated communities. Ann Rev Microbiol 56:187–209

Sun JL, Zhang SK, Chen JY et al (2012) Efficacy of acidic and basic electrolyzed water in eradicating *Staphylococcus aureus* biofilm. Can J Microbiol 58(4):448–454

Sun JP, Zhao L, Yu WY et al (2017) Effect of acidic electrolyzed water on polyphenoloxidase from *Litopenaeus vannamei*. Food Sci 39(9):7–12

Vázquez-Sánchez D, Cabo ML, Ibusquiza PS et al (2014) Biofilm-forming ability and resistance to industrial disinfectants of *Staphylococcus aureus* isolated from fishery products. Food Control 39:8–16

Wang JJ, Zhang ZH, Li JB et al (2014a) Modeling *Vibrio parahaemolyticus* inactivation by acidic electrolyzed water on cooked shrimp using response surface methodology. Food Control 36(1):273–279

Wang JJ, Sun WS, Jin MT et al (2014b) Fate of *Vibrio parahaemolyticus* on shrimp after acidic electrolyzed water treatment. Int J Food Microbiol 179(22):50–56

Wang JJ, Lin T, Li JB et al (2014c) Effect of acidic electrolyzed water ice on quality of shrimp in dark condition. Food Control 35(1):207–212

Wang M, Wang JJ, Sun XH et al (2015a) Preliminary mechanism of acidic electrolyzed water ice on improving the quality and safety of shrimp. Food Chem 176:333–341

Wang XL, Li YQ, Li QG et al (2015b) Research on spatial-temporal aggregation of China's environment pollution and food safety: the interaction between emergency environmental accidents and foodborne disease outbreaks. China's Population, Res Environ 25(12):53–61

Wolcott RD, Rhoads DD, Dowd SE (2008) Biofilms and chronic wound inflammation. J wound care 17(8):333–341.

Wu Y, Wen J, Ma Y et al (2014) Epidemiology of foodborne disease outbreaks caused by *Vibrio parahaemolyticus*, china, 2003–2008. Food Control 46(46):197–202

Xiao J, Klein MI, Falsetta ML et al (2012) The exopolysaccharide matrix modulates the interaction between 3D architecture and virulence of a mixed-species oral biofilm. PLoS Pathogen 8(4):e1002623

Xie J, Sun XH, Pan YJ et al (2012a) Combining basic electrolyzed water pretreatment and mild heat greatly enhanced the efficacy of acidic electrolyzed water against *Vibrio parahaemolyticus*, on shrimp. Food Control 23(2):320–324

Xie J, Sun XH, Pan YJ et al (2012b) Physicochemical properties and bactericidal activities of acidic electrolyzed water used or stored at different temperatures on shrimp. Food Res Int 47(2):331–336

Xuan XT, Fan YF, Ling JG et al (2017) Preservation of squid by slightly acidic electrolyzed water ice. Food Control 73:1483–1489

Yao X, Zhao AJ, Du SP et al (2017) The effect of acid electrolysis water ice on the quality of small yellow croaker and the change of enzyme activity in muscle tissue. Food Sci 38(13):244–250

Zhang CL, Chen X, Xia XD et al (2017) Viability assay of *E. coli* o157: h7 treated with electrolyzed oxidizing water using flow cytometry. Food Control 88:47–53

Zeng XP, Tang WW, Ye GQ et al (2010) Studies on disinfection mechanism of electrolyzed oxidizing water on *E. coli* and *Staphylococcus aureus*. J Food Sci 75(5):M253–M260

Zhang X, Wang H, Li N et al (2015) High CO_2-modified atmosphere packaging for extension of shelf-life of chilled yellow-feather broiler meat: a special breed in asia. LWT—Food Sci Technol 64(2):1123–1129

Zhao AJ, Wang M, Zhao F et al (2016) Application of acidic electrolyzed water ice in sterilization and preservation of shrimp. Modern Food Sci Technol 32(3):126–131, 245

Zhou R, Liu Y, Xie J et al (2011) Effects of combined treatment of electrolysed water and chitosan on the quality attributes and myofibril degradation in farmed obscure puffer fish (*takifugu obscurus*) during refrigerated storage. Food Chem 129(4):1660–1666

Chapter 7
Application of Electrolyzed Water on Environment Sterilization

Charles Nkufi Tango, Mohammed Shakhawat Hussain and Deog-Hwan Oh

7.1 Introduction

Food processing environments play major roles in propagating and transmitting of foodborne pathogens (Holah 2014; Møretrø et al. 2012; Shaughnessy et al. 2011; Srey et al. 2013). Contamination of fresh food may occur at any stage food supply chain; from food production to consumption (Srey et al. 2013; Tango et al. 2015). In the field, the fresh produce has frequent contact with environmental surfaces, including the polluted water, soil, or air. In food industries, the major source of food contamination is through the food processing facilities and worker's hands. Many outbreaks of foodborne pathogens have been found to be associated with food environmental surfaces (Tango et al. 2014). There is vast evidence in scientific literature that contaminated environmental surfaces play an important role in the propagation of several foodborne pathogens and diseases (Bagge-Ravn et al. 2003; Kim et al. 1999). The food industry is in search for effective cleaning and disinfection (sanitization) methods that can reduce common spoilage and pathogenic bacteria and prevent economic loss and foodborne outbreaks.

Cleaning of food processing facilities is considered as the primary measure for the reduction and prevention of economic lose and foodborne outbreaks. However, cleaning without disinfection is not efficient. The use of any disinfectant solution in cleaning promotes a break up or dissolution of organic materials associated with biofilm and makes the microbial cells susceptible to killing by the disinfectant. Currently, several studies have demonstrated that environmental cleaning and disinfection methods can improve the reduction and prevent the propagation of foodborne

C. N. Tango (✉)
Department of Chemistry and Agricultural Industries, Faculty of Agronomy, University of Kinshasa, Kinshasa, Democratic Republic of the Congo
e-mail: charlynkufi2@yahoo.fr

M. S. Hussain · D.-H. Oh
Department of Bioconvergence Science and Technology, College of Agriculture and Life Science, Kangwon National University, Chuncheon, Republic of Korea

T. Ding et al. (eds.), *Electrolyzed Water in Food: Fundamentals and Applications*,
https://doi.org/10.1007/978-981-13-3807-6_7

pathogens in food industry (Pascual et al. 2007). Food deposits on processing plants can harbor potentially foodborne pathogens which if left to grow can present a serious risk to the public health. To control such bacterial growth the food deposits must initially be removed using an effective cleaning technique. An efficient cleaning and disinfection strategy consists of a sequence of rinses with water (containing a detergent) and disinfection applied in various combinations of temperatures, time, and concentrations. Therefore, the food contact surface should be properly cleaned with an alkaline solution to remove all substances that can support microbial growth. Subsequently, a good disinfectant should be applied for killing microorganisms. EW can be applied as a cleaner (EW containing sodium hydroxide) and a disinfectant (EW containing hypochlorous acid) (Rahman et al. 2016). In general, EW is produced by electrolysis of a brine solution and/or hydrochloric acid to alter the molecular structure of water. This reaction is induced by electrical current inside an electrolytic cell in which the electrodes are separated or not by a membrane. In cells separated by membrane, the cathode area generates an alkaline electrolyzed water (AlEW) with a high pH of 10–11 and a low ORP of −800 to −900 mV while the anode produces AEW at low pH of 2.7 and ORP of >1100 mV (Hricova et al. 2008; Huang et al. 2008; Rahman et al. 2016). SAEW with a pH of 5.0–6.5 and ORP between 750 and 890 mV is produced by electrolysis in a non-membrane cell (Cao et al. 2009; Koide et al. 2011; Tango et al. 2017). Due to the properties of AlEW such as high pH, negative and low ORP, and sodium hydroxide and H_2 contents (Al-Haq et al. 2005), it could be used as an alternative for trisodium phosphates for removing organic or inorganic materials from surfaces (Hao et al. 2015; Kim et al. 2005). Therefore, SAEW could be used to interrupt microbial environmental spread (Kim et al. 2005; Koseki et al. 2004).

The importance of using EW as an emerging method for improving cleaning and disinfection processes is based on its good biocidal potential, low by-products regeneration, wide microbial spectrum, and its ability to be generated in situ. Moreover, it also has a significant advantage of being an environmental friendly method that reduces industrial cost and facilitates their compliance with statutory obligations. These reasons not only make EW a good alternative to traditional chlorine but also to other chemical disinfectants, including quaternary ammonium, hydrogen peroxide, phenol, formaldehyde, and iodine compounds which are extensively used in cleaning and disinfection in the food industry. Such compounds can cause discoloration and irritation of skin, cancer, respiration problems, and corrosion of food processing equipment (Fasenko et al. 2009; Hricova et al. 2008; Ni et al. 2016; Zheng et al. 2013) as well as bacterial resistance (Ni et al. 2016; Zamora 1986).

It has been reported that the different types of EW have successfully demonstrated excellent cleaning and disinfection abilities in removing airborne dust and microorganisms from environmental surfaces in the food industry. Scientific and industrial literature showed that spraying AEW and AlEW were used to clean and reduce the dust levels and bacterial aerosols in layer breeding houses (Zhao et al. 2014; Zheng et al. 2012, 2013). AlEW has been reported effective in eliminating spoilage and pathogens bacteria on stainless steel surfaces in animal transport vehicles (Ni et al. 2016). AEW is reported to reduce foodborne pathogens contamination in seafood processing surfaces (Liu and Su 2006; Phuvasate and Su 2010), on ceramic platform surfaces in fish markets (Huang et al. 2006). SAEW water has also been used in reducing foodborne pathogens on the equipment surfaces in swine barns (Hao et al. 2013), on different surfaces and equipment in food processing operations (Guentzel et al. 2008), on animal houses (Hao et al. 2013a, 2014; Ni et al. 2015), on the surface of plastic poultry transport cages (Zang et al. 2015), and in removing biofilms and pathogenic bacteria in dental equipment and clinical environment (Fertelli et al. 2013; Komachiya et al. 2014; Stewart et al. 2014). Moreover, EW was applied in agriculture as an alternative to conventional fungicides for the control of foliar pathogenic fungi in greenhouse (Buck et al. 2003; Mueller et al. 2003). Although several researches on the application of EW have been performed among academic researchers and food processors, but the application of EW as cleaner and disinfectant agent is not typically and frequently used in agriculture and food industry. Therefore, this chapter reviews the advantage of use of EW for improving the cleaning and disinfection in environmental surface in food industry and hospital and summarizes the major factors that influence the sanitization potency of EW as an alternative technology.

7.2 Definition of Terms (Disinfectant and Sanitizer)

Cleaning is a physio-chemical process aiming at the elimination of soils, debris, food residues, and other materials which may enhance bacteria proliferation and biofilm formation (Simões et al. 2010; Srey et al. 2013). It is the interaction of the chemical and physical reactions that determine the efficiency and rate of the cleaning method. The main chemical reaction during cleaning processing is hydrolysis, saponification, chelation, and oxidation. The physical reaction during cleaning processing induces different actions, including wetting, penetration, emulsification, dispersion, and solubilization. The disinfection process involves the killing of spoilage and pathogenic microorganisms which can present a severe risk to the economic loss and the health of the consumer. Disinfection solution provokes a reaction with microbial cell and facilities the penetration of cell membranes to produce a biocidal or biostatic action (Holah 2014). Another notion important for understanding microbial control is the sanitization. Sansebastiano et al. (2007) defined the sanitization as a reduction of pathogenic bacteria population colonizing a given substrate in order to meet the safety level for public health. Detergents are often chemical compounds used to dissolve grease and remove dirt and soil attached on any surface. Appropriate acidic or alkaline solution can also be used. EW meets all of these functions and properties and if it is used effectively it can be a powerful cleaning and disinfection in food industry and other public health-related industries.

7.3 Mode of Application of Electrolyzed Water on Environment

Most currently, EW treatments have been performed using dipping or submersion method. However, regarding the complexity of food processing equipment, the dipping and submersion method may not be a convenient technique to clean and disinfect the food processing plants. Several procedures can be used for environmental disinfection with EW. The common among these methods are rinse, spraying, and clean-in-place. There are a number of factors, including type of factory, food, and materials used, that have been demonstrated to drive the choice of method applying EW cleaning and disinfection on food process environment and production equipment.

7.3.1 Spraying EW Cleaning and Disinfecting

The spraying disinfectants are often used generally in modern intensive poultry and livestock production. Spraying EW is an approach to reduce airborne dust and bacteria is an essential part of sanitization purpose in animal houses. Spraying EW during production can reduce dust and other harmful particles on surface and improve the air quality inside food plant and animal housing (Zheng et al. 2012, 2013). Its application seems to be doubled because it permits to avoid the application of antibiotic additives in feedstuffs and to minimizing the necessity to use therapeutic drugs. The utilization of AEW has limited potential for spraying in poultry houses. Since it can easily dissolve to Cl_2 gas due to volatilization, which causes chlorine loss, therefore, decreasing acidic EW sanitization potency with time. SAEW with a neutral (pH 6.0–6.5) pH is considered as a novel disinfectant for spraying in food plant and breeding houses. The AEW and SAEW spraying treatment showed expected performance to reduce the airborne dust and bacterial aerosols in animal and agricultural production buildings (Hao et al. 2014; Mueller et al. 2003; Zhao et al. 2014; Zheng et al. 2012).

One of the earliest EW spraying investigations studied the effects of spraying methods and conditions on properties of EW and their antimicrobial effects against *Listeria monocytogenes*. Experiment data demonstrated that high air pressure spray retained more chlorine and gave a higher ORP than low air pressure. EW containing high initial chlorine concentration (about of 88 mg/L) achieved at least a 3–4 log cfu/mL reduction in *L. monocytogenes* populations when sprayed using the spray gun, while spraying treatment using a bigger commercial sprayers (backpack sprayer or a poly-tank) inactivated *L. monocytogenes* population (9.4 log cfu/mL reductions) completely (Hsu et al. 2004). The results in this report showed that the sprayer size, type, power are important factors to control for using EW spraying has potential mode to inactivate microorganisms on food processing equipment. Application of EW spray treatment onto the wall of poultry breeding was performed to decrease the pathogenic microorganisms. Samples for microbiological analysis were collected at the time sequence of 20–60 min after spraying treatment using smear method. In this study, the staphylococci, streptococci, and enterococci were completed inactive after 60 min of treatment. There was a significant (82%) decrease in fungi populations after 60 min spraying treatment (Jirotková et al. 2013). EW spraying treatment has been demonstrated to decontaminate the greenhouse-grown floricultural crops threatened by foliar plant pathogenic fungi. In this experiment, the authors used the spraying mode and reported that EW spraying treatment caused minor damage to some plant species but, overall, it appears to be safe to use as a foliar spray on a wide variety of bedding plants grown under greenhouse conditions (Buck et al. 2003). Another study is applied.

EW spraying treatment to evaluate the management of powdery mildew on gerbera daisy and the compatibility of EW with different traditional fungicides and pesticides. The authors stated that EW has a potential option for controlling powdery mildew on gerbera daisies and the EW spray treatment offers to growers an excellent treatment to decrease the use of traditional fungicides in greenhouses. EW was compatible with five different fungicides, while it was not compatible with thiophanate methyl. Therefore, EW spraying treatment can be used in integrated management system of greenhouse-grown floricultural crops (Holah 2014). Spraying treatment was used to compare the effectiveness of SAEW and traditional environmental sanitizers, including benzalkonium chloride (BC) and povidone-iodine (PI) solutions, on the reduction of airborne microorganisms in a layer breeding house. The results showed that there is no significant reduction of airborne bacteria among these sanitizers. However, SAEW has been showed to be more effective in airborne fungi compared to BC and PI treatments (Hao et al. 2014).

EW uses in spraying treatment contains usually a high available chlorine concentration because the spraying treatment significantly decreased the initial properties of EW, including chlorine concentration and ORP. The reduction is important for chlorine concentration, which can vary between 20 and 90% (Hsu et al. 2004). In addition, the loss of the chlorine component during spraying treatment depends on sprayer type, nozzle size, air pressure, and temperature (Chuang et al. 2013b; Hao et al. 2014; Hsu et al. 2004; Zhao et al. 2014). Zhao et al. (2014) evaluated the changes in SAEW properties during praying treatment under various air conditions. Their results showed that the chlorine concentration loss can be categorized into initial and traveling step of EW fine particles. Loss rate at the initial step was low, in short period of time, and probably related to sudden aerosolization near the nozzle. While the loss from nozzle to target was more important and associated with air temperature since aerosols evaporated quicker at higher temperatures. Evaporation of aerosols increases the relative contact surface, and releases and decomposes the sanitizing components (Koide et al. 2011). Therefore, air temperature becomes an important factor to control during EW spraying process. While the use of smallest which can accelerate the interfacial mass transfer of chlorine gas, resulting in moderate chlorine loss (Ovissipour et al. 2015). Chuang et al. (2013a) evaluated the inactivation efficacy of neutral EW to *Bacillus subtilis* and *Escherichia coli* using different types of sprayer. The experimental results demonstrated that number of

Fig. 7.1 Schematic representative of sequential steps used in cleaning and disinfecting

nozzle on sprayer device also significantly influences the sanitization of spraying processing.

7.3.2 *Cleaning-in-Place*

Cleaning-in-place (CIP) is a process of cleaning or disinfecting of stationary equipment, including vessels, tankers pipes, and filters without opening or disassembling the equipment. The process involves usually a minimal manual intervention on the part of the operator and is automatically controlled. CIP consisting of a sequentially four-step process: pumping or circulating through or on surface of plant the water following by washing with cleaning (alkaline solution) and disinfection (acid sanitizer) solutions under conditions of increased mechanical force of turbulence and flow velocity, and finally the sanitizing rinse before the next food production. The cleaning step is usually performed at very high temperature and followed by rinsing with proper water before disinfection step, which should be also followed by water rinsing to remove effectively the excess of disinfection solution (Fig. 7.1). The treatment parameters such as alkaline and acid treatment time, concentration, and temperature were significant parameters (Walker et al. 2005a). The milk industry and breweries are the largest users of CIP method since they include pipeline and some of the most difficult to clean processing plant. It is known that the safety of beer and raw milk essentially depends on using clean processing equipment during the production. While EW can be a cleaning effective and environmental friendly sanitizer for improving CIP times and saving potential water and energy.

EW was successfully employed for CIP of food processing equipment to remove the food depot and microorganisms. The results demonstrated that EW can maintain a level of safety and comply with microbiological integrity and sensory testing requirements (Chen et al. 2012). A pilot-scale test system was studied to evaluate the effectiveness of EW for CIP procedure of dairy processing plant, specifically a refrigerated milk storage tank and a tank used for sterilization of milk. A response surface model was used to optimize temperatures and times for both AlEW and acid EW treatments to clean the vessel soiled by liquid milk. Parameters for four-step CIP procedures using EW were applied: washing with AlEW at 54.6 °C for 20.5 min and sanitize with acid EW water at 25 °C for 10 min. The efficacy of the CIP procedure was assessed using a microbiological enrichment method, ATP bioluminescence, and residual protein detection assays. The authors reported that a complete CIP procedure using EW with optimal operational temperatures and time was able to return the surface of the test vessel to a satisfactory clean condition with non-detectable residual ATP and protein. Another work has been performed to compare the bacterial activity and corrosion assessment of AEW and peracetic acid using both laboratory and industrial scales. The results showed that AEW, containing a free chlorine concentration of above 17 mg/L, was satisfactory and an effective alternative to peracetic acid (2%) in terms of safety. The corrosion rate of the AEW was similar to that of distilled water and it was significantly lower than that of peracetic acid (Chen et al. 2012). Jiménez-Pichardo et al. (2016) studied and developed an optimal design of experiment to clean and disinfect the milk spoilage bacterial strain such as *Pseudomonas aeruginosa*, *Enterococcus faecalis,* and *Micrococcus luteus* on stainless steel surface as a model in the dairy industry. The results of this study showed that washing with AlEW (100 mg/L NaOH) followed by neutral EW (40 mg/L total chlorine concentration) at 30 °C for 10 min was able to removed 3.90 ± 0.25 and 3.20 ± 0.20 log CFU/cm^2 of milk spoilage bacteria on stainless steel with and without electro-polishing, respectively. They concluded that the cleaning and sanitization procedure of stainless steel plates with a combination of AlEW and neutral EW represents an excellent alternative for the dairy industry (Jiménez-Pichardo et al. 2016).

Several researches developed the mathematical models to describe the depot and microorganism removal from food processing plants during CIP processing using a pilot-scale system. A successful mathematical model was used to describe the milk deposit removal process during the CIP process with combined alkaline and AEW through a stainless steel straight pipe as a simulator pipeline. The mass of milk depot on the pipeline was measured using a high precision balance after the initial soiling and the end of each CIP cycle was evaluated using ATP bioluminescence method. Experimental results demonstrated that a satisfactory CIP performance for the simulator pipeline was achieved with the time duration of the CIP process that was shortened by 55% (10 s warm water rinse, 3 min AlEW wash, and 6 min acid EW wash) and yielded a deposit removal of 0.28 mg/mg m^2 at the end of the washing processing (Zheng et al. 2013). Another model was developed to optimize the temperature of EWs during CIP processing for milk depot removal using a pilot-scale pipeline milking system. The cleaning efficiency of the AlEW (45–75 °C) washing was estimated by ATP bioluminescence test and safety effectiveness of AEW (25–45 °C) was evaluated by microbiological analysis through enrichment culture. Response surface methodology showed that the optimal logarithmic mean temperatures of 58.8 and 39.3 °C for the AlEW and AEW, respectively, were effective for milk depot removal during CIP of pipeline. Moreover, combination of AlEW and AEW CIP outperformed the conventional CIP technique both in terms of bacterial inactivation and percentage Relative Light Units reduction (Dev et al. 2014). A similar result was observed when the effective treatment time and temperature combinations of both AlEW and AEW were evaluated by response surface modeling of a temperature range of 25–60 °C and a time range of 5–20 min. These results showed that most of the treatments at 60 °C successfully removed all detectable bacteria and ATP on different materials used in dairy processing (Walker et al. 2005b). Overall, the results described above demonstrate that AlEW and AEW have the potential to be used as a cleaning and sanitizing agent for materials used in brewery and dairy system.

7.3.3 *Manual EW Cleaning and Disinfection*

Food processing environment is recognized as an important source of recontamination. The microbial pathogens may access food processing environment through the raw materials, workers, pests, and mobile equipment, including forklifts. Some bacteria may survive and grow in food processing environment through biofilm form in floor, walls, interfaces between floor and equipment. Cleaning and disinfection of these environments require a high amount of cleaning and disinfecting solutions and washing time. EW is on-site produced sanitizer, less toxic and effective against a large spectrum of microorganisms, and can be positively used for cleaning and disinfecting these parts of food processing environment as well as household kitchen. Manual cleaning and disinfection are generally applied in cleaning and disinfecting the floor, rail, and wall in food factory, service, or hospital (Hao et al. 2013b). In this chapter, manual EW cleaning or disinfection is considered as a method in which the food processing equipment or surface is soaked, rinsed, or washed with liquid disinfectant. It is often associated with manual or mechanical brushing and EW is spread manually using polyvinyl chloride horse pipe but during spraying treatment EW is spread by mechanical or electrical pumps. Only a few studies have investigated the effectiveness of EW on cleaning and disinfecting food factory and hospital floor, wall, and mobile equipment in detail. While SAEW, containing an ACC of 250 mg/l and pH value of 6.19, was flushed and spread for 5 min to disinfect the microbial population on floor and wall in layer breeding house after cleaning with tap water. A reduction of 0.79 and 1.06 log cfu/cm^2 was observed, respectively, for floor and wall (Hao et al. 2013a). An increased microbial reduction of 1.07 and 3.18 log cfu/cm^2, respectively, on surface of floor and wall on swine barns was found when the chlorine concentration and time treatment were increased at 300 mg/L and 30 min, respectively (Hao et al. 2013b). A high reduction could be obtained if the cleaning was performed with AlEW solution. The sequential cleaning with AlEW should remove debris and residues so that microorganisms can be more susceptible to disinfecting with AEW (Fig. 7.1, Hricova et al. 2008). AlEW solution contain an amount of free hydroxide ions which promote a rapid hydrolysis of organic food residues, including protein, carbohydrate, and effective saponification of fats and oils, principally at high temperatures.

Application of good EW cleaning and disinfecting mode should reduce the cross contamination in food processing and medical plants. Due to zoning, limitations of movements of personnel, and lack of good cleaning and disinfection programs, and mop and bucket washing technique appear to risk leaving residual microbial contamination of surfaces, a careful application of a spraying and manual (associated with physical force) methods that may be able to permeate into difficult-to-reach places becomes a potential method to clean and sanitize the food processing and medical environment.

7.4 Factors Affecting the Efficacy of EW on Environmental Cleaning and Disinfection

In general, the rate of EW-induced inactivation of microbial population depend markedly on the chlorine concentration, pH, and ORP of EW and temperature and time treatment (Hricova et al. 2008; Rahman et al. 2016). This chapter is concerned with the factors that impact specifically on action of EW during cleaning and disinfecting environmental surfaces. The sanitization efficacy of EW on surface contact in food industry and hospital is affected by a number of factors related to environmental conditions and by the sensitivity of the target microorganisms (Hricova et al. 2008; Møretrø et al. 2012; Rahman et al. 2016). The main factors influencing the effectiveness of EW cleaning and disinfection are method of EW application, size and location of microorganism, microbial form (planktonic or biofilm), and the presence of organic or inorganic material on the environment of food industry.

7.4.1 Number and Location of Microbial Population

When the microbial populations colonize an environment, the cell number is initially low. They become higher over time and are overlapped in environment in which these populations survive. The microbial cells located on outer layers are directly in contact with the sanitizer agent and are consequently inactivated, while those that located underlying layers have more chance to survive the treatment (Gómez-López et al. 2005; Vázquez-Sánchez et al. 2014), even though a long exposure time allows to penetrate deeper. In addition, during the cleaning and disinfecting of large-scale production plant, the sanitizer solution cannot reach all parts of the plant (Russell 2008) and microbial population located at different locations such as wall, air, and feed debris inside of animal layer house may react differently to EW solutions. Because the populations grown on the debris and wall are in more affinity to substrate compared to that fluctuating on air. Consequently, location of microbial population can reduce the effectiveness of EW cleaning and disinfection. Interestingly, one study investigated efficacy of SAEW on the inactivation of bacterial and fungal population located at different solid samples (dust, feces, feathers, and feed) inside layer breeding house. The results showed that the initial population of bacteria and fungi were approximately same in feces (9.02 log cfu/mL) and dust (9.55 log cfu/mL), however, the inactivation was higher in feces (>8.02 log cfu/mL) compared to dust (2.78 log cfu/mL) after 5 min SAEW spraying disinfection (Hao et al. 2013a).

Number of microbial population also should be considered during improving disinfection approach on environmental surface. The disinfection of higher cells density in any medium requests high concentration and more time as well as high warm temperature. The effectiveness of chlorine from neutral EW and traditional chlorine from sodium hypochlorite was investigated from the values of the overall load during washing tanks. Results demonstrated that the disinfection was more dependent on the initial microbial density, which was directly associated with the type of fresh produce washed (Machado et al. 2016).

7.4.2 *Microbial Form*

Microbial cells associated with biofilms are much more resistant to disinfection methods than their planktonic counterparts. Increased resistance to disinfection methods is achieved due to the changes in physiology when the cells are attached to surfaces, moreover, to the production of extracellular polysaccharides as a protective barrier. Resistance to disinfection methods is also due to cell-to-cell signaling, genetic exchange, and amplifying the production of degradative enzymes by attached cells (Bredholt et al. 1999; Russell 2008). Carpentier and Cerf (2011) described the microbial resistance as a situation in which the extent of killing by an antimicrobial agent applied at a bactericidal concentration is less than what is expected. Deza et al. (2005) investigated the efficacy of neutral EW (NEW, ACC: 63 mg/L, pH: 6.05, ORP: 849 mV) in reducing TSB cultivated *L. monocytogenes*, *E. coli* O157:H7, and *P. aeruginosa* populations on stainless steel surfaces. They reported that washing inoculated stainless steel surfaces for 1 min in NEW solution decreased the pathogens of all strains by more than 6 log. However, the biofilm developed on LB broth of the same pathogenic bacteria was treated for 5 min with NEW (ACC: 70 mg/L, pH: 6.5, and ORP: +1015 mV), showed a reduction of about 2.0, 2.5, and 4.0 log, respectively, for *L. monocytogenes, E. coli* O157:H7, and *P. aeruginosa* on stainless steel surface (Moradi and Tajik 2017). *S. aureus cells* on stainless steel have been demonstrated to be more sensitive to SAEW treatment than staphylococcal biofilm formed on stainless steel (Ni et al. 2016; Tango et al. 2014).

It is known that the killing mechanism of EW is mainly due to chlorine concentration and acidic pH plays effective role in planktonic cells inactivation. Several studies investigated the pH effect on EW removing biofilm of different bacteria and reported that the decreasing pH did not influence the sanitization activity of EW against different bacteria (Han et al. 2017; Sun et al. 2012; Vázquez-Sánchez et al. 2014). The lack of pH effect is probably due to the presence of the extracellular matrix, which acts as a bacterial protective barrier that is not sensitive to pH effect and limits the diffusion of HOCl into the biofilm deepest layers (Han et al. 2017). Therefore, the application of SAEW for removing biofilm seems effective for controlling pathogenic biofilm formation in any environmental surfaces, since SAEW with pH between 5.5 and 6.5 contains approximately 98% of HOCl of available chlorine concentration, which is 80 times more efficacy as a disinfecting agent than as hypochlorite ion at an equivalent concentration (Cao et al. 2009; Tango et al. 2014). The sequential cleaning and disinfecting with AlEW and SAEW have been found more effective in killing cells and biofilm removing from different environmental surfaces of food processing and hospital (Jiménez-Pichardo et al. 2016; Nakano et al. 2016; Vázquez-Sánchez et al. 2014).

7.4.3 Presence of Organic and Inorganic Material in Food Processing and Medical Environment

Failure in cleaning before disinfection may detriment significantly the action of AEW solution since food processing and hospital environment contain numerous forms of organic materials from food residue, milkstone, soil, fecal material, blood, and serum (Russell 2008). The presence of organic materials in sanitization processing of environmental surfaces is an important consideration of any disinfection process. Organic material interferes with the sanitization activity of disinfectant, especially chlorine-based disinfectants. This interference commonly takes the form of a reaction between the highly reactive compounds such HOCl and the ammonium radical of protein compounds (Jo et al. 2018), consequently change appears on the form of active chlorine component and reduces the effectiveness of EW. The reaction leads to the formation of some cancerigenic compounds such as trihalomethanes (Gómez-López et al. 2005). Jo et al. (2018) investigated the detrimental effect of different organic materials from lipid, carbohydrate, and protein on the sanitization effect of SAEW. Experimental results showed that the change rate of free chlorine to combined chlorine was proportional to increasing rate of peptone and tryptone into SAEW. Similar experimental data have also been reported using AEW by Oomori et al. (2000). An alternative possibility is that organic material protects microbial population from HOCl solution attack (Russell 2008). Organic protein has been incorporated into SAEW testing inactivation of foodborne pathogens. Experiment results demonstrated that the inactivation of foodborne pathogens by SAEW decreased with increasing the peptone or tryptone. Similar results were found when protein compounds were added in AEW (Oomori et al. 2000). The change of free to combined chlorine was slightly observed when corn oil was added into solution but did not affect the sanitization effect of EW. Lipid material did not impact the effectiveness of EW against foodborne pathogens because the free chlorine may react with carbon–carbon double bonds in the presence of corn oil since it contains a high number of unsaturated fatty acid chains (Jo et al. 2018; Oomori et al. 2000).

The high contents of certain specific inorganic component in food are sometimes used to inhibit the growth of certain microorganisms. However, some process in food industry, including pyrolysis, transform organic materials into gaseous, liquid component, and solid residue containing carbon and ash. Certain biomass types also carry considerable proportion of inorganic materials (Arvanitoyannis 2010). Inorganic materials deposited on the surface from the fluid stream after production could impact on the cleaning of certain food processing such as brewery and dairy industries. The effect of inorganic depots on the disinfection process has been studied and showed the protection of microorganisms during disinfection by inorganic residue (Al-Haq et al. 2005; Al-Qadiri et al. 2016a). Still now no study has investigated the effect of inorganic contaminants on the effectiveness of EW solutions in cleaning and disinfecting process on environmental food processing and medical equipment. The knowledge of such a process may improve the safety approach on food industry and hospital.

7.5 Application of EW on Different Environments

Different types of EW solutions can be used as an effective combined cleaning and disinfection method, which have several advantages in the food and healthcare environments to control the contamination of food spoilage and pathogens or transmission of disease. The cleaning and disinfection aimed at environmental surfaces resulted in significant improvement in disinfection practices, which are different regarding environments include food manufacture, farms, agriculture, and hospital. The range of published EW applications for disinfection of environment surfaces is summarized in Table 7.1 and the description of these applications in more detail is given in the following sections.

7.5.1 Food Processing Environments

EW has been successfully used in the food processing equipment as a potential disinfectant. Cutting boards can be disinfected using neutralized electrolyzed water (NEW). Al-Qadiri et al. (2016b) have shown that combination of NEW, quaternary ammonium, and lactic acid based solutions are effective against *Salmonella Typhimurium*, *Escherichia coli* O157:H7, *Campylobacter jejuni*, *Listeria monocytogenes,* and *Staphylococcus aureus* from the surface of polypropylene and wooden food cutting boards. An approximate 5 log reductions have obtained using NEW for 5 min. Moreover, Deza et al. (2005) have shown similar effectiveness of NEW to eliminate *E. coli*, *Pseudomonas aeruginosa*, *S. aureus,* and *L. monocytogenes* from plastic and wooden kitchen cutting boards. Importantly, in this study, it has

Table 7.1 Application of electrolyzed water for cleaning and disinfection method in food processing, agriculture, and health care associated environments

Category	Objective	Mode of application	Type of environment	Microorganism	References
Food processing surface	Combined effect using AlEW and NEW as cleaning disinfection procedure	Manual with agitation on stainless steel plates	Milking processing equipment	*P. aeruginosa* *E. faecalis* *M. luteus*	Jiménez-Pichardo et al. (2016)
	Efficacy of NEW on biofilm at different temperature and type of milk.	Manual rinsing on stainless steel coupon	Milk processing surface	*L. monocytogenes E. coli* *P. aeruginosa*	Moradi and Tajik (2017)
	Efficacy of NEW on viral inactivation at different food processing surfaces	Manual application Stainless steel surfaces	Food processing equipment	Human norovirus	Moorman et al. (2017)
	Comparison efficacy of NEW with sodium hypochlorite at different free chlorine concentrations and pHs	Cleaning-in-place of stainless steel flow cell system	Fresh-cut vegetables plant	*E. coli*	Machado et al. (2016)
	Comparison efficacy of AEW, SAEW, and AlEW to control the biofilm	Manual rinsing on stainless steel and plastic coupon	Surfaces of processing equipment	*B. cereus*	Hussain et al. (2018)
	Synergistic effect of AlEW and AEW for washing milking system	Cleaning-in-place of stainless steel straight pipe	Milk processing surface	Organic materials	Wang et al (2016)

(continued)

Table 7.1 (continued)

Category	Objective	Mode of application	Type of environment	Microorganism	References
	Optimization of the temperature of AlEW and AEW for maximum cleaning of the milking system	Clean-in-place of pipeline	Milking system equipment	*P. fluorescens* B2 *M. luteus* *E. faecalis* *E. coli*	Dev et al. (2014)
	Comparison of efficiencies of SAEW and AEW with sodium hypochlorite and chlorine dioxide	Manual rinsing of stainless steel plates	Meat processing surface	*P. fluorescens* H5	Wang et al. (2018)
	Effectiveness of AEW on different food processing surfaces	Pumping and spraying method	Slaughterhouse equipment surfaces	*S. aureus* *S.* Typhimurium E. coli O157:H7 L. monocytogenes MBS	Serraino et al. (2010)
	Comparison of efficacy of AlEW+SAEW and AlEW+AEW with composite phenol solution	Spraying of inoculated stainless steel plates	Animal transport vehicles	*E. coli* *S. typhimurium* *S. aureus*	Ni et al. (2016)
	Comparison of efficacy of NEW with NaClO solution at same conditions of pH, ORP, and active chlorine	Manual rinsing of plastic and Wooden	Kitchen cutting boards	*E. coli* *P. aeruginosa* *S. aureus* *L. monocytogenes*	Deza et al. (2007)

(continued)

Table 7.1 (continued)

Category	Objective	Mode of application	Type of environment	Microorganism	References
	Effective of EW at different concentrations and pressurized spray pressures	Spray application on slicing blades	Meat processing surfaces	*L. monocytogenes E. coli* O157:H7 *Salmonella* sp.	Veasey and Muriana (2016)
	Sequential application of EW solution and BAC and PAA at different washing times	Manual rinsing of stainless steel coupons	Fishery processing surfaces	*S. aureus*	Vázquez-Sánchez et al. (2014)
	Efficacy of AEW (50 mg/L) at different food processing surfaces	Spraying application conveyor belt coupons	Seafood processing plants	*L. monocytogenes* *M. morganii*	Mccarthy and Burkhardt (2012)
	Comparison efficiency of NEW, BAC, and lactic-acid-based solutions	Manual spraying technique on wooden and polypropylene	Food cutting boards	*S. Typhimurium, E. coli O157:H7, C. jejuni* *L. monocytogenes S. aureus*	Al-Qadiri et al. (2016b)
	Efficacy of SAEW at different treatment times and washing speeds	Automated produce washer	Fresh fruits and vegetable service environment	*E. coli* O157:H7 *S.* Typhimurium	Afari et al. (2015)
	Effective and corrosive effects of AEW in comparison with PAA	Cleaning-in-place on bright beer tank	Brewery equipment	*L. brevis* *S. diastaticus* *B. subtilis*	Chen et al. (2012)

(continued)

Table 7.1 (continued)

Category	Objective	Mode of application	Type of environment	Microorganism	References
Poultry and Farm	Efficacy of SAEW at different concentrations and times	Spray application on plastic cages	Poultry transport cages	*S. enteritids*	Zang et al. (2015)
	Comparison efficiency of SAEW with benzalkonium chloride and povidone-iodine solution	Spray application on hen house equipment	Egg-producing house	*Salmonella* *E. coli*	Hao et al. (2013a)
	Effectiveness of SAEW at different spatial locations	Spraying in animal house surface	Layer breeding house	airborne microorganisms	Hao et al. (2014)
	Comparison efficiency of SAEW with povidone-iodine	Spraying in the surface of floors, walls, feed troughs, and egg conveyor belts	Surfaces of layer houses	TAB Coliforms, *Staphylococci* Yeasts and molds	Ni et al. (2015)
	Effectiveness of NEW as a drinking water addictive for chickens	Supplementation in drinking water	Drinking lines in chicken houses	TVC *E. coli*	Bügener et al. (2014)
	Comparison efficiency of SAEW with EDP and povidone-iodine	Spraying technique	Swine barns environment	*Salmonella* *S. aureus* Coliforms Fungi	Hao et al. (2013b)

(continued)

Table 7.1 (continued)

Category	Objective	Mode of application	Type of environment	Microorganism	References
	Comparison efficiency of AlEW and AEW with iodophor Mikroklene at processing plants	Pressurized spraying technique	Abattoir	Total coliforms *E. coli*	Bach et al. (2006)
	Free chlorine loss from MLAEW at different air temperatures	Spraying process using different nozzle sizes	Poultry house surface	TAB	Zhao et al. (2014)
	Effectiveness of NEW on dust reduction	Spraying method	Laying breeding cages	Dust	Zheng et al. (2012)
Restaurants/Feeding facilities	Comparison efficiency of AEW with quaternary ammonium compound, sodium hypochlorite	Automatic and manual washing process	Food service facilities	*E. coli* K12 *L. innocua*	Sigua et al. (2011)
	Comparison efficiency of AEW with UC–light, ozone	Mechanical and manual ware washing	Food service facilities	*E. coli K-12* *S. epidermidis*	Handojo et al. (2009)
Healthcare environments	Efficiency of MLEW on bioaerosols at different available chlorine and ventilation rates	Pumping and spraying method	Indoor healthcare environment	Airborne bacteria *S. aureus* λ virus	Chuang et al. (2013b)
	Efficacy of SAEW against flow and static biofilm	Manual rinsing	Hospital pipeline system	*E. faecalis*	Cheng et al. (2016)

(continued)

Table 7.1 (continued)

Category	Objective	Mode of application	Type of environment	Microorganism	References
	Comparison efficiency of AEW with glutaraldehyde (2%)	Manual washing	Endoscopes equipment	Hepatitis B virus	Lee et al. (2004)
	Effectiveness of SAEW for preventing microbial contamination	Manual washing	Dental unit water system	*S. aureus* *C. albicans* *P. aeruginosa*	Komachiya et al. (2014)
	Efficacy of applied AEW biocidal action on bioaerosol	Spraying method	Radiology unit equipment	TAB	Pintaric et al. (2015)
	Effects of sequential AlEW and AEW washing to remove or inactivate microbial pathogens	Manual washing	Hospital metal surfaces	*S. aureus* *P. aeruginosa* *C. albicans* Fungus	Nakano et al. (2016)
	Comparison efficiency of AEW with glutaraldehyde (2%)	Manual immersion	Dental unit equipment	*S. aureus*	Jnanadev et al. (2011)
	Effectiveness of AEW for environmental decontamination	Fogging spraying application	Hospital hard surfaces	MRSA *A. baumannii*	Clark et al. (2006)
	Effectiveness of AEW	Pumping and spraying method	Dental unit waterline	VBC	Kohno et al. (2004)

Slightly acidic electrolyzed water (SAEW), membrane-less electrolyzed water (MLEW), acidic electrolysed water (AEW), peracetic acid (PAA), disinfection powder (EDP), total aerobic bacteria (TAB), total viable cell (TVC), meticillin-resistant staphylococcus aureus (MRSA), viable bacteria count (VBC), and mesophilic bacterial suspension (MBS) also revealed that the disinfection efficacy of NEW is comparable to that of NaOCl as well has the advantage of having a larger storage time. EW has used to control biofilm from conveyor belt coupons as well as and raw fish surfaces. For instance, *L. monocytogenes* and *M. morganii* biofilms formed on conveyer belt coupons can reduce up to 7 log cfu after 5 min of EW treatment (Mccarthy and Burkhardt 2012). Recently, Hussain et al. (2018) have found that BEW can remove *B. cereus* from stainless steel and plastic coupons which are frequently using in the food processing industry. Moreover, EW has reported to be used in the industrial food processing equipment. For instance, EW can be used effectively to disinfectant slicing blades in the meat processing industry to eliminate *L. monocytogenes*, *E. coli* O157:H7, and *Salmonella* spp. (Veasey and Muriana 2016). In this study, inoculated slicing blades have shown reductions of 3.6 and 5.7-log reduction for clean, 0.6, and 3.3-log reduction for dirty slicing blades. Several combination treatments have been applied with EW to disinfectant food processing surfaces. For example, stainless steel surfaces have been disinfected for *E. coli* ATCC 25922, *Pichia pastoris* GS115, and *Aureobasidium pullulans* 2012 using NEW combined with ultrasound (Zhao et al. 2017). Moreover, Jiménez-Pichardo et al. (2016) have reported that AlEW and NEW can be used as an alternative to clean and disinfect stainless steel plate surfaces, particularly for the dairy industry. In addition to reduction of bacterial load, EW can also be used to disinfect viruses. For example, NEW is shown to be effective in reducing human norovirus (NoV) on stainless steel surfaces (Moorman et al. 2017). Another study demonstrated that EW could be used as a surface sanitizer to remove histamine-producing bacterial contamination. Washing those bacteria inoculated ceramic tile and stainless steel with EW (pH: 2.7, chlorine: 50 ppm, ORP: 1160 mV) for 5 min removed the histamine-producing bacteria on both surfaces. The reductions of histamine-producing bacteria ranged from >0.92 log CFU/cm^2 for *Klebsiella pneumoniae* on ceramic tile to >5.41 log CFU/cm^2 for *Enterobacter aerogenes* on stainless steel depending on the populations survived on chips after inoculation (Al-Qadiri et al. 2016b). AlEW has been also applied in electrical industry to improve the surface cleanliness of electroplated nickel plates (Takenouchi and Wakabayashi 2006). The nickel plates were efficiently rinsed by reducing by half the amount of residual sulfate ion compared to deionized water. The sulfate ions probably were selectively absorbed on the nanosize colloidal hydrogen bubbles, or substituted for anions absorbed on the hydrogen bubbles present in AlEW. Therefore, the sulfate ions were easily detached from nickel plates.

7.5.2 Poultry and Farm Environments

Facing the air quality problem caused by airborne microorganisms in the poultry and farm environments, scientists have shown interest in the use of EW to inactivate airborne microorganisms without causing harmful effects to workers and animals (Bach et al. 2006; Bügener et al. 2014; Hao et al. 2013a; Zhao et al. 2014; Zheng et al. 2012). Animal houses often faces problems with microbial contaminates which can cause adverse effects in environmental and public health (Weichao et al. 2016). EW proves to be effective in reducing the risk of pathogen transmission in the animal farm such as poultry and other animal transport systems. Wang et al. (2018) have reported the potential use of SAEW in poultry cabinet to reduce bacteria in chicken carcasses. In addition to eliminate bacterial load from different surfaces, EW can also be used to disinfect viruses. For example, EW has been successfully used to disinfect norovirus and hepatitis A virus on stainless steel coupons (Fang et al. 2016). Human norovirus can be inactivated using NEW in both suspension and stainless steel surfaces, which indicates that this treatment could control potential viral epidemic in the poultry industry (Moorman et al. 2017). Recently, NEW or SAEW treatment is introduced to reduce microbial contaminants from the surfaces and air of animal houses (Zang et al. 2015).

The transportation of animals from farms could pose a risk of enteric pathogens transmission. SAEW is effectively used for disinfection of plastic poultry transport cages to eliminate *S. Enteritidis* (Zang et al. 2015). A reduction of 3.12 log 10 cfu/cm^2 for *S. Enteritidis* is found for transport cages at tap water cleaning (for 15 s) followed by SAEW treatment (for 40 s at an ACC of 50 mg/l). The SAEW treatment time and ACC showed effect on the reductions of *S. Enteritidis*. Air of animal houses is usually contaminated with airborne microorganisms, in particular, during the day, due to the continuous movement of the birds. Spraying SAEW is potential in reducing these airborne microorganisms in layer breeding houses (Hao et al. 2014). SAEW spraying in the floor, wall, feed trough, and egg conveyor belt surfaces of layer houses has significant effect in reducing total aerobic bacteria, *Staphylococci*, coliforms, yeasts, and molds (Ni et al. 2015). It has shown that SAEW is more effective than povidone-iodine at reducing total aerobic bacteria, coliforms, yeasts, or molds on the wall surface. In addition to SAEW, NEW spray can also remove significant amount of dust levels in a layer breeding house (Afari et al. 2015). EW is effectively used in footbath solution on digital dermatitis incidence of dairy cows (Nakano et al. 2016).

7.5.3 Food Service Facilities

Foodborne diseases continue to be a public health concern, especially in the restaurants and food service outlets (Steffen et al. 2010). In the US, the high incidence of foodborne outbreaks could be associated to food service facilities due to poor personal hygiene of food handlers, incorrect storage conditions, and cross contamination (Wernersson et al. 2004). EW has successfully been used as a disinfectant for food service establishments, especially restaurant and hygiene maintenance in factory (Handojo et al. 2009; Sigua et al. 2011; Steffen et al. 2010). Sigua et al. (2011) compared the effectiveness of NEW with sodium hypochlorite and quaternary ammonium compound on ware washing operations in restaurants. The comparison was done by the number of ware washing cycles that each disinfectant could reach a reduction of 5 log in *E. coli* K12 and *L. innocua* populations. The authors reported that even though, the sanitizing effectiveness was reduced with number of washing cycle, due to increased organic materials in the cleaning solutions, however, NEW was significantly effective in cleaning and disinfection than the traditional sanitizers. The similar results were also observed when NEW was compared with quaternary ammonium compound, sodium hypochlorite, and PAA acid for inactivation of *E. coli* K-12 and *S. epidermidis* population during tableware washing process (Handojo et al. 2009). The above results showed that the application of EW during ware washing is efficacy means to inactivate the contamination with *E. coli* K12, *L. innocua,* and *S. epidermidis* on the food facility surfaces. NEW has the advantages of chlorine off-gassing concerns and oxidant/disinfectant residuals that are stable over longer periods of time. Additionally, the presence of hydrochloric acid and free radicals, which possess also sporicidal activity, highly contribute to improve the efficacy of washing. Solutions containing higher concentrations of available chlorine (for instance, 100 mg/L) are useful for washing by means of high-pressure sprayers (Fertelli et al. 2013; Steffen et al. 2010). The potential advantages of this EW for food facility surfaces disinfection include simplicity, efficiency, and ability to maintain sufficient hygienic condition during service when the tableware was thoroughly washed (Fertelli et al. 2013).

7.5.4 Healthcare Environment

The health care associated infection remains a major source of morbidity and mortality (Rutala and Weber 2013). The important source of nosocomial pathogens, include methicillin-resistant *S. aureus* (MRSA), *Clostridium difficile*, norovirus, *Acinetobacter*, and vancomycin-resistant *Enterococcus* (VRE) (Fertelli et al. 2013; Rutala and Weber 2013). The literatures supported the role of contaminated hard surface environments in the transmission of healthcare associated infections (Rutala and Weber 2013). The common use of EW is in the food processing and agricultural systems. Beside, EW can also be used in healthcare environments, especially dental water pipeline and radiology and endoscopes equipment (Cheng et al. 2016; Clark et al. 2006; Komachiya et al. 2014; Lee et al. 2004). Pintaric et al. (2015) investigated the effectiveness of AEW for microbial reduction on diagnostics equipment including patient couches of computer tomography (CT) and magnetic resonance imaging (MRI) scanners. AEW aerosolization demonstrated a reduction of 80.19–92.14% (from 1.14 to 2.54 CFU/m3) in the total number of microorganisms. EW demonstrated a potential as a biocide aerosolization for establishing a biosecurity between operational and diagnostic interventions. EW has been also used to disinfect external medical treatment on movement or shipment of animals, vehicles, and people around foot-and-mouth disease virus (FMDV) infected farms. For example, the moderate effect of SAEW in inactivation of FMDV is shown in a study of Harada et al. (2015). Furthermore, EW is shown to be effective in the disinfection of wet tissue and to inhibit in human cervical carcinoma HeLa cell proliferation (Nakamura and Muraoka 2017; Seo et al. 2015). Disinfection process using EW was reported in several studies including to disinfecting water system and inactivating the planktonic and biofilm bacteria from metal surfaces using dental arch and reprocessed surgical instruments. As endodontic root canal irrigant disinfection simulation, Cheng et al. (2016) studied the cleaning and disinfection effect of SAEW to remove the planktonic and biofilm of *E. faecalis* on microfluidic system. The viable cell count and morphological evaluation demonstrated that SAEW had a sanitization effect similar observed with 5.25% NaOCl against both the flow and static *E. faecalis* biofilms. Another work investigated the possibility of cleaning and disinfecting the dental unit water system by injecting SAEW for 64 h (Komachiya et al. 2014). The polymerase chain reaction showed that no viable cell of *S. aureus* and *P. aeruginosa* was detected and biofilm formation was four times lower with SAEW than that observed with tap water injection. A previous study showed that AEW exerted a similar disinfecting activity on dental waterline in a short time (Kohno et al. 2004). The above-discussed results demonstrated that EW solutions aid in reducing hospital equipment and hard surface contamination after effective cleaning and disinfection process.

7.6 Potential Advantages of EW on Environment: Comparison of EW with Other Environmental Surfaces

The cleaning and disinfection program aimed at environmental surfaces should be carefully developed because microorganisms attached to environmental surfaces are more resistant to sanitizing agents than free-living counterparts since disinfectant agents have a limited ability to penetrate the protective stratum of microbial polymers (Al-Haq et al. 2005). Potential advantage of EW solutions is enormous as they can in principle meet numerous time food industry and healthcare safety program. Alternative cleaning and disinfecting agent offers a reduced cleaning time and increased food plant running time. EW cleaning and disinfecting solutions can be used in food processing plants and healthcare facilities with dramatically reducing cost and energy consumption for the process. EW solutions increase the safety level and prevent the foodborne outbreaks and infection transmission. Since it has been found to provide a greater cleaning and disinfection results than traditional sanitizers, including benzalkonium chloride and povidone-iodine solutions (Hao et al. 2014; Hao et al. 2013a). It has also proved that EW as disinfectant plays a great role in maintaining the drinking line system in anima layer houses and seemed to minimize the drug use without showing undesirable impact on performance parameters and mortality rates (Bügener et al. 2014). The on-site production system electrolyzed water can actually generate enough solution for use in numerous facility types and applications.

7.7 Impact of EW in Disinfection on Environment

Sustainable food processing and healthcare facility management require technologies and tools to evaluate the environmental impact, therefore, going green these days, and for changes toward environmental improvements. Green solutions and practices deliver occupational and environmental safety benefits and at the same time can often create operational efficiencies. EW solutions are considered currently as green cleaning and sanitizing agents and optimal use of green solutions generate cost savings and energy, provide greater suitability in producing solutions, and eliminate the use of harsh chemical products at the facility, which often have an adverse impact on the environment. The traditional cleaning and disinfecting agents are often purchased in bulk and transported to the facilities, creating transportation costs, storage space requirements, and the effort needed to continually reorder chemicals. Application of EW solutions is benefit because it can be used as a foliar spray on a large species of bedding plants grown under greenhouse conditions to promote the fungal disease control without serious phytotoxicity problems (Buck et al. 2003; Mueller et al. 2003). The on-site production of EW solutions requests an ease and accessible ingredients to produce EW, resulting in environmental benefit of greenhouse gas emission and global warming during chemicals transportation and application. It produces minimum by-product wastes and is sustainable based on current and future economic and safety needs.

7.8 Conclusions

The cleaning and disinfection of environmental surfaces in food processing and medical plants are factors of foodborne outbreaks and infection prevention program. EW has a potential to be used as cleaning and disinfecting sanitizer in washing and sanitizing processing. Contrary to the food disinfection, it is difficult to achieve a requested quality and safety level during EW cleaning and disinfecting environmental surfaces since many factors and parameters limit the effectiveness of EW solutions. Both the understanding of affected effectiveness of EW and the application of appropriate instrument and method for specific case are the key of an excellent EW cleaning and disinfection processing. EW is an alternative technology for not only food safety and suitable healthcare facilities but also mitigation of carbon emission and reducing the secondary wastes with negative effect on environment through substituting traditional technologies.

References

Afari GK, Hung YC, King CH (2015) Efficacy of neutral pH electrolyzed water in reducing *Escherichia coli* O157: H7 and *Salmonella Typhimurium* DT 104 on fresh produce items using an automated washer at simulated food service conditions. J Food Sci 80:M1815–M1822

Al-Haq MI, Sugiyama J, Isobe S (2005) Applications of electrolyzed water in agriculture and food industries. Food Sci Technol Res 11:135–150

Al-Qadiri HM, Al-Holy MA, Shiroodi SG et al (2016a) Effect of acidic electrolyzed water-induced bacterial inhibition and injury in live clam (Venerupis philippinarum) and mussel (Mytilus edulis). Int J Food Microbiol 231:48–53

Al-Qadiri HM, Ovissipour M, Al-Alami N et al (2016b) Efficacy of neutral electrolyzed water, quaternary ammonium and lactic acid-based solutions in controlling microbial contamination of food cutting boards using a manual spraying technique. J Food Sci 81:M1177–M1183

Arvanitoyannis IS (2010) Waste management for the food industries. Academic Press

Bach S, Jones S, Stanford K et al (2006) Electrolyzed oxidizing anode water as a sanitizer for use in abattoirs. J Food Prot 69:1616–1622

Bagge-Ravn D, Ng Y, Hjelm M et al (2003) The microbial ecology of processing equipment in different fish industries—analysis of the microflora during processing and following cleaning and disinfection. Int J Food Microbiol 87:239–250

Bredholt S, Maukonen J, Kujanpää K et al (1999) Microbial methods for assessment of cleaning and disinfection of food-processing surfaces cleaned in a low-pressure system. Eur Food Res Technol 209:145–152

Buck J, Van Iersel M, Oetting R et al (2003) Evaluation of acidic electrolyzed water for phytotoxic symptoms on foliage and flowers of bedding plants. Crop Protect 22:73–77

Bügener E, Kump AW-S, Casteel M et al (2014) Benefits of neutral electrolyzed oxidizing water as a drinking water additive for broiler chickens. Poultry Sci 93:2320–2326

Cao W, Zhu ZW, Shi ZX et al (2009) Efficiency of slightly acidic electrolyzed water for inactivation of Salmonella enteritidis and its contaminated shell eggs. Int J Food Microbiol 130:88–93

Carpentier B, Cerf O (2011) Persistence of Listeria monocytogenes in food industry equipment and premises. Int J Food Microbiol 145:1–8

Chen L, Chen R, Yin H et al (2012) Cleaning in place with onsite-generated electrolysed oxidizing water for water-saving disinfection in breweries. J I Brewing 118:401–405

Cheng X, Tian Y, Zhao C et al (2016) Bactericidal effect of strong acid electrolyzed water against flow Enterococcus faecalis biofilms. J Endodont 42:1120–1125

Chuang CY, Yang S, Chang MY et al (2013a) Inactivation efficiency to Bacillus subtilis and *Escherichia coli* bacterial aerosols of spraying neutral electrolyzed water. J Air Waste Manage 63:1447–1456

Chuang CY, Yang S, Huang HC et al (2013b) Applying the membrane-less electrolyzed water spraying for inactivating bioaerosols. Aerosol Air Qual Res 13:350–359

Clark J, Barrett S, Rogers M et al (2006) Efficacy of super-oxidized water fogging in environmental decontamination. J Hosp Infect 64:386–390

Dev SR, Demirci A, Graves RE et al (2014) Optimization and modeling of an electrolyzed oxidizing water based clean-in-place technique for farm milking systems using a pilot-scale milking system. J Food Eng 135:1–10

Deza M, Araujo M, Garrido M (2005) Inactivation of *Escherichia coli*, *Listeria monocytogenes*, *Pseudomonas aeruginosa* and *Staphylococcus aureus* on stainless steel and glass surfaces by neutral electrolysed water. Lett Appl Microbiol 40:341–346

Deza M, Araujo M, Garrido M (2007) Efficacy of neutral electrolyzed water to inactivate *Escherichia coli*, *Listeria monocytogenes*, *Pseudomonas aeruginosa*, and *Staphylococcus aureus* on plastic and wooden kitchen cutting boards. J Food Protect 70:102–108

Fang J, Cannon JL, Hung YC (2016) The efficacy of EO waters on inactivating norovirus and hepatitis A virus in the presence of organic matter. Food Control 61:13–19

Fasenko G, O'dea Christopher E, Mcmullen L (2009) Spraying hatching eggs with electrolyzed oxidizing water reduces eggshell microbial load without compromising broiler production parameters. Poultry Sci 88:1121–1127

Fertelli D, Cadnum JL, Nerandzic MM et al (2013) Effectiveness of an electrochemically activated saline solution for disinfection of hospital equipment. Infect Cont Hosp Ep 34:543–544

Gómez-López V, Devlieghere F, Bonduelle V et al (2005) Factors affecting the inactivation of micro-organisms by intense light pulses. J Appl Microbiol 99:460–470

Guentzel JL, Lam KL, Callan MA et al (2008) Reduction of bacteria on spinach, lettuce, and surfaces in food service areas using neutral electrolyzed oxidizing water. Food Microbiol 25:36–41

Han Q, Song X, Zhang Z et al (2017) Removal of foodborne pathogen biofilms by acidic electrolyzed water. Front Microbiol 8:988

Handojo A, Lee J, Hipp J et al (2009) Efficacy of electrolyzed water and an acidic formulation compared with regularly used chemical sanitizers for tableware sanitization during mechanical and manual ware-washing protocols. J Food Protect 72:1315–1320

Hao X, Li B, Wang C et al (2013a) Application of slightly acidic electrolyzed water for inactivating microbes in a layer breeding house. Poultry Sci 92:2560–2566

Hao X, Li B, Zhang Q et al (2013b) Disinfection effectiveness of slightly acidic electrolysed water in swine barns. J Appl Microbiol 115:703–710

Hao X, Cao W, Li B et al (2014) Slightly acidic electrolyzed water for reducing airborne microorganisms in a layer breeding house. J Air Waste Manage 64:494–500

Hao J, Li H, Wan Y et al (2015) Combined effect of acidic electrolyzed water (AEW) and alkaline electrolyzed water (AlEW) on the microbial reduction of fresh-cut cilantro. Food Control 50:699–704

Harada Y, Lekcharoensuk P, Furuta T et al (2015) Inactivation of foot-and-mouth disease virus by commercially available disinfectants and cleaners. Biocontrol Sci 20:205–208

Holah J (2014) Cleaning and disinfection practices in food processing. In: Hygiene in Food Processing, 2nd edn. Elsevier, pp 259–304

Hricova D, Stephan R, Zweifel C (2008) Electrolyzed water and its application in the food industry. J Food Protect 71:1934–1947

Hsu SY, Kim C, Hung YC et al (2004) Effect of spraying on chemical properties and bactericidal efficacy of electrolysed oxidizing water. Int J Food Sci Technol 39:157–165

Huang YR, Hsieh HS, Lin SY et al (2006) Application of electrolyzed oxidizing water on the reduction of bacterial contamination for seafood. Food Control 17:987–993

Huang YR, Hung YC, Hsu SY et al (2008) Application of electrolyzed water in the food industry. Food Control 19:329–345

Hussain MS, Kwon M, Tango CN et al (2018) Effect of electrolyzed water on the disinfection of bacillus cereus biofilms: the mechanism of enhanced resistance of sessile cells in the biofilm matrix. J Food Prot 81:860–869

Jiménez-Pichardo R, Regalado C, Castaño-Tostado E et al (2016) Evaluation of electrolyzed water as cleaning and disinfection agent on stainless steel as a model surface in the dairy industry. Food Control 60:320–328

Jirotková D, Šoch M, Kernerová N et al (2013) Possibilities of the use of electrolyzed water in poultry breeding. Acta Universitatis Cibiniensis. Series E: Food Technol 17:129-136

Jnanadev K, Babu CS, Shetty SS et al (2011) Disinfecting the acrylic resin plate using electrolyzed acid water and 2% glutaraldehyde: a comparative microbiological study. J Indian Prosthodont Soc 11:36–44

Jo HY, Tango CN, Oh DH (2018) Influence of different organic materials on chlorine concentration and sanitization of slightly acidic electrolyzed water. LWT 92:187–194

Kim JG, Yousef AE, Dave S (1999) Application of ozone for enhancing the microbiological safety and quality of foods: a review. J Food Protect 62:1071–1087

Kim C, Hung YC, Russell S (2005) Efficacy of electrolyzed water in the prevention and removal of fecal material attachment and its microbicidal effectiveness during simulated industrial poultry processing. Poultry Sci 84:1778–1784

Kohno S, Kawata T, Kaku M et al (2004) Bactericidal Effects of Acidic Electrolyzed Water. Jpn J Infect Dis 57:52–54
Koide S, Shitanda D, Note M et al (2011) Effects of mildly heated, slightly acidic electrolyzed water on the disinfection and physicochemical properties of sliced carrot. Food Control 22:452–456
Komachiya M, Yamaguchi A, Hirai K et al (2014) Antiseptic effect of slightly acidic electrolyzed water on dental unit water systems. The Bulletin of Tokyo Dental College 55:77–86
Koseki S, Yoshida K, Kamitani Y et al (2004) Effect of mild heat pre-treatment with alkaline electrolyzed water on the efficacy of acidic electrolyzed water against *Escherichia coli* O157: H7 and Salmonella on lettuce. Food Microbiol 21:559–566
Lee JH, Rhee PL, Kim JH et al (2004) Efficacy of electrolyzed acid water in reprocessing patient-used flexible upper endoscopes: Comparison with 2% alkaline glutaraldehyde. J Gastroen Hepatol 19:897–903
Liu C, Su YC (2006) Efficiency of electrolyzed oxidizing water on reducing Listeria monocytogenes contamination on seafood processing gloves. Int J Food Microbiol 110:149–154
Machado I, Meireles A, Fulgêncio R et al (2016) Disinfection with neutral electrolyzed oxidizing water to reduce microbial load and to prevent biofilm regrowth in the processing of fresh-cut vegetables. Food Bioprod Process 98:333–340
Mccarthy S, Burkhardt IW (2012) Efficacy of electrolyzed oxidizing water against *Listeria monocytogenes* and *Morganella morganii* on conveyor belt and raw fish surfaces. Food Control 24:214–219
Moorman E, Montazeri N, Jaykus LA (2017) Efficacy of neutral electrolyzed water for inactivation of human norovirus. Appl Environ Microbiol 00653–00617
Moradi M, Tajik H (2017) Biofilm removal potential of neutral electrolysed water on pathogen and spoilage bacteria in dairy model systems. J Appl Microbiol 123:1429–1437
Møretrø T, Heir E, Nesse LL et al (2012) Control of Salmonella in food related environments by chemical disinfection. Food Res Int 45:532–544
Mueller D, Hung YC, Oetting R et al (2003) Evaluation of electrolyzed oxidizing water for management of powdery mildew on gerbera daisy. Plant Dis 87:965–969
Nakamura K, Muraoka O (2017) Effect of electrolyzed water produced using carbon electrodes on HeLa cell proliferation. Biosci Trends 11:688–693
Nakano Y, Akamatsu N, Mori T et al (2016) Sequential Washing with Electrolyzed Alkaline and Acidic Water Effectively Removes Pathogens from Metal Surfaces. PLoS ONE 11:e0156058
Ni L, Cao W, Zheng W et al (2015) Reduction of microbial contamination on the surfaces of layer houses using slightly acidic electrolyzed water. Poultry Sci 94:2838–2848
Ni L, Zheng W, Zhang Q et al (2016) Application of slightly acidic electrolyzed water for decontamination of stainless steel surfaces in animal transport vehicles. Prev Vet Med 133:42–51
Oomori T, Oka T, Inuta T et al (2000) The efficiency of disinfection of acidic electrolyzed water in the presence of organic materials. Anal Sci 16:365–369
Ovissipour M, Al-Qadiri HM, Sablani SS et al (2015) Efficacy of acidic and alkaline electrolyzed water for inactivating *Escherichia coli* O104: H4, *Listeria monocytogenes*, *Campylobacter jejuni*, *Aeromonas hydrophila*, and *Vibrio parahaemolyticus* in cell suspensions. Food Control 53:117–123
Pascual A, Llorca I, Canut A (2007) Use of ozone in food industries for reducing the environmental impact of cleaning and disinfection activities. Trends Food Sci Tech 18:S29–S35
Phuvasate S, Su YC (2010) Effects of electrolyzed oxidizing water and ice treatments on reducing histamine-producing bacteria on fish skin and food contact surface. Food Control 21:286–291
Pintaric R, Matela J, Pintaric S (2015) Suitability of electrolyzed oxidizing water for the disinfection of hard surfaces and equipment in radiology. J Environ Health Sci Eng 13:6
Rahman S, Khan I, Oh DH (2016) Electrolyzed water as a novel sanitizer in the food industry: Current trends and future perspectives. Compr Rev Food Sci F 15:471–490
Russell AD (2008) Factors influencing the efficacy of antimicrobial agents. Principles and practice of disinfection, preservation and sterilization. Blackwell, London:98–127

Rutala WA, Weber DJ (2013) Disinfectants used for environmental disinfection and new room decontamination technology. Am J Infect Control 41:S36–S41
Sansebastiano G, Zoni R, Bigliardi L (2007) Cleaning and disinfection procedures in the food industry general aspects and practical applications. Food Safety: 253–280
Seo JH, Lee DJ, Lee MK et al (2015) Studies on the antibacterial activity of wet-tissue saturated with electrolytic water of NaCl solution. J Korea Tech Assoc Pulp Paper Ind 47:147–153
Serraino A, Veronese G, Alonso S et al (2010) Bactericidal activity of electrolyzed oxidizing water on food processing surfaces. Ital J Food Sci 22
Shaughnessy MK, Micielli RL, Depestel DD et al (2011) Evaluation of hospital room assignment and acquisition of Clostridium difficile infection. Infect Cont Hosp Ep 32:201–206
Sigua G, Lee YH, Lee J et al (2011) Comparative efficacies of various chemical sanitizers for warewashing operations in restaurants. Food Control 22:13–19
Simões M, Simoes LC, Vieira MJ (2010) A review of current and emergent biofilm control strategies. LWT-Food Sci Technol 43:573–583
Srey S, Jahid IK, Ha SD (2013) Biofilm formation in food industries: a food safety concern. Food Control 31:572–585
Steffen H, Duerst M, Rice RG (2010) User experiences with ozone, electrolytic water (active water) and UV-c light (ventafresh technology) in production processes and for hygiene maintenance in a Swiss sushi factory. Ozone: Sci Eng 32:71–78
Stewart M, Bogusz A, Hunter J et al (2014) Evaluating use of neutral electrolyzed water for cleaning near-patient surfaces. Infect Cont Hosp Ep 35:1505–1510
Sun JL, Zhang SK, Chen JY et al (2012) Efficacy of acidic and basic electrolyzed water in eradicating *Staphylococcus aureus* biofilm. Can J Microbiol 58:448–454
Takenouchi T, Wakabayashi S-I (2006) Rinsing effect of alkaline electrolyzed water on nickel surfaces. J Appl Electrochem 36:1127–1132
Tango CN, Wang J, Oh DH (2014) Modeling of *Bacillus cereus* growth in brown rice submitted to a combination of ultrasonication and slightly acidic electrolyzed water treatment. J Food Protect 77:2043–2053
Tango CN, Hong SS, Wang J et al (2015) Assessment of enterotoxin production and cross-contamination of *Staphylococcus aureus* between food processing materials and ready-to-eat cooked fish paste. J Food Sci 80
Tango CN, Khan I, Kounkeu PFN et al (2017) Slightly acidic electrolyzed water combined with chemical and physical treatments to decontaminate bacteria on fresh fruits. Food Microbiol 67:97–105
Vázquez-Sánchez D, Cabo M, Rodríguez-Herrera J (2014) Single and sequential application of electrolyzed water with benzalkonium chloride or peracetic acid for removal of *Staphylococcus aureus* biofilms. J Food Safety 34:199–210
Veasey S, Muriana PM (2016) Evaluation of electrolytically-generated hypochlorous acid ('electrolyzed water') for sanitation of meat and meat-contact surfaces. Foods 5:42
Walker SP, Demirci A, Graves R et al (2005a) Cleaning milking systems using electrolyzed oxidizing water. Transactions of the ASAE 48:1827–1833
Walker SP, Demirci A, Graves RE et al (2005b) Response surface modelling for cleaning and disinfecting materials used in milking systems with electrolysed oxidizing water. Int J Dairy Technol 58:65–73
Wang X, Puri VM, Demirci A et al (2016) Mathematical modeling and cycle time reduction of deposit removal from stainless steel pipeline during cleaning-in-place of milking system with electrolyzed oxidizing water. J Food Eng 170:144–159
Wang H, Cai L, Li Y et al (2018) Biofilm formation by meat-borne Pseudomonas fluorescens on stainless steel and its resistance to disinfectants. Food Control 91:397–403
Weichao Z, Li N, Xue H et al (2016) Optimization of slightly acidic electrolyzed water spray for airborne culturable bacteria reduction in animal housing. Int J Agr Biol Eng 9:185–191
Wernersson ES, Johansson E, Håkanson H (2004) Cross-contamination in dishwashers. J Hosl Infect 56:312–317

Zamora JL (1986) Chemical and microbiologic characteristics and toxicity of povidone-iodine solutions. Am J Surg 151:400–406

Zang Y, Li B, Bing S et al (2015) Modeling disinfection of plastic poultry transport cages inoculated with Salmonella enteritids by slightly acidic electrolyzed water using response surface methodology. Poultry Sci 94:2059–2065

Zhao Y, Xin H, Zhao D et al (2014) Free chlorine loss during spraying of membraneless acidic electrolyzed water and its antimicrobial effect on airborne bacteria from poultry house. Ann Agr Env Med 21

Zhao L, Zhang Y, Yang H (2017) Efficacy of low concentration neutralised electrolysed water and ultrasound combination for inactivating Escherichia coli ATCC 25922, Pichia pastoris GS115 and Aureobasidium pullulans 2012 on stainless steel coupons. Food Control 73:889–899

Zheng W, Li B, Cao W et al (2012) Application of neutral electrolyzed water spray for reducing dust levels in a layer breeding house. J Air Waste Manage 62:1329–1334

Zheng W, Kang R, Wang H et al (2013) Airborne bacterial reduction by spraying slightly acidic electrolyzed water in a laying-hen house. J Air Waste Manage 63:1205–1211

Chapter 8
Application of Electrolyzed Water on Livestock

S. M. E. Rahman and H. M. Murshed

8.1 Introduction

Each year an estimated 76 million Americans become ill from consuming foods contaminated with pathogenic microbes and their toxins (Mead et al. 1999). Since many of the reported foodborne outbreaks have been linked to meat products or to contact with food animals or their waste, enteric pathogens in livestock and poultry are of concern. In terms of those specific pathogens that contribute substantially to foodborne illnesses, *Campylobacter jejuni* and *Salmonella* species are prevalent in poultry, cattle, swine, and sheep, whereas enterohemorrhagic *Escherichia coli* O157:H7 is a major concern in cattle and sheep and *Yersinia enterocolitica* in swine (Doyle and Erickson 2006). Livestock and poultry production facilities are associated with much higher concentrations of airborne microorganisms compared to the ambient environment (Zhao et al. 2014). The environment within animal houses is often contaminated with pathogenic microorganisms and contaminated surfaces in such facilities may act as reservoirs for pathogenic microorganisms (Hao et al. 2013a, b; Ni et al. 2015). Exposure to high levels of airborne microbes in animal houses can have negative impacts on the health of both the animals and the workers (Radon et al. 2002; Zheng et al. 2016). Microorganisms contaminating animal houses are also responsible for disease infection among animals (Zhao et al. 2011) and can even enter the food chain (Leach et al. 1999). Provision of a healthy environment for animal production is receiving increased attention.

Disinfection is a commonly recommended approach for disease prevention in animal houses (Schmidt et al. 2004; Rodríguez Ferri et al. 2010; Zheng et al. 2013). This can help to lower the potential for disease infection and transmission in animal houses by reducing the population of pathogenic microorganisms on the surfaces or in the air. Numerous chemical disinfectants such as benzalkonium chloride, formaldehyde,

S. M. E. Rahman (✉) · H. M. Murshed
Department of Animal Science, Bangladesh Agricultural University, Mymensingh 2202, Bangladesh
e-mail: ehsan_bau@yahoo.com

T. Ding et al. (eds.), *Electrolyzed Water in Food: Fundamentals and Applications*,
https://doi.org/10.1007/978-981-13-3807-6_8

and glutaraldehyde are used for disinfection against bacterial infections in animal houses (Sundheim et al. 1998). However, the use of these chemical disinfectants has limited potential due to their toxicity, corrosiveness, and/or volatility (Lewis and McIndoe 2004). Also, the resistance and cross-resistance of pathogens to chemical disinfectants have been reported (Aarnisalo et al. 2000; Sidhu et al. 2001). Therefore, it is essential to develop alternative disinfectants for decontamination in animal houses. Over the last decade, EW has become increasingly popular as an alternative disinfectant for decontamination in animal industries. Acidic electrolyzed water (AEW, $pH<2.7$), slightly acidic electrolyzed water (SAEW pH 5.0–6.5) and neutral electrolyzed water (NEW, 6.5–8.5) are the three main types of EWs reported as alternative disinfectants for decontamination. AEW has been reported to be an effective antimicrobial agent in the food industry (Huang et al. 2008; Hricova et al. 2008; Rahman et al. 2016). However, AEW can easily release Cl_2 gas due to its volatility, which causes chlorine loss, thus decreasing AEW bactericidal activity over time (Rahman et al. 2012a; Cui et al. 2009). The strong acidity ($pH<2.7$) of AEW can also cause corrosion of equipment (Guentzel et al. 2008). These disadvantages potentially limit the use of AEW in some applications such as animal houses. In contrast, SAEW and NEW are near neutral pH and more stable than AEW (Cui et al. 2009). They have been increasingly used for the prevention and control of microorganisms (Rahman et al. 2016; Deza et al. 2005; Abadias et al. 2008; Koide et al. 2011). Studies have shown EW as a decontaminant on eggshell (Fasenko et al. 2009; Bialka et al. 2004), meat and meat contact surfaces (Arya et al. 2018; Veasey and Muriana 2016; Rahman et al. 2012b, 2013), livestock and poultry (Doyle and Erickson 2006), layer houses (Zhao et al. 2014; Hao et al. 2013a, 2014; Ni et al. 2015; Zheng et al. 2013), swine barns (Hao et al. 2013b), animal houses (Zheng et al. 2016), abattoirs (Bach et al. 2006), dairy industry (Himmelmann et al. 2017; Kalit et al. 2015; Pichardo et al. 2016), and broiler processing (Rasschaert et al. 2013). However, little work has been done on the application of electrolyzed water in disinfection of livestock facilities. The aim of this work is to revise the potential of electrolyzed water (EW) as a cleaning agent in animal production systems. EW has been successfully used in disinfection processes, as agent for sanitizing equipment and drinking water. Its effects on animal physiology, health, and performance are also revised.

8.2 EW for Livestock Farm House Sanitation and Biosecurity

Animal housing and transportation equipment can also harbor pathogens and contribute to contamination of animals; however, a primary source of enteric pathogens transmitted to livestock and poultry is manure (Doyle and Erickson 2006). Many intervention technologies including antibiotics, vaccination, cleaning, and wiping have been introduced to reduce or prevent diseases in animal houses. Spraying or soaking is extensively used in advanced livestock breeding houses (Rodríguez

Ferri et al. 2010). Spray application of membrane-less acidic electrolyzed water (MLAEW) is a novel technique for disinfection in livestock houses. A study (Zhao et al. 2014) investigated the loss of free chlorine (FC—the major germicidal component in MLAEW) over distance during spraying, as affected by air temperature and initial FC concentration. The antimicrobial effect of MLAEW on airborne bacteria from an aviary laying hen house was examined. Recently, a MLAEW spray has been applied in swine and poultry houses to inactivate airborne microorganisms. Chuang et al. (2011) reported that the level of total airborne bacteria was reduced by 70% by spraying MLAEW in a cage hen house. A considerable number of studies have reported the increased use of SAEW and NEW for controlling contamination in animal houses, including facilities for swine, poultry, and dairy. Hao et al. (2013a) reported the cleaning effectiveness of SAEW in the pH range of 5.0–6.5 in layer houses. Treatment with SAEW effectively decreased the survival rates of *Salmonella* spp. and *E. coli* by 21 and 16%, respectively. Hao et al. (2013b) also reported on the potential applications of SAEW in swine barns. SAEW containing 300 ppm of active chlorine was flushed onto surfaces and sprayed within the whole swine barn. Spraying with SAEW significantly ($P < 0.05$) reduced the microbial count on the rail, floor, and walls of the swine barns. In another study, Hao et al. (2014) reported the inactivation of airborne bacteria in a commercial layer house in northern China by treatment with SAEW. The results showed that the airborne microorganism and fungi counts were reduced by 4.85 and 3.45 log CFU/m^3, respectively, after 30 min of exposure to SAEW. Recently, Zheng et al. (2014) studied the efficacy of SAEW in reducing airborne culturable bacteria and particulate matter levels in hen houses. The results of the study showed the inactivation of airborne culturable bacteria attached to particulate matter. In summary, research has revealed EW to be a potential antimicrobial agent for reducing microbial presence in layer houses, swine barns, slaughterhouse, and animal breeding houses.

8.3 EW for Slaughterhouse/Abattoir

The hides of cattle are the primary source of pathogens such as *E. coli* O157:H7 that contaminate pre-evisceration carcasses during commercial beef processing. Treatment using ozonated water reduced hide aerobic plate counts by 2.1 log CFU/100 cm^2 and reduced *Enterobacteriaceae* counts by 3.4 log CFU/100 cm^2. Therefore, ozonated and EO waters can be used to decontaminate hides during processing and may be viable treatments for significantly reducing pathogen loads on beef hides, thereby reducing pathogens on beef carcasses (Bosilevac et al. 2005). Improper handling of raw meat during slaughtering leads to the contamination of meat with *Listeria monocytogenes*, *Salmonella*, and some strains of *E. coli* resulting in foodborne illnesses (Priyanka et al. 2016). According to the Centers for Disease Control and Prevention (CDC 2015), foodborne illnesses from pathogens have caused 15,202 million illnesses, 950 hospitalizations, and 15 deaths in the United States in 2015. The contamination of beef carcasses during slaughter and processing is undesirable but unavoidable in the conversion of live animals to meat for consumption. Live animal contamination and transference during the carcass breaking and dressing processes contribute to the microbial contamination of beef. Inadequately cleaned equipment in meat processing facilities has been identified in many studies as contributing to the transfer of microorganisms to beef carcass surfaces, thus potentially having serious implications for food safety (Bach et al. 2006). They also reported electrolyzed oxidizing anode (EOA) water may be a suitable alternative or complement to iodophor (IOD) as a sanitizer in small-to-medium-sized abattoirs. In another study, Rasschaert et al. (2013) reported EO water seems to be a promising technique to reduce the number of *Campylobacter* on broiler carcasses during processing, although the technology needs improvement before large-scale application. Combination of ultrasound and SAEW might be an effective decontamination technique in meat industry (Flores et al. 2017). Park et al. (2002) recently evaluated the effectiveness of EO water as a sanitizer for decontaminating surfaces such as glass, stainless steel, ceramic tile, and vitreous china. The effectiveness in reducing microbial contamination on such surfaces suggests that EO water may be an effective and appropriate sanitizer for use in the meat processing industry.

8.4 EW for Processing Plant (Equipment/Surfaces)

Bacterial cross-contamination can occur from the preparation equipment and tableware during food processing. Improper cleaning and sanitization of the tools used in the food industry were reported as a serious problem by the U.S. Food and Drug Administration (FDA 2009). Thus, cleaning and sanitization of these tools should be optimized to ensure food safety. In this context, EW has been employed as a novel sanitizing agent to reduce the bacterial count on food contact surfaces to acceptable levels. The effectiveness of NEW or AEW has been examined by several researchers and they have been recommended as novel food contact surface sanitizers (Izumi 1999; Deza et al. 2005; Deza et al. 2007). The use of AEW solution effectively reduced the bacterial populations on metal and plastic surfaces, as well as on disposable fabric wipes (Lee et al. 2007a, b). Handojo et al. (2009) used NEW and AEW to achieve antimicrobial effect against *E. coli* K-12 and *Staphylococcus epidermidis* inoculated onto stainless steel cutlery, ceramic plates, and drinking glasses. Their results indicated that treatment with AEW and NEW reduced the bacterial populations by more than 5 log CFU per tableware item for all the treatment conditions. In another study, the disinfection efficacy of NEW on cutting boards (hard and bamboo boards) inoculated with *E. coli* K-12 and *Listeria innocua* was examined. Significant reduction in foodborne pathogens was recorded, regardless of the type of cutting board sample treated (Monnin et al. 2012).

8.5 EW for Hatchery

Research has demonstrated that the hatchery is one of the most important areas, within a vertically integrated poultry company, in which *Salmonella* control is essential. Studies have associated the presence of *Salmonella* in the hatchery with contamination among broiler flocks. The ability of EO water to reduce eggshell microbial load without negatively affecting hatchability or chick quality may make it a useful product for hatching egg sanitation. Reducing microbial contamination of eggshells may help to decrease the incidence of bacterial infections in developing embryos and newly hatched chicks. Microorganisms can penetrate the eggshell through shell pores or cracks and can kill the developing embryo, reduce hatchability, and negatively affect the chick post hatching (Fasenko et al. 2009). They also reported that total aerobic bacteria counts were significantly lower on eggs that were sprayed with acidic EO water compared with eggs left untreated. In another study, Ni et al. (2014) demonstrated that SAEW had an equivalent or higher efficiency to reduce *Salmonella enteritidis*, *E. coli* O157:H7, *Staphylococcus aureus,* and indigenous microbiota present on shell eggs compared to chlorine dioxide and NaClO solution, and has similar bactericidal activities with AEW at the same available chlorine concentration. The combination treatment of SAEW with AlEW had better reduction than SAEW or AlEW alone. Therefore, SAEW shows the potential to be used for sanitization of eggshells as an environmental friendly disinfection agent. Further studies should be elucidated to determine the levels of bacterial inoculation and the synergistic effects of combining technologies.

8.6 EW for Dairy Industry

Microbial contamination is an important issue in dairy farms. According to regulations, the dairy farm must be cleaned and sanitized regularly. Good cleaning practices in dairy farms help to reduce the incidence of disease like mastitis and ensure the production of high-quality milk. Mastitis is commonly a severe problem of dairy farmers worldwide, which is a result of the interaction between the cow, pathogen, and the environment (Huton et al. 1990). The best way to prevent mastitis occurrence is to keep the animals clean (WHO 1984). The disinfection effect of SAEW on cow's teats, milking cups, towels, and hands of the workers was investigated. The prevention of mastitis in dairy cows by SAEW was also examined. Results show the percentage reduction of aerobic plate count (APC) on the cow's teats, towels, and the milking cups treated by SAEW with ACC of 40 mg/L achieved 91.2, 97.1, and 93.3%, respectively. SAEW with a mild pH 5.6–6.5 is not only effective to reduce the population of APC on the cow teats, towels, hands of workers, and milking cups but also could prevent the occurrence of subclinical mastitis. The findings from the current study imply that SAEW with environment friendly, low cost, no corrosion, high stability and bactericidal activity, and a less potential health hazard to the worker and environment could be an attractive candidate for application in dairy farms. Pichardo et al. (2016) also reported that the cleaning and sanitization procedure of stainless steel plates (SSP) with a combination of AlEW and NEW represents an excellent alternative for the dairy industry. A commercial disinfectant based on EO water was used, and as a material for disinfecting milk sample with a known initial number of microorganisms. Disinfectant, taken from the original factory-sealed bottles, was diluted with double-distilled water in the three ratios, 1: 10, 1: 20, and 1: 30. The significant decrease (P <0.05) of the number of aerobic mesophilic bacteria was observed in the cases of disinfectant's stronger concentration (1: 10). Weaker concentration of disinfectant (1: 20 and 1: 30) showed less antibacterial activity, and in that case, it is necessary to carry out rinsing of the equipment before using disinfectant, to remove organic milk matter which decreases the disinfection effect of free chlorine. The results confirmed that EO disinfectant is acceptable and sufficiently effective in the dairy industry, but it is recommended to use higher concentration, especially if traces of organic compounds retain on the treated surface (kalit et al. 2015). The safety of raw milk largely depends on using a clean milking system during the milk production. The milking system cleaning process widely used on dairy farms is a highly automated process called cleaning-in-place (CIP), which comprises four cycles: (i) warm water rinse, (ii) alkaline wash, (iii) acid wash, and (iv) sanitizing rinse before the next milking event. Electrolyzed oxidizing (EO) water is an emerging technology, which consists of acidic and alkaline solutions by the electrodialysis of dilute sodium chloride solution. Deposit removal data from the simulator formed the basis for developing mathematical models to describe the deposit removal process during the CIP process with EO water. Stainless steel straight pipe specimens were placed at the end of undisturbed entrance length along the simulator pipeline. The mass of milk deposits on the inner surfaces of the specimens were measured using a high-precision balance after the initial soiling, and then after certain time durations within the warm water rinse, alkaline wash, and acid wash cycles. A unified first-order deposit removal rate model dependent on nth power of remaining deposit mass was used for all three cycles. ATP bioluminescence method was also used as a validation approach at the end of each CIP cycle. Experimental results showed that the milk deposit on the inner surfaces of the specimens was removed rapidly by the warm water rinse within 10 s of rinse time. For the alkaline and acid wash cycles, the co-existence of a fast deposit removal at the beginning of the wash cycle and a slow deposit removal throughout the entire wash cycle were inferred. The proposed models matched the experimental data with small root mean square errors (0.23 mg/[(mg)(m^2)] and 0.07 mg/[(mg)(m^2)] for the upstream and downstream locations, respectively) and satisfactory percent error differences (3.67% and 0.93% for the upstream and downstream locations, respectively). Based on the experimental data and the proposed models, the time duration of the CIP process was shortened by 55% (10 s warm water rinse, 3 min alkaline wash, and 6 min acid wash) and validated, which yielded an average deposit of 0.28 mg/[(mg)(m^2)] at the end of the CIP as compared with that of 0.29 mg/[(mg)(m^2)] at the end of the original CIP, to achieve a satisfactory CIP performance for the simulator (Wang et al. 2016). In the dairy industry, cleaning and disinfection of surfaces are important issues and development of innovative strategies may improve food safety.

8.7 EW for Drinking Water in Poultry

The use of EO water as a supplementary measure can be considered to permanently achieve a better drinking water quality and had no negative effects on bird health and performance. Furthermore, the use of EO water contributes to current European Union requirements regarding the reduction of using antibiotic substances (Bügener et al. 2014). Ionized alkaline water is the potential candidate to improve some physiological traits for Ross broiler chickens. However, treated water has no effect on some blood parameters and liver enzymes, despite positive effect on some biochemical parameters in serum of animal, especially on low sugar and cholesterol and triglyceride (Jassim and Aqeel 2017). Tap water alone (TW) or treated with 3% of SAEW was used in this experiment to study its effect on water quality, blood biochemical parameters, and milk yield and composition. SAEW can be used at 3% rate as a powerful and economic agent for sanitizing drinking water for dairy ewes with no effects on animal performance (Bodas et al. 2013).

8.8 EW for Transport

Contaminated poultry and livestock transport vehicles have been reported to be sources of many disease pathogens including *Salmonella*, *Campylobacter*, *E. coli*, *Streptococcus suis*, *Mycoplasma hyopneumoniae*, *Actinobacillus pleuropneumoniae*, classic swine fever virus, porcine epidemic diarrhea virus, porcine circovirus, and porcine reproductive and respiratory syndrome virus under field and experimental conditions (Dee and Corey 1993; Fedorka-Cray et al. 1997; Fussing et al. 1998; Rajkowski et al. 1998; Gebreyes et al. 2004; Dee and Deen 2006; Amass et al. 2007; Patterson et al. 2011; Hernández et al. 2013; Lowe et al. 2014). The effectiveness of SAEW in reducing *E. coli*, Salmonella typhimurim, *S. aureus,* or bacterial mixtures on stainless steel surfaces was evaluated and compared its efficacy with composite phenol solution for reducing total aerobic bacteria in animal transport vehicles by Ni et al. (2016). The results suggest that the bactericidal efficiency of SAEW was higher than or equivalent to that of composite phenol and SAEW may be used as an effective alternative for reducing microbial contamination of animal transport vehicles. In order to reduce the risk of enteric pathogens transmission in animal farms, the disinfection effectiveness of slightly acidic electrolyzed water (SAEW, pH 5.85–6.53) for inactivating *S. enteritidis* on the surface of plastic poultry transport cages was evaluated by Zang et al. (2015). The coupled effects of the tap water cleaning time (5–15 s), SAEW treatment time (20–40 s), and available chlorine concentrations (ACCs) of 30–70 mg/L on the reductions of *S. enteritidis* on chick cages were investigated using a central composite design of the response surface methodology (RSM). The established RS model had a goodness of fit quantified by the parameter R2 (0.971), as well as a lack of fit test ($P > 0.05$). The maximum reduction of 3.12 log 10 CFU/cm^2 for *S. enteritidis* was obtained for the cage treated with tap water cleaning for 15 s followed by SAEW treatment for 40 s at an ACC of 50 mg/L. Results indicate that the established RS model has shown the potential of SAEW in disinfection of bacteria on cages.

8.9 EW for Disease Prevention and Control

Respiratory symptoms resulting from disease states such as PRRS and streptococcus infections can be alleviated by administration of electrolyzed water via fog or mist delivery (US) in an animal confinement. The delivery may be in doses spaced evenly throughout the day and may include delivery of between about 1 and about 6 gallons per dose. In one embodiment, the dose is delivered via a distribution system of nozzles over a time period of about 1 min. In others, delivery may require 5, 10, or 20 min and may include components in addition to the electrolyzed water. It is understood that the mist is inhaled, and is also distributed onto the surfaces in the confinement and on the animals themselves (Watson 2015). Electrolyzed reduced water (ERW) stimulated glucose uptake into muscle and adipocyte cells as well as insulin. It also stimulated the secretion of insulin from pancreatic beta cells and improved the sugar tolerance damage in type 2 diabetes model mice. Reduced water impaired the tumor phenotypes such as rapid growth, anchorage-independent growth in a soft agar, morphology, telomere maintenance, and abilities of invasion, metastasis, and angiogenesis. It activated the cancer immune systems, suppressing the tumor growth in vivo. Reduced water is expected for utilization for prevention and therapy of various diseases (Shirahata 2002). Wound healing may be accelerated by applying a hydrocolloid occlusive dressing on burn surfaces after they are cleaned with EOW (Xin et al. 2003). Foot-and-mouth disease (FMD) is a highly transmissible and 37 economically important viral disease of domestic and wild cloven-hoofed animals. This disease results in devastating consequences worldwide. Bui et al. (2017) first report the virucidal effects of acidic (pH 2.6) and alkaline (pH > 11.7) EW solutions against FMDV. The prominent effect of alkaline EW on viral genome integrity and its virucidal mechanism, which is thought not to rely on FAC, emphasizes its potential utility as a disinfectant for field application. The disinfection effect of SAEW use in a farm where Pseudomonas mastitis has spread was evaluated. Procedural changes and equipment modifications did not improve environmental contamination or the incidence of Pseudomonas mastitis. When all the equipment and the parlor environment were sterilized with SAEW (pH 5–6.5, available chlorine 12 parts per million) before and during milking time, the incidence of clinical and subclinical Pseudomonas mastitis cases decreased significantly ($P < 0.0001$) and disappeared. These findings suggest that SAEW effectively reduced the incidence of mastitis in a herd contaminated by *Pseudomonas* species. This is the first report to demonstrate the effectiveness of disinfection by SAEW against mastitis pathogens in the environment (Kawai et al. 2017). Slightly acidic electrolyzed water (SAEW, pH 5.0–6.5) is a novel disinfectant with environmental friendly broad spectrum microbial decontamination properties which could have significant utility on farm. SAEW shows promise as a disinfectant for use on pig farms to reduce the spread of both PRRSV and PRV, and to limit the morbidity associated with those viruses (Hao et al. 2013c). An experiment was carried out to investigate the disinfection efficiency of SlAEW on teat dipping of milking cows, and the effect was compared with that using 0.5% povidone-iodine (PVP-I). The results showed that the bactericidal index of SlAEW increased as the increase of ACC and bath time of teat dipping. The average bactericidal index to total aerobic bacteria reached up to 2.21 log CFU/cm^2 using SlAEW with ACC of 60 mg/L and a bath time of 15 s, while the average bactericidal index of 0.5% PVP-I was approximately 1.52 log CFU/cm^2. Holding-up time is critical for the teat-dipping bath to achieve better disinfection efficiency. The bactericidal index of direct teat flushing using SlAEW with a low ACC (20 mg/L) was 1.47 log CFU/cm^2. After a 20-day test, the subclinical mastitis incidence of the dairy cows, by application of the SlAEW on the teats before and after milking decreased by 12.9% (Nan et al. 2010).

8.10 EW for Livestock Production

Electrolyzed water is a new type disinfectant agent which has been recognized as instantaneous, universally applicable, effective, cheapness, safe, without chemical residues, and no pollutions. Based on the electrolysis form and degrees, it can be divided into five types, and every type used in different areas. It was widely used in food processing industry. In recent years, researchers start to use it in the disinfection of livestock production, and achieved sound effects (Chen et al. 2012). Use of EO is an eventual alternative way of disinfection of breeding halls resulting in reduction of ammonia emissions up in the air from the stable environment and in suppressing pathogenic microorganisms, too. Usage of this system thus enables save usage and significant saving on cost on disinfection (Jirotková et al. 2012).

Laying hen digestive microflora and litter contamination levels are represented by the bacteriological quality of water, feedstuffs and mixed diets. The three EW solutions used in experiments had a total bactericidal effect on the total Coliforms. In the same time, the total number of germ decreased until 30 times against the control. The experiment highlights the role of active chlorine on the microbiological load in the water, feedstuffs, mixed diets, eggshells, and accommodation space. As the concentration of active chlorine is much higher as more significant is the bactericidal effect of the EW solution. These experiments proved that the EW solution could be used as disinfectant agent to control the microorganisms from water, feedstuffs, laying hens living spaces, and on the surface of eggshells (Surdu et al. 2009).

Table 8.1 Use of electrolyzed water as cleaning and disinfection method in different livestock-related fields

Category	Objective	Mode of application	Type of environment	Microorganism	References
Livestock house	Investigation of the FC loss from MLAEW aerosols over distances from spray origin at different air temperatures	MLAEW was sprayed using a spray gun (PILOT Mini, 0.5 mm nozzle, Walther Pilot NA, Chesterfield, MI, USA) connected to an air compressor	Poultry house	*Airborne bacteria*	Zhao et al. (2014)
	Evaluation of the inactivation efficacy of spraying SAEW to inhibit microorganisms on solid materials and on facility surfaces and the effect of SAEW on survival rates of *Salmonella* and *E. coli* in an egg-producing house	The house equipment was sprayed with SAEW having an ACC of 250 mg/L, pH value of 6.19, and ORP of 974 mV	Layer breeding house	*Salmonella* and *E. coli*	Hao et al. (2013a)
	Effectiveness of NEW as a drinking water addictive for chickens	Supplementation in drinking water	Drinking lines in chicken houses	TVC, *E. coli*	Bügener et al. (2014)
	Comparison efficiency of SAEW with EDP and povidone-iodine	Spraying technique	Swine barns environment	*Salmonella*, *S. aureus*, Coliforms, fungi	Hao et al. (2013b)
	Effectiveness of NEW on dust reduction	Spraying method	Laying breeding cages	Dust	Zheng et al. (2012)

(continued)

Table 8.1 (continued)

Category	Objective	Mode of application	Type of environment	Microorganism	References
Transportation of Animals	Evaluation of the efficacy of SAEW in reducing populations of bacterial mixtures on stainless steel surfaces typically used in transport vehicles, and to compare its efficacy with composite phenol solution for reducing total aerobic bacteria in animal transport vehicles	Disinfectant solutions were sprayed using a manual sprayer	Animal transport vehicles	*E. coli*, *Salmonella* typhimurium, *S. aureus* or bacterial mixtures	Ni et al. (2016)
	Effective of SAEW at different concentrations and times	Spray application on plastic cages	Poultry transport cages	*S. enteritids*	Zang et al. (2015)
Dairy industry	Synergistic effect of AlEW and AEW for washing milking system	cleaning-in-place of stainless steel straight pipe	Milk processing surface	Organic materials	Wang et al. (2016)
	Optimization of the temperature of AlEW and AEW for maximum cleaning of the milking system	Clean-in-place of pipeline	Milking system equipment	*P. fluorescens* B2, *M. luteus*, *E. faecalis*, *E. coli*	Dev et al. (2014)
	Investigation of the efficiency of EO water as a disinfectant in the dairy industry	Dilution with milk	Raw milk	*APC*	Kalit et al. (2015)

(continued)

Table 8.1 (continued)

Category	Objective	Mode of application	Type of environment	Microorganism	References
Slaughterhouse and meat processing plant	Effectiveness of AEW on different food processing surfaces	Pumping and spraying methods	Slaughterhouse equipment surfaces	*S. aureus*, *S.* Typhimurium, *E. coli* O157:H7, *L. monocytogenes*, MBS	Serraino et al. (2010)
	Effectiveness of EW at different concentrations and pressurized spray pressures	Spray application on slicing blades	Meat processing surfaces	*L. monocytogenes*, *E. coli* O157:H7 *Salmonella* sp.	Veasey and Muriana (2016)
	Comparison efficiency of AlEW and AEW with iodophor Mikroklene at processing plants	Pressurized spraying technique	Abattoir	Total coliforms, *E. coli*	Bach et al. (2006)
	Evaluation of the use of ozonated or EO waters in wash steps to reduce hide contamination in experiments using a model hide washing system	Sprayed with a nozzle	Slaughterhouse	APC, EBC	Bosilevac et al. (2005)
Disease prevention and control	Evaluation of the virucidal effects of EW against FMDV	Mixture was made	In vitro	*FMD virus*	Bui et al. (2017)
	Efficacy of electrolyzed oxidizing water (EOW) and hydrocolloid occlusive dressings in the acceleration of epithelialization in excised burn wounds in rats	Irrigation	–	*Wound*	Xin et al. (2003)

(continued)

Table 8.1 (continued)

Category	Objective	Mode of application	Type of environment	Microorganism	References
	Sterilization the equipment and the environment of the milking parlor at milking time to control mastitis in a dairy herd with a high incidence of Pseudomonas infection	SAEW sprayed from drop hoses during and after milking	Dairy farm	*Pseudomonas mastitis*	Kawai et al. (2017)
	Understanding the virucidal activity of SAEW is important for its practical application as a prophylactic disinfectant on pig farms	SAEW were mixed with the cell culture solution	Swine farm	*PRRSV and PRV Bartha K-61*	Hao et al. (2013c)
	To explore electrolyzed water as a possible footbath additive along with copper sulfate on the prevention of digital dermatitis	Mixed with footbath solutions	Dairy farm	*Digital dermatitis*	Himmelmann et al. (2017)
	Evaluation of the decontamination efficiency of SAEW on cow's teats and feet, milking cups, towels and hands of the workers, and examine the prevention of mastitis in dairy cows	The towels usually used to swab the teats, milking cups and workers' hands were cleaned with SAEW	Dairy farm	Mastitis	Nan et al. (2010)

8.11 Conclusion

EW exhibits strong antimicrobial activity against a broad spectrum of microorganisms. The definitions of SAEW and NEW are clarified in this review. SAEW and NEW can be produced by different systems, affecting their active components and thereby their antimicrobial activity. HOCl and OCl^- are responsible for the antimicrobial activity of SAEW and NEW. Spraying SAEW is considered as an alternative approach for reducing the microbial populations in animal houses. Increasing the ACC of SAEW and the spraying volume could improve its antimicrobial effectiveness on the surfaces in animal houses (Table 8.1). The airborne microbial population in animal houses can be greatly reduced by spraying SAEW. The airborne microbe reduction by spraying SAEW is influenced by ACC, spray volume, aerosols size, and ventilation. Moreover, while using EW as a disinfectant does not leave residues in the product after its application, it could be considered as a better choice in the production of dairy products for sensitive group of consumers such as children, pregnant women, elderly people, and people with chronic diseases. Additional research is required in order to optimize the application (concentration, time, temperature, and method/s) of EW and to further evaluate the effectiveness of EW in combination with other intervention/s.

References

Aarnisalo K, Salo S, Miettinen H et al (2000) Bactericidal efficiencies of commercial disinfectants against *Listeria monocytogenes* on surfaces. J Food Saf 20(4):237–250

Abadias M, Usall J, Oliveira M et al (2008) Efficacy of neutral electrolyzed water (NEW) for reducing microbial contamination on minimally-processed vegetables. Int J Food Microbiol 123(1–2):151–158

Amass SF, Thompson B, Dimmich KM et al (2007) Impact of down time on reducing aerobic bacterial counts in cleaned and disinfected trailers. J Swine Health Prod 15:37–41

Arya R, Bryant M, Degala HL et al (2018) Effectiveness of a low-cost household electrolyzed water generator in reducing the populations of *Escherichia coli* K12 on inoculated beef, chevon, and pork surfaces. J Food Process Preserv. https://doi.org/10.1111/jfpp.13636

Bach SJ, Jones S, Stanford K et al (2006) Electrolyzed oxidizing anode water as a sanitizer for use in abattoirs. J Food Protect 69(7):1616–1622

Bialka KL, Demirci A, Knabel SJ et al (2004) Efficacy of electrolyzed oxidizing water for the microbial safety and quality of eggs. Poult Sci 83:2071–2078

Bodas R, Bartolomé DJ, Tabernero MJ et al (2013) Electrolyzed water as novel technology to improve hygiene of drinking water for dairy ewes. Res Vet Sci 95:1169–1170

Bosilevac JM, Shackelford SD, Brichta DM et al (2005) Efficacy of ozonated and electrolyzed oxidative waters to decontaminate hides of cattle before slaughter. J Food Protect 68(7):1393–1398

Bügener E, Kump AW, Casteel M et al (2014) Benefits of neutral electrolyzed oxidizing water as a drinking water additive for broiler chickens. Poult Sci 93(9):2320–2326

Bui VN, Nguyen KV, Pham NT et al (2017) Potential of electrolyzed water for disinfection of foot-and-mouth disease virus. J Vet Med Sci 79(4):726–729

CDC (Center for Disease Control and Prevention) (2015) Foodborne outbreaks. Retrieved from https://www.cdc.gov/foodsafety/outbreaks/. Assessed on 21 Sept 2018

Chen HY, Zhou LY, Li FD et al (2012) The application of electrolyzed water in livestock production. China Anim Husb Vet Med 39(2):74–78

Chuang C, Fang W, Yang S (2011) A Study of membrane-less electrolyzed water fogging-spread for airborne bacteria and fungus decontamination in hen house. Proceedings of International Conference on Agricultural and Biosystems Engineering; Feb 20–21 2011; Hongkong, China; IERI, 2011

Cui X, Shang Y, Shi Z et al (2009) Physicochemical properties and bactericidal efficiency of neutral and acidic electrolyzed water under different storage conditions. J Food Eng 91(4):582–586

Dee SA, Corey MM (1993) The survival of *Streptococcus suis* on farm and veterinary equipment. J Swine Health Prod 1:17–20

Dee SA, Deen J (2006) Evaluation of an industry-based sanitation protocol for transport vehicles contaminated with porcine reproductive and respiratory syndrome virus. J Swine Health Prod 14:126–132

Dev SR, Demirci A, Graves RE (2014) Optimization and modeling of an electrolyzed oxidizing water based clean-in-place technique for farm milking systems using a pilot-scale milking system. J Food Eng 135:1–10

Deza MA, Araujo M, Garrido MJ (2005) Inactivation of *Escherichia coli*, *Listeria monocytogenes*, *Pseudomonas aeruginosa* and *Staphylococcus aureus* on stainless steel and glass surfaces by neutral electrolyzed water. Lett Appl Microbiol 40(5):341–346

Deza MA, Araujo M, GarridoMJ (2007) Efficacy of neutral electrolyzed water to inactivate Escherichia coli, Listeria monocytogenes, Pseudomonas aeruginosa, and Staphylococcus aureus on plastic and wooden kitchen cutting boards. J Food Prot 70(1):102–108

Doyle MP, Erickson MC (2006) Reducing the carriage of foodborne pathogens in livestock and poultry. Poul Sci 85:960–973

Fasenko GM, O'Dea Christopher EE, McMullen LM (2009) Spraying hatching eggs with electrolyzed oxidizing water reduces eggshell microbial load without compromising broiler production parameters. Poul Sci 88:1121–1127

FDA (2009) Food and Drug Administration. Report on the occurrence of foodborne illness risk factors in selected institutional foodservice, restaurant, and retail food store facility types. US Food and Drug Administration, Silver Spring, MD

Fedorka-Cray PJ, Hogg A, Gray JT et al (1997) Feed and feed truck as sources of *Salmonella* contamination in swine. Swine Health Prod 5:189–193

Flores DRM, Athayde DR, Silva MS et al (2017) The use of ultrasound and slightly acidic electrolyzed water as alternative technologies in the meat industry. Food Res 1(5):140–146

Fussing V, Barfod K, Nielsen R et al (1998) Evaluation and application of ribotyping for epidemiological studies of *Actinobacillus pleuropneumoniae* in Denmark. Vet Microbiol 62:145–162

Gebreyes WA, Davies PR, Turkson PK et al (2004) *Salmonella enterica* serovars from pigs on farms and after slaughter and validity of using bacteriologic data to define herd *Salmonella* status. J Food Protect 67:691–697

Guentzel JL, Liang Lam K, Callan MA et al (2008) Reduction of bacteria on spinach, lettuce, and surfaces in food service areas using neutral electrolyzed oxidizing water. Food Microbiol 25(1):36–41

Handojo A, Lee J, Hipp J, Pascall MA (2009) Efficacy of electrolyzed water and an acidic formulation compared with regularly used chemical sanitizers for tableware sanitization during mechanical and manual ware-washing protocols. J Food Prot 72(6):1315–1320

Hao X, Li B, Wang C et al (2013a) Application of slightly acidic electrolyzed water for inactivating microbes in a layer breeding house. Poult Sci 92(10):2560–2566

Hao X, Li B, Zhang Q et al (2013b) Disinfection effectiveness of slightly acidic electrolyzed water in swine barns. J Appl Microbiol 115(3):703–710

Hao X, Shen Z, Wang J et al (2013c) In vitro inactivation of porcine reproductive and respiratory syndrome virus and pseudorabies virus by slightly acidic electrolyzed water. Vet J 197(2):297–301

Hao X, Cao W, Li B et al (2014) Slightly acidic electrolyzed water for reducing airborne microorganisms in a layer breeding house. J Air Waste Manag 64(4):494–500
Hernández M, Gómez-Laguna J, Luque I et al (2013) Salmonella prevalence and characterization in a free-range pig processing plant: tracking in trucks, lairage, slaughter line and quartering. Int J Food Microbiot 162:48–54
Himmelmann HK, Jones BW, Bewley JM (2017) Effects of the addition of electrolyzed water to a footbath solution on digital dermatitis incidence. J Dairy Vet Anim Res 6(1):00168
Hricova D, Stephan R, Zweifel C (2008) Electrolyzed water and its application in the food industry. J Food Prot 71(9):1934–1947
Huang YR, Hung YC, Hsu SY et al (2008) Application of electrolyzed water in the food industry. Food Control 19(4):329–345
Huton CT, Fox LK, Hancock DD (1990) Mastitis control practices: differences between herds with high and low milk somatic cell counts. J Dairy Sci 73(4):1135–1143
Izumi H (1999) Electrolyzed water as a disinfectant for fresh-cut vegetables. J Food Sci 64(3):536–539
Jassim EQ, Aqeel CH (2017) Effect of alkaline water and/or magnetic water on some physiological characteristic in broiler chicken. JEZS 5(5):1643–1647
Jirotková D, Šoch M, Kernerová N et al (2012) Use of electrolyzed water in animal production. JMBFS 2(2):477–483
Kalit S, Kos T, Kalit MT et al (2015) The efficacy of electrolysed oxidizing water as a disinfectant in the dairy industry. J Hyg Eng Des 12:28–32
Kawai K, Shinozuka Y, Uchida I (2017) Control of *Pseudomonas* mastitis on a large dairy farm by using slightly acidic electrolyzed water. Anim Sci J 88(10):1601–1605
Koide S, Shitanda D, Note M et al (2011) Effects of mildly heated, slightly acidic electrolyzed water on the disinfection and physicochemical properties of sliced carrot. Food Control 22(3–4):452–456
Leach SA, Williams A, Davies AC et al (1999) Aerosol route enhances the contamination of intact eggs and muscle of experimentally infected laying hens by *Salmonella typhimurium* DT 104. FEMS Microbiol Lett 171(2):203–207
Lee J, Gupta M, Lopes J, Pascall M (2007a) Comparative efficacies of hydrogen peroxide and acidic sanitizers for microbial reduction on metal can and low density polyethylene film surfaces. J Food Sci 72:M335–M339
Lee J, Lopes J, Pascall M (2007b) Development of a sanitizing fabric wipe for use on food contact surfaces. J Food Sci 72(9):M375–M381
Lewis S, McIndoe AK (2004) Cleaning, disinfection and sterilization of equipment. Anaesth Intensiv Care Med 5(11):360–363
Lowe J, Gauger P, Harmon K et al (2014) Role of transportation in spread of porcine epidemic diarrhea virus infection. Emerg Infect Dis 20:872–874
Mead PS, Slutsker L, Dietz V et al (1999) Food-related illness and death in the United States. Emerg Infect Dis 5:607–625
Monnin A, Lee J, Pascall MA (2012) Efficacy of neutral electrolyzed water for sanitization of cutting boards used in the preparation of foods. J Food Engin 110(4):541–546
Nan S, Wang C, Li B et al (2010) Disinfection efficiency and prevention of mastitis by slightly acidic electrolyzed water in a dairy farm. CIGR XVIIth world congress, Québec City, Canada, 13–17 June 2010
Ni L, Cao W, Zheng W et al (2014) Efficacy of slightly acidic electrolyzed water for reduction of foodborne pathogens and natural microflora on shell eggs. Food Sci Technol Res 20(1):93–100
Ni L, Cao W, Zheng W et al (2015) Reduction of microbial contamination on the surfaces of layer houses using slightly acidic electrolyzed water. Poult Sci 94(11):2838–2848
Ni L, Zheng W, Zhang Q et al (2016) Application of slightly acidic electrolyzed water for decontamination of stainless steel surfaces in animal transport vehicles. Prev Vet Med 133:42–51
Park H, Hung YC, Kim C (2002) Effectiveness of electrolyzed water as a sanitizer for treating different surfaces. J Food Prot 65:1276–1280

Patterson AR, Baker RB, Madson DM et al (2011) Disinfection protocols reduce the amount of porcine circovirus type 2 in contaminated 1:61 scale model livestock transport vehicles. J Swine Health Prod 19:156–164
Pichardo RJ, Regalado C, Tostado EC et al (2016) Evaluation of electrolyzed water as cleaning and disinfection agent on stainless steel as a model surface in the dairy industry. Food Control 60:320–328
Priyanka B, Patil RK, Dwarakanath S (2016) A review on detection methods used for foodborne pathogens. The Indian J Med Res 144(3):327–338
Radon K, Danuser B, Iversen M et al (2002) Air contaminants in different European farming environments. Ann Agric Environ Med 9(1):41–48
Rahman SME, Park JH, Wang J et al (2012a) Stability of low concentration electrolyzed water and its sanitization potential against foodborne pathogens. J Food Eng 113:548–553
Rahman SME, Park J, Song KB et al (2012b) Effects of slightly acidic low concentration electrolyzed water on microbiological, physicochemical, and sensory quality of fresh chicken breast meat. J Food Sci 77(1):M35–M41
Rahman SME, Wang J, Oh DH (2013) Synergistic effect of low concentration electrolyzed water and calcium lactate to ensure microbial safety, shelf life and sensory quality of fresh pork. Food Control 30(1):176–183
Rahman SME, Khan I, Oh DH (2016) Electrolyzed water as a novel sanitizer in the food Industry: current trends and future perspectives. Compr Rev Food Sci Food Saf 15(3):471–490
Rajkowski KT, Eblen S, Laubauch C (1998) Efficacy of washing and sanitizing trailers used for swine transport in reduction of *Salmonella* and *Escherichia coli*. J Food Protect 61:31–35
Rasschaert G, Piessens V, Scheldeman P et al (2013) Efficacy of electrolyzed oxidizing water and lactic acid on the reduction of *Campylobacter* on naturally contaminated broiler carcasses during processing. Poult Sci 92:1077–1084
Rodríguez Ferri EF, Martínez S, Frandoloso R et al (2010) Comparative efficacy of several disinfectants in suspension and carrier tests against *Haemophilus parasuis* serovars 1 and 5. Res Vet Sci 88(3):385–389
Schmidt PL, O'Connor AM, McKean JD et al (2004) The association between cleaning and disinfection of lairage pens and the prevalence of Salmonella enterica in swine at harvest. J Food Protect 67(7):1384–1388
Serraino A, Veronese G, Alonso S (2010) Bactericidal activity of electrolyzed oxidizing water on food processing surfaces. Italian J Food Sci 22(2):222–228
Shirahata S (2002) Reduced water for prevention of diseases. In: Shirahata S, Teruya K, Katakura Y (eds) Animal cell technology: basic and applied aspects. Animal cell technology: basic and applied aspects, vol 12. Springer, Dordrecht
Sidhu MS, Langsrud S, Holck A (2001) Disinfectant and antibiotic resistance of lactic acid bacteria isolated from the food industry. Microb Drug Resist 7(1):73–83
Sundheim G, Langsrud S, Heir E et al (1998) Bacterial resistance to disinfectants containing quaternary ammonium compounds. Int Biodeterior Biodegrad 41(3–4):235–239
Surdu I, Vatuiu I, Jurcoane S et al (2009) The sanitation effect of electrolyzed water (neutral anolyte-ank) on pathogen agents from living space and feedstuffs used in laying hens nutrition. In: 8th international symposium of animal nutrition and biology, Balotesti, Rumania, 24–25 Sept. Disponible en: http://metaviac.siat.ro/Default_engl.aspx (01/04/2012)
Veasey S, Muriana PM (2016) Evaluation of electrolytically-generated hypochlorous acid ('electrolyzed water') for sanitation of meat and meat-contact surfaces. Foods 5:42. https://doi.org/10.3390/foods5020042
Wang X, Puri VM, Demirci A et al (2016) Mathematical modeling and cycle time reduction of deposit removal from stainless steel pipeline during cleaning-in-place of milking system with electrolyzed oxidizing water. J Food Eng 170:144–159
Watson RD (2015) Use of electrolyzed water for treatment and prevention of common livestock diseases and conditions, including PRRS. US Patent App.14/180,046, 2015, free patents online.com

World Health Organization (WHO) (1984) Guidelines on disinfection in animal husbandry for prevention and control of zoonotic diseases. In: Russell AD, Yarnych VS, Koulikovskii AV (eds) Elaboration of the guidelines was coordinated by the veterinary public health unit of the World Health Organization, Geneve, WHO/VPH/84.4

Xin H, Zheng YJ, Hajime N et al (2003) Effect of electrolyzed oxidizing water and hydrocolloid occlusive dressings on excised burn-wounds in rats. Chin J Traumatol 6(4):234–237

Zang Y, Li B, Bing S et al (2015) Modeling disinfection of plastic poultry transport cages inoculated with *Salmonella* enteritids by slightly acidic electrolyzed water using response surface methodology. Poult Sci 94:2059–2065

Zhao Y, Aarnink AJA, Koerkamp PWGG et al (2011) Detection of airborne Campylobacter with three bioaerosol samplers for alarming bacteria transmission in broilers. Biol Eng 3(4):177–186

Zhao Y, Xin H, Zhao D et al (2014) Free chlorine loss during spraying of membrane less acidic electrolyzed water and its antimicrobial effect on airborne bacteria from poultry house. Ann Agr Env Med 21(2):249–255

Zheng W, Li B, Cao W et al (2012) Application of neutral electrolyzed water spray for reducing dust levels in a layer breeding house. J Air Waste Manag Assoc 62(11):1329–1334

Zheng W, Kang R, Wang H et al (2013) Airborne bacterial reduction by spraying slightly acidic electrolyzed water in a laying-hen house. J Air Waste Manag 63(10):1205–1211

Zheng W, Zhao Y, Xin H et al (2014) Airborne particulate matter and culturable bacteria reduction from spraying slightly acidic electrolyzed water in an experimental aviary laying-hen housing chamber. Trans ASABE 57(1):229–236

Zheng W, Ni L, Li B (2016) Electrolyzed water and its application in animal houses. Front Agric Sci Eng 3(3):195–205

Chapter 9
Application of Electrolyzed Water in Agriculture

Fereidoun Forghani

9.1 Introduction

The need for technologies to enhance the safety of food and produce which cause less risk as well as being environmentally friendly approaches for management of plant pathogens is increasing as concerns on the impact of pesticides on environment and human health are now a great topic of interest to the public. Despite the extensive advancements in food safety that mankind has reached as well as in many other fields, foodborne diseases, food-related outbreaks, and different types of losses during the food supply chain still happen and many plant and animal pathogens pose serious concerns within the agricultural community.

The best way to reduce the incidence of foodborne disease, in particular, would be to secure a safe food supply. Today, the food supply chain is much more complicated than before in production, handling, processing, transportation, storage, and retail, making it a challenging task to deliver food with the optimum safety to the consumers. Thus, improvement of the whole supply chain safety would not be possible except if every single step's safety is improved to a maximum extent. Clearly, one of the major parts of the supply chain to achieve such goal is in agriculture.

Improvement of production phase safety, which is mainly taking place in the agriculture section, is of great importance where improvement in the practices and interventions as well as a continuous development of new interventions is a must. In this context, there is an ever growing list of interventions being developed to reach a higher level of final product safety, efficiency, cost, and final product quality. When it comes to microbial safety of the produce and food products, electrolyzed water as a relatively new and environmentally friendly product is one of the interventions applied. The suitability of electrolyzed water in agriculture has been reported

F. Forghani (✉)
Center for Food Safety, College of Agricultural and Environmental Sciences, University of Georgia, Griffin, GA, USA
e-mail: fereidoun.forghani@uga.edu

T. Ding et al. (eds.), *Electrolyzed Water in Food: Fundamentals and Applications*,
https://doi.org/10.1007/978-981-13-3807-6_9

(Huang et al. 2008). Hence, in this chapter, we will briefly discuss its applications, advantages, disadvantages, and future application possibilities in more detail.

In general, considering the variety of needs in food production including agriculture and several types of electrolyzed water available today with a wide range of properties, it seems that electrolyzed water may be a candidate for current and future use in agriculture in a good number of applications. These applications include but are not limited to dairy, hog, poultry, equine, and bovine applications. Silage making and sterilization of agricultural devices and equipment are two other fields with numerous possibilities for the application of electrolyzed water.

In addition to animal husbandry and related applications, there are several areas in crop production that express great potential for electrolyzed water application such as pesticidal and insecticidal reduction on plants, potato blight removal, antifungal application on rice or similar cereals, vegetable seed germination, plant growth acceleration, and possibly control. These are only a few examples and upon further research and case studies in future, there will probably be a long list added to these applications.

Electrolyzed water in agriculture can be used both as a pre-harvest and post-harvest intervention not only against pathogenic bacteria but also to control product loss, for shelf-life extension, against undesirable fungi, for chemical removals such as pesticide inactivation and toxin inactivation or mitigation procedures (Al-Haq et al. 2005). Many of these applications have been discussed in the other chapters of this book.

In recent years, applications of electrolyzed water in agriculture have expanded to new areas such as product enhancement (Liu et al. 2013). Since 1931 that electrolyzed water was first introduced as electrolyzed reduced water and applied in agriculture as well as other sectors, different types of it such as acidic, slightly acidic, and neutral electrolyzed water have been applied in agriculture (Rahman et al. 2016), which will be further discussed in this chapter as follows.

The possibilities for the application of electrolyzed water as a pre-harvest technology include but are not limited to its use for seed treatment, product yield improvement, product quality improvement, pathogen suppression, and plant disease control. All these advantages beside the possibility of reducing water consumption by different methods such as reusing the municipal sewage water are among the possibilities available for the application of electrolyzed water in the pre-harvest phase of agriculture sector, which will each and/or in combination eventually result in the product safety, consumers' better health, and farmers profits to increase.

9.2 Electrolyzed Water Application in Plant Irrigation

It has been reported that acidic electrolyzed water can be a good antifungal agent against the foliar diseases in plants. Also, reports have shown the suitability of using electrolyzed water for disinfecting strawberry plants against *Botrytis cinerea* (Rahman et al. 2016). *B. cinerea* is a necrotrophic fungus that affects many plant species, although its most notable hosts may be wine grapes. One of these plants heavily impacted is strawberry in which electrolyzed water has been effective in reducing the product damage.

Another study reported that application of neutral electrolyzed water as a foliar spray through an overhead irrigation system was a suitable approach for controlling *Colletotrichum fructicola* on strawberry (Hirayama et al. 2016). Results obtained in this study showed that the overhead irrigation using neutral electrolyzed water either alone or in combination with fungicides not only enhanced the anthracnose disease control on strawberry but also did not result in any visible phytotoxicity. This is an important advantage for electrolyzed water in comparison with many other soluble treatments due to the longevity of treatment during crop production, which lasts for weeks.

Another study assessed the ability of electrolyzed water in cultivation of Chinese cabbage and other leafy vegetables and its effects on properties such as photosynthesis, growth, yield, and nutritional quality. Results obtained showed that electrolyzed water could enhance the leaf net photosynthetic rate, leaf number, and yield. More interestingly, alkaline electrolyzed water irrigation on the root could increase the vitamin C content of Chinese cabbage (Hou et al. 2011).

A study in 2011 showed the potential of electrolyzed water to be applied in sprout production. In this study, electrolyzed water was replaced for tap water in producing mung bean sprouts. Results obtained showed that it could not only reduce the microbial numbers on the surface of mung bean sprouts but could also promote the sprouts growth (Rui et al. 2011). Application of slightly acidic electrolyzed water in this study resulted in at least 1 log CFU/g lower microbial counts on the sprouts compared to the ones that were treated by tap water. This result is promising for future efforts since 1 log difference in case of microbial contamination of produce is an acceptable achievement. However, it can still be further improved.

Despite the few reports mentioned above, application of electrolyzed water for plant irrigation is a relatively new concept and brings a diverse pool of opportunities for future studies to find suitable types of electrolyzed water to be used based on the plants type, production environment, and efficiency as well as the procedures that can be used. One of the possibilities in this regard can be the hydroponic systems that are increasingly being used in agriculture and are known for their many advantages including enhanced microbial safety. An attribute can be even further enhanced if electrolyzed water is added to the system.

9.3 Application in Plant Disease Control

Fungal infections cause significant losses to growers and retailers in many types of produce. On the other hand, application of fungicides during the crop season or prior to harvest forces growers to spend millions of dollars only within the US per year (Pimentel et al. 2000). In addition, fungicides have other disadvantages such as negative environmental impacts and occupational exposure danger during fungicide application on field. Also, there is a growing demand by consumers for limiting application of chemicals such as fungicides in the agricultural production of all crops. Hence, alternative methods with similar efficiency and less negative effects are necessary and electrolyzed water can be one of these new generations of antifungal treatments.

A study using electrolyzed water at near neutral pH for the management of gray mold disease on strawberry plants showed that not only the electrolyzed water treatment was effective in prevention of the disease but also did not show phytotoxic effects on the plant (Guentzel et al. 2011). The results from this study suggested that electrolyzed water solutions could be used to manage the infection of the fungi *Botrytis cinerea* causing disease in grapes, strawberries, and stone fruits and *Monilinia fructicola* which is responsible for brown rot disease in peaches, apples, and other stone fruits (Al-Haq et al. 2002; Imran et al. 2001).

In another study, effects of spraying electrolyzed water on the severity of powdery mildew infection caused by *sphaerotheca fuliginea* Pollacci fungus was investigated on cucumbers (Fujiwara et al. 2000). This fungus is widespread on cucurbits, especially during dry and hot periods, and severely affects cucumber stems and leaves leading to even death. Results obtained in this study showed that electrolyzed water would be an alternative to chemical fungicides for the control of powdery mildew in cucumber.

However, it also emphasized that further efforts are necessary in order to minimize the occasional physiological disorder caused to the cucumber leaves after a period of electrolyzed water application (Fujiwara et al. 2000). In another study, these scientists used acidic electrolyzed water for the same purpose in order to control the powdery mildew infection on cucumber and reported the same results as their previous study (Fujiwara et al. 2009).

Another promising result regarding the application of electrolyzed water for the mitigation of plant disease was reported on tomato (Abbasi and Lazarovitz 2006). The study investigated the effects of acidic electrolyzed water on the viability of bacterial and fungal plant pathogens and on bacterial spot disease of tomato. Results obtained showed that foliar sprays of acidic electrolyzed water under greenhouse and field conditions on tomato reduced *Xanthomonas campestris* population and leaf spot severity on tomato foliage in the greenhouse. Furthermore, field study showed that multiple sprays of electrolyzed water reduced bacterial spot severity on tomato foliage. In addition, fruit yield was either enhanced or not affected by the electrolyzed water sprays. These results also indicate a potential use of acidic electrolyzed water as a contact disinfectant in agriculture (Abbasi and Lazarovitz 2006).

9.4 Application in Seeds Decontamination and Germination

Application of electrolyzed water in seed decontamination and germination improvement seems to be another promising area. A study investigating the effects of acidic electrolyzed water treatment of wheat grain revealed that electrolyzed water not only had disinfecting effect on the wheat kernels but also improved their germination (Bonde et al. 1999). The study reported that acidic electrolyzed water has certain advantages over NaOCl in this regard. Considering the increasing demand for safer dry foods especially different types of flour produced, application of electrolyzed water as an irrigation component prior to harvest and milling can be a suitable strategy to improve microbial safety of such products considering the fact that to date there are very limited interventions available during the milling process for microbial decontamination.

In another study, acidic electrolyzed water in comparison with other treatments was evaluated for treating alfalfa and broccoli seeds (Kim et al. 2006). The study included assessment of antimicrobial effects of treatments against *Escherichia coli* O157:H7 too. Results obtained revealed that treatment at 55 °C for 10 min reduced *E. coli* O157:H7 population by 3.4 and 3.3 log CFU/g for the alfalfa and broccoli seeds, respectively. In addition, the germination percentages for seeds after treatment were approximately 90% which is an acceptable rate for the viability of the seeds after treatment.

The effect of soaking Chinese cabbage seeds with electrolyzed water was studied under laboratory conditions. The results showed that pH value of the electrolyzed water was an important factor that affected the Chinese cabbage seeds germination. The acidic electrolyzed water of pH 3.30 could accelerate seeds germinating and raise fresh weight of shoots significantly. On the other hand, strong acidic electrolyzed water and alkaline electrolyzed water would slow down the speed of seeds germination and inhibit both germination rate and fresh weight of shoots.

Interestingly, neutral electrolyzed water had no clear effect on germination potential, germination rate, and fresh weight of shoots. Thus, overall results showed the suitability of acidic electrolyzed water for soaking cabbage seeds for a faster germination. However, another important influencing factor was the soaking time which had a positive effect if kept less than 2 h. Longer treatment times resulted in slower germination (Deng et al. 2010).

9.5 Application in Foliage and Flower Production

In 2003, Buck et al. performed a study evaluating acidic electrolyzed water for phytotoxic symptoms on foliage and flowers of bedding plants as well as its fungicidal effects (Buck et al. 2003). Results obtained from their study revealed that electrolyzed water was effective against a variety of fungi and showed a promising broad-spectrum fungicidal ability for the control of foliar diseases of greenhouse-grown ornamentals. They also reported that the composition of the salts used for electrolyzed water production would affect their phytotoxic effects.

Primarily in their study, Buck et al. confirmed that the optimal electrolyzed water if used appropriately did not have excessive phytotoxic effect on the plants. For this purpose, they used a total of 15 species of bedding plants and demonstrated that electrolyzed water did not produce any visible phytotoxicity in most of them while in some other there were minor white spots observed on flowers over a foliar spray period of 7 weeks. However, overall electrolyzed water appeared to be safe to use as a foliar spray treatment on a wide variety of bedding plants grown under greenhouse conditions.

In another study, Mueller and colleagues reported electrolyzed water to be a viable option for controlling powdery mildew on gerbera daisies providing growers an additional tool to reduce the use of traditional fungicides in greenhouse (Mueller et al. 2003). Powdery mildew is a fungal disease and has been a major concern for greenhouse growers.

In brief, Mueller and colleagues work showed that when used in an integrated system, electrolyzed water could be useful for reducing powdery mildew. In addition, it was also effective when combined with most of the conventional pesticides showing a synergistic potential. These results clearly show the potential of electrolyzed water to be used as a new fungicide by the flower greenhouse growers. This would be not only financially helpful for the growers but also in favor of the environment due to the safer nature and/or lower cost of electrolyzed water compared to current treatments for mitigating powdery mildew such as sulfur, lime sulfur, neem oil, and potassium bicarbonate, although there is a lot of room for further study in this regard.

9.6 Conclusions

Although the application of electrolyzed water in different fields has been around for decades, there are always new reports being published on a new aspect of its use. The same is true for electrolyzed water application in agriculture. Therefore, there is always new possibilities and new opportunities. For example, application of electrolyzed water for decontamination of avocados has been reported and adding a new product to the category of products electrolyzed water has successfully been used for their treatment (Rodríguez-Garcia et al. 2011).

In another study, electrolyzed water was reported as a useful tool for preventing enzymatic browning in Chinese Yam (Jia et al. 2015). Hence, the list of new applications for electrolyzed water has yet to become longer. This, in addition to technology improvements and application of advanced hurdle treatments, promises a bright future for electrolyzed water as an environmentally friendly and efficient tool in agriculture for maximizing produce quality and consumers safety.

Furthermore, as described in Chap. 4 of this book, electrolyzed water brings endless possibilities for using it with other technologies to develop improved hurdle approaches. This ability can also be extended to agriculture. A few examples for future exploration which might be well fit for agriculture sector can be a combination of electrolyzed water and ozonated water, production of microbubbled water from electrolyzed water and also multistep application of electrolyzed water with other conventional sanitizers or pesticides currently in use.

Natural products and organic produce are other possible candidates for application of electrolyzed water in agriculture in the future. With the increase in peoples' interest in consuming such products, chemicals being unacceptable to be used and the very limited antimicrobial treatments and pesticides approved to be used in organic production, environmentally friendly, and safe treatments such as electrolyzed water are of great need.

References

Abbasi PA, Lazarovitz G (2006) Effect of acidic electrolyzed water on the viability of bacterial and fungal plant pathogens and on bacterial spot disease of tomato. Can J Microbiol 52:915–923

Al-Haq MI, Seo Y, Oshita S et al (2002) Disinfection effects of electrolyzed oxidizing water on suppressing fruit rot of pear caused by *Botryosphaeria berengeriana*. Food Res Int 35:657–664

Al-Haq MI, Sugiyama J, Isobe S (2005) Applications of electrolyzed water in agriculture and food industries. Food Sci Technol Res 11:135–150

Bonde MR, Nester SE, Khayat A et al (1999) Comparison of effects of acidic electrolyzed water and NaOCl on *Tilletia indica* teliospore germination. Plant Dis 83:627–632

Buck JW, Van Iersel MW, Oetting RD et al (2003) Evaluation of acidic electrolyzed water for phytotoxic symptoms on foliage and flowers of bedding plants. Crop Prot 22:73–77

Deng L, Huang C, Zhao S et al (2010) The effect of electrolyzed water on Chinese cabbage seeds germination. J Agric Mech Res 2:133–136

Fujiwara K, Doi R, Iimoto M et al (2000) Effects of spraying electrolyzed anode-side water and pH-available chlorine concentration-regulated water on the severity of powdery mildew infection and percentage of leaves with a leaf burn-like physiological disorder on cucumber leaves. Environ Control Biol 38:33–38

Fujiwara K, Fujii T, Park JS (2009) Comparison of foliar spray efficacy of electrolytically ozonated water and acidic electrolyzed oxidizing water for controlling powdery mildew infection on cucumber leaves. Ozone Sci Eng 31:10–14

Guentzel JL, Callan MA, Liang LK et al (2011) Evaluation of electrolyzed oxidizing water for phytotoxic effects and pre-harvest management of gray mold disease on strawberry plants. Crop Prot 30:1274–1279

Hirayama Y, Asano S, Watanabe K et al (2016) Control of *Colletotrichum fructicola* on strawberry with a foliar spray of neutral electrolyzed water through an overhead irrigation system. J Gen Plant Pathol 82:186–189

Hou M, Gao J, Deng L et al (2011) Effect of electrolyzed water on growth and development, nutritional quality of Chinese cabbage. Hubei Agric Sci 7(50):1342–1346

Huang YR, Hung YC, Hsu SY et al (2008) Application of electrolyzed water in the food industry. Food Control 19:329–345

Imran AM, Seo Y, Oshita S et al (2001) Fungicidal effectiveness of electrolyzed oxidizing water on postharvest brown rot of peach. HortScience 36:1310–1314

Jia GL, Shi JY, Song ZH et al (2015) Prevention of enzymatic browning of Chinese Yam (*Dioscorea* spp.) using electrolyzed oxidizing water. J Food Sci 80:C718–C728

Kim HJ, Feng H, Kushad MM et al (2006) Effects of ultrasound, irradiation, and acidic electrolyzed water on germination of alfalfa and broccoli seeds and *Escherichia coli* O157:H7. J Food Sci 71:168–173

Liu R, He X, Shi J et al (2013) The effect of electrolyzed water on decontamination, germination and aminobutyric acid accumulation of brown rice. Food Control 33:1–5

Mueller DS, Hung YC, Oetting RD et al (2003) Evaluation of electrolyzed oxidizing water for management of powdery mildew on gerbera daisy. Plant Dis 87:965–969

Pimentel D, Lack L, Zuniga R et al (2000) Environmental and economic costs of nonindigenous species in the United States. Bioscience 50:53–65

Rahman S, Khan I, Oh DH (2016) Electrolyzed water as a novel sanitizer in the food industry: current trends and future perspectives. Compr Rev Food Sci Food Saf 15:471–490

Rodríguez-Garcia O, González-Romero VM, Fernández-Escartín E (2011) Reduction of *Salmonella enterica*, *Escherichia coli* O157:H7, and *Listeria monocytogenes* with electrolyzed oxidizing water on inoculated hass avocados (*Persea americana* var. Hass). J Food Prot 74:1552–1557

Rui L, Jianxiong H, Haijie L et al (2011) Application of electrolyzed functional water on producing mung bean sprouts. Food Control 22:1311–1315

Chapter 10
Hurdle Enhancement of Electrolyzed Water with Other Techniques

Deog-Hwan Oh, Imran Khan and Charles Nkufi Tango

10.1 Introduction

EW has shown promising potential as a broad-spectrum sanitizer of foods. It offers a number of applications from on-the-spot and green production to low cost and labor. However, the individual sanitization efficacy of EW was found to be insufficient to completely inactivate or decontaminate many food products (Mikš-Krajnik et al. 2017; Tango et al. 2017). In the food industry, the novel approach of hurdle technology has been introduced to guarantee microbial safety, nutritional quality, and the economic viability of food products (Luo and Oh 2016). Hurdle technology, also known as combination preservation, combined methods, barrier technology, combined processes, and combination techniques, is the application of two or more basic food preservation techniques to reduce the extreme conditions of individual treatments and enhance their effectiveness (Khan et al. 2017). The use of hurdle technology in the food industry, without clear scientific knowledge, can be traced back to early civilization. Hurdle technology provides safe, stable, and improved nutritional quality. Though an individual food preservation treatment, for instance, electrolyzed water, could be used for acceptable food safety, the recent trend in the food industry is to maintain an improved quality of food without compromising food safety (Rostami et al. 2016). However, recently, the use of hurdle technology has been employed in the food industry to control the main factors such as water activity, pH, temperature, and their impact on microbes in food products (Mukhopadhyay and Gorris 2014). The applications of hurdle technology are well suited to the increased consumer demand for minimally processed food (Guerrero et al. 2017).

D.-H. Oh (✉) · I. Khan · C. N. Tango
Department of Bioconvergence Science and Technology, College of Agriculture and Life Science, Kangwon National University, Chuncheon, Republic of Korea
e-mail: deoghwa@kangwon.ac.kr

C. N. Tango
Department of Chemistry and Agricultural Industries, Faculty of Agronomy, University of Kinshasa, Kinshasa, Democratic Republic of the Congo

T. Ding et al. (eds.), *Electrolyzed Water in Food: Fundamentals and Applications*,
https://doi.org/10.1007/978-981-13-3807-6_10

The advanced steps towards the successful development of hurdle technology to preserve foods need diverse knowledge and a profound understanding of the influence of individual hurdle factors and the combined treatment on microorganisms at the cellular level. Hurdle technology offers the best framework for the integration of optimal individual hurdles that in combination achieve improved food quality while maintaining food safety and stability (Mukhopadhyay and Gorris 2014).

EW in combination with other technologies has been utilized in the food industry for more than 15 years. Therefore, hurdle enhancement of EW simultaneously reduces the microbial count and enhances the food quality and safety. Hurdle enhancement of EW has gained immense attention due to its synergistic effectiveness at inhibiting the targeted microorganisms. We have reported a number of various combinations of EW and other technologies to enhance food quality and safety (Forghani and Oh 2013; Luo and Oh 2016; Mansur et al. 2015; Ngnitcho et al. 2017). The hurdle enhancement of low-concentration EW was studied in combination with calcium lactate against foodborne pathogens inoculated on fresh pork, and the results showed synergistic effects compared to those of the individual treatments. Interestingly, the shelf life of fresh pork was extended up to 6 days when stored at 4 °C. Moreover, the synergistic effects for of calcium lactate and EW were obtained at very low concentrations (Rahman et al. 2013). In another study, we examined the hurdle enhancement of EW with fumaric acid and mild heat treatment against foodborne pathogens inoculated on fresh pork. Similar synergistic effects for EW and fumaric acid at 40 °C were obtained (Mansur et al. 2015).

10.1.1 Definition and Concept of Hurdle Technology

Hurdle techniques have been unknowingly utilized in the traditional foods of many developing countries. However, in the late 1970s, the concept was reinvented in the meat industry by Leistner by applying hurdle techniques to extend the shelf life of sausage (Leistner and Gorris 1995). Hurdle technology was defined as "the deliberate combination of existing and novel preservation techniques in order to establish a series of preservative factors (hurdles) that any microorganisms present should not be able to overcome". Different hurdles are explained in Table 10.1.

The primary mechanism of hurdle technology is to disrupt the homeostasis condition of targeted microorganisms. This can be achieved by keeping microorganisms under a continuous stress, which may be chemical, physical or environmental. Microorganisms have the capability of responding to stress conditions and building their stability against it, while hurdles techniques overcome the stability wall of the microorganisms and disrupt homeostasis, leading to the inhibition of microorganisms multiplying and hence resulting in death (Mukhopadhyay and Gorris 2014). The hurdle concept relies on different factors that promote the growth of microorganisms. These factors constantly interact with each other and become an issue when these materials (physical, chemical, or environmental) support the growth of microorganisms. These factors must be addressed properly so that the growth of microorganisms is inhibited. All the hurdles applied in the food industry must be above the critical limit. For example, in the case of pH and water activity, if the hurdles are below the critical limits, the growth of microorganisms will persist and cause problems. When these individual hurdles are present in combination, the growth of microorganisms may be inhibited at levels above the individual critical limits for microorganisms (Leistner 1992, 2000; Mukhopadhyay and Gorris 2014). The expert application of hurdles in food is important because of individual critical limits and interactions with the food matrix. Some hurdles have the ability to interact with food material and in turn reduce the availability of hurdles to microorganisms, e.g., chlorine-based hurdles can form secondary products such as chloramines, haloketones, trihalomethanes, haloacetic acids, and chloropicrins (Bilek and Turantaş 2013). In such cases, all the possible interactions and the optimal application of each hurdle could be addressed.

Currently, the modern food industry has a number of processes and preservative technologies that assist the acquisition of safe and stable foods. Some of these technologies (e.g., freezing, chilling, drying, modified atmosphere packing, vacuum packing, coating, antimicrobial packaging, curing, fermenting, acidifying, and adding preservatives, bacteriophages, or lysozymes) operate to inhibit or prevent the microbial growth, while some technologies (e.g., sterilization, pasteurization, ultrahigh pressurization, ohmic heating, electrolyzed water, organic acid treatment and irradiation) inactivate the microorganisms or reduce its access to food (Block 2001). The food industry has been engaged in finding a "golden bullet" technology to overcome all the issues associated with the current technologies and address all consumers' demands. Unfortunately, none of these preservative techniques have

Table 10.1 Different hurdles applied in foods

Category	Types
Physical hurdles	Sterilization, pasteurization, blanching, chilling and freezing, radiations, electromagnetic energy, photodynamic inactivation, ultrahigh pressure, ohmic heating, ultrasonication, packaging films, modified atmosphere packaging, aseptic packaging, food microstructure, antimicrobial coating, antimicrobial packaging
Chemical hurdles	Nitrite, nitrate, ozone, carbon dioxide, oxygen, organic acids, lactic acid, lactate, acetic acid, acetate, ascorbic acid, benzoic acid, sulphite, phosphates, glucono-δ-lactone, propylene glycol, ethanol, levulinic acid, chlorine, hypochlorous acid, monolaurin, calcium lactate, essential oil
Physiochemical hurdles	Low pH, low water activity, low redox potential, Millard reaction products, spices, herbs, lactoperoxidase, lysozyme, chelators, surface treatment agents, smoking, electrolyzed water
Hurdles derived from microbes	Probiotics (competitive hurdles), bacteriocins, antibiotics, bacteriophages
Other hurdles	Free fatty acids, chitosan and its derivatives, nisin

yet proven to be that technology (Mukhopadhyay and Gorris 2014). Many of the food processing technologies achieve some of the desired targets by inactivating the microorganisms and maintaining the food quality and safety. For instance, the use of irradiation technology has shown long-lasting and promising results in a variety of foods. However, it is a known fact that consumers as well as regulators are not persuaded of its safe application (Farkas and Mohácsi-Farkas 2011).

10.1.2 Mode of EW Either Alone or in Combination

The effect of EW on the inactivation of microorganisms is greatly influenced by a number of factors such as pH, ORP, flow rate of electrolytes, temperature, etc. The killing mechanism of EW can be attributed to the active chlorine species including Cl_2, HOCl, and $^{-}$OCl. Apart from active chlorine species, others such as reactive oxygen species (ROS) also contribute to the killing of microorganisms. The most active form of EW is HOCl, where HOCl attacks the cell wall, cell membrane, DNA, mitochondria and enzymes of microbial cells, which leads to cell death. However, there is little effect of ROS and $^{-}$OCl on microbial cells. Yet, the action mechanisms of EW underlying the microbial inactivation are still not fully understood. Ding et al. (2016) reported that EW caused injury to the cell wall, outer membranes and nucleus, which lead to the leakage of intracellular materials such as DNA, K+, and proteins and lower the dehydrogenase activities of *Staphylococcus aureus*. Liao et al. (2007) reported in a study that the high ORP of EW could remarkably influence the EW disinfection activity by allowing penetration of the outer and inner membranes. In addition to ORP, a reduced pH also significantly influences the disinfection power of EW (Waters and Hung 2014). When EW was applied alone to study the mechanism of action in *S. aureus*, the results indicated that EW caused leakage of intracellular potassium, 2,3,5-triphenyl tetrazolium chloride-dehydrogenase activity inhibition and ultrastructure disruption of the cell (Ding et al. 2016). The use of hurdle technology to enhance the antimicrobial activity of EW has already been reported by a number of researchers (Fig. 10.1), for instance, the use of fumaric acid in combination with EW against foodborne pathogens in pork (Mansur et al. 2015). The authors reported that the available chlorine concentration was the main factor in microbial inactivation over that of ORP and pH and further suggested microscopic tests and mathematical modeling to understand the basics of the interaction of the hurdles with bacterial cells and the mode of action of each hurdle.

Different hurdles have different modes of action against microbes. The organic acids become dissociated after penetration resulting in the reduction of pH, which impairs or inhibits the growth of microorganisms. Moreover, the anionic part of the organic acids accumulates and disrupts many metabolic functions and leads to death (Giannattasio et al. 2013). EW alone can eliminate or inhibit a portion of microbes in food, but it cannot satisfy the food safety and stability requirements under a low intensity condition. The hurdle enhancement of EW in combination with organic acid could eliminate or inhibit a large portion of microbes by using different mechanisms. The use of a single hurdle (chilling, preservative, or EW) may be compromised by some microbes; however, the use of many hurdles could provide a different mechanism of action that is not possible for a microbe to overcome. Manas and Pagán (2005) reported that the damage to microbial cells induced by EW is not necessarily sufficient to cause death, and sublethal injury might happen. Later, when the conditions become appropriate, these injured cells might revive, start growing and become pathogenic again, posing potential threats to food safety and public health. Interestingly, once the microorganism becomes vulnerable by one hurdle and loses the capacity to further resist, the second hurdle can easily neutralize the microbe. However, the selection of hurdles to accompany EW mainly depends on the composition of the food product and the expected variations in the level of the target microorganism. EW is mainly applied in the beginning of food processing, followed by other hurdle technologies along the food processing chain. When pork is washed with EW + fumaric acid, chilling and preservatives could be applied in the food chain to inhibit or kill the microbes present after the initial treatment to maintain the food quality and extend the shelf life.

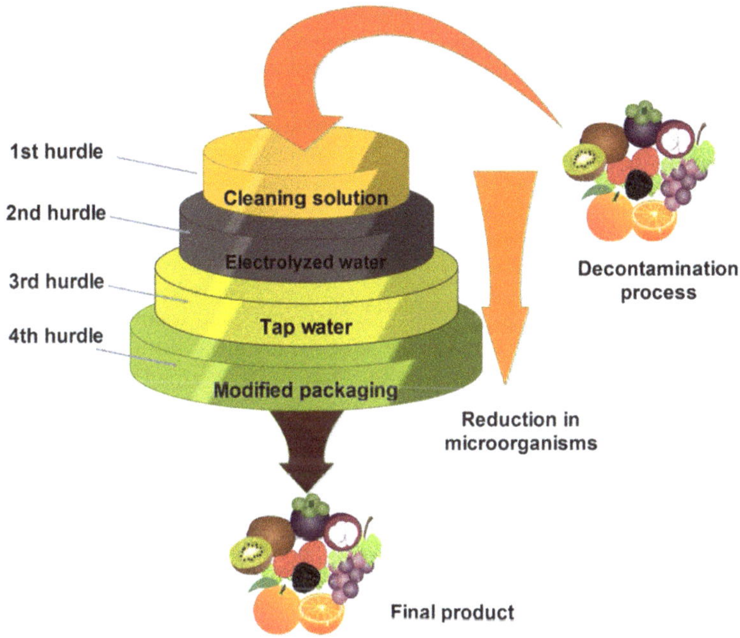

Fig. 10.1 Hurdle enhancement of EW with other techniques

10.1.3 Applications in Foods

It is well established that hurdle technology has been exploited by the food industry for microbial stability and safety of sustenance for many centuries. This concept, however, has unwittingly been traditionally used for food preservation, especially in developing countries (Leistner and Gorris 1995). The continuously increasing demand for fresh and improved-quality food reinvented the use of hurdle technology some 38 years ago for the preservation of meat and is considered highly favorable when compared to other technologies. EW, alone and in combination with other techniques, has shown promising results in controlling microbial growth in food and enhancing the shelf life. It is known that AEW significantly reduces the microbial count in food, but at the same time, adversely affects the organoleptic properties of food products. Therefore, NEW was developed to resolve the issue, but at the same time, the effectiveness of EW became reduced. Afterwards, scientists started using EW in combination with other decontamination technologies under optimal conditions to improve the microbial stability and safety of food along with an enhanced shelf life. Forghani et al. (2015) reported the use of EW alone and in combination with ultrasound against foodborne pathogens in kashk (a low-fat dried yogurt). The results indicated that a 2/1 EW/kashk ratio exhibited 1.42, 1.13, 1.24, and 1.37 log CFU/mL microbial reductions in *S. aureus*, *Bacillus cereus*, *Escherichia coli*, and *Aspergillus fumigatus*, respectively, at room temperature. A synergistic microbial reduction was observed when EW was combined with ultrasonication and resulted in 1.87, 1.67, 1.71, and 1.91 log CFU/mL reductions in *S. aureus*, *B. cereus*, *E. coli*, and *A. fumigatus*, respectively. It was suggested that the developed hurdle approach could be utilized for the sanitization of kashk and other similar products. Similarly, ultrasound improved the bactericidal activity of EW, which resulted in 1.77 (1.29) log reductions on total aerobic bacteria and 1.50 (1.29) log reductions on yeasts and molds for cherry tomatoes (strawberries). All the tested qualities of the cherry tomatoes and strawberries remained the same except the firmness of the cherry tomatoes decreased (Ding et al. 2015). In another study, Rahman et al. (2011) studied the combined treatment of EW and citric acid with mild heat for shredded carrots. The results indicated that at 50 °C, the 1% citric acid and EW treatment showed a reduction of 3.7 log CFU/g for the total bacterial count and the yeast and mold, as well as effective reductions in *Listeria monocytogenes* (3.97 log CFU/g) and *E. coli* O157:H7 (4 log CFU/g). The effectiveness of EW was improved with the use of citric acid and could be used in the minimally processed food industry, as it maintained the sensory and microbial quality of the fresh-cut carrots and enhanced the overall shelf life of the produce.

10.1.4 Types of Combination Processes of EW

Each inactivation treatment induces physiological or structural changes to microbial cells, independent of whether the cells are injured or completely destroyed. In the case that the microbial cells are not finally destroyed, the recovery of these changes may depend on whether the combined treatment is simultaneous or sequential. The combination effect is frequently relative to cellular recovery, damage, and hemostasis maintenance (13). The significant progress in the knowledge of the sanitizer's mechanisms of inactivation and damage and of microbial recovery and stress adaptation offer the possibility of planning novel hurdle treatments to improve the antimicrobial quality and safety. These concepts are helpful for determining the type of combination (simultaneous or sequential) that should be applied for a specific case. In the case of sequential hurdles, the application of the appropriate order is a challenge to effective hurdle treatment. However, it has been proven that the application of two sanitizers in the inverse order could render an antagonistic effect (Lee et al. 2003). EW has been applied as a combination in different ways including simultaneous and sequential treatments. Biological and chemical agents are often simultaneously combined with EW, since they can interfere with the properties of EW. While the sequential treatments in EW hurdle technology are commonly associated with physical and chemical agents, those are susceptible to affecting the properties of EW. An effective synergistic effect was observed when organic acids, chitosan, and nisin were simultaneously applied in a washing process with EW against foodborne pathogens on food products (Arevalos-Sánchez et al. 2012; Mansur et al. 2015; Tango et al. 2014a, b; Xu et al. 2014). Ultraviolet irradiation, calcium oxide, and multipulsed high pressure were successfully used in subsequent treatment with EW against foodborne pathogens on food products (Mikš-Krajnik et al. 2017; Tango et al. 2017; Wang et al. 2018a; b).

10.2 EW in Combination with Thermal Treatment

It has commonly been observed that preservative heat treatments cause a modification of the nutritional and sensory properties of food, and the EW inactivation effect at ambient temperature has often been scarcely reported (Condón et al. 2011; Tango et al. 2014a, b). Mild heat treatment has been reported to accelerate the sanitizer rate of microbial inactivation, enhance the physiological and sensory quality of food, and to be energy saving (Koseki et al. 2004; Liu et al. 2017; Tango et al. 2014a, b; Xie et al. 2012a, b). Several studies have been performed to understand the additional effect of heat on the potency of EW (Liu et al. 2017; Tango et al. 2014a, b), which has been explained. Mild temperatures (40–60 °C) induce an increase in the vapor pressure of the medium and disperse cell aggregates (Condón et al. 2011), and therefore, the cell surface area exposed to the EW solution increases and so does the efficiency of the combined treatment (Tango et al. 2017; Tango et al. 2014a, b). It has been demonstrated that the potential of EW is highly expressed at mild temperatures (40–60 °C). The mild temperatures can enhance the inactivation of EW-resistant microorganisms such as *S. aureus* and *L. monocytogenes* (Tango et al. 2014a, b; Tango et al. 2015), and the high chlorine concentration in EW can inactivate heat-resistant microorganisms such as *Alicyclobacillus acidoterrestris* and *B. cereus* spores (Park et al. 2009; Torlak 2014; Yamaner et al. 2016). Koseki et al. (2004) demonstrated the effect of temperature on the effectiveness of EW towards foodborne pathogen reductions using heads of lettuce inoculated with *Salmonella* and *E. coli* O157:H7 and washed with EW (pH 2.6, chlorine concentration of 40 ppm) at different temperatures for 1 min. A reduction of 0.6–0.9 log 10 CFU/g for *E. coli* O157:H7 and Salmonella was observed for 1 min EW exposures at 4 and 20 °C, respectively, results that were not significantly different from that obtained with DW. However, when the temperature was increased to 50 °C, a significant bacterial reduction of 2.7–3.0 log CFU/g for both pathogens was found after washing with EW. The heated EW treatment was also evaluated for natural microflora on the fresh produce. The low-concentration EW (4 mg/L and pH 2.3) combined with heat (50 °C) resulted in a higher (2 log CFU/g) reduction in the count of aerobic mesophilic bacteria, yeasts, and molds on fresh broccoli (Liu et al. 2017).

Several studies have demonstrated the effect of heated EW in the inactivation or decontamination of spoilage and pathogenic bacteria on fresh produce (Koide et al. 2011; Mansur et al. 2015; Ngnitcho et al. 2018; Tango et al. 2014a, b; Tango et al. 2015; Xie et al. 2012a, b). The scientific investigation of these results showed that treatment with mildly heated SAEW was more effective than that obtained with AEW. Treatments involving mild-temperature AEW have the severe problem that chlorine gas easily volatilizes at higher temperatures, which might lead to a reduction in the effectiveness of EW during washing/processing produce. AEW must be applied at the lowest possible temperatures to prevent the volatilization of chlorine gases in EW treatments (Koseki et al. 2004). Moreover, an increase in temperature is more effective than an increase in chlorine concentration. For instance, increasing the concentration from 51 ppm (1.00 log CFU/g) to 78 ppm (1.02 log CFU/g) did not significantly affect the reduction of *Vibro parahaemolyticus* inoculated on shrimp at similar pH values of 2.44 and 2.39, respectively. However, an increase in temperature from 20 to 50 °C showed a significant reduction of 1.00–3.11 log CFU/g (Xie et al. 2012a, b).

10.3 EW in Combination with Chemical Treatments

Chemical decontamination techniques, aiming at improving the microbiological safety and quality of food, are usually implemented at the postharvest level and at different points of the food processing chain alone or in combination with other intervention technologies. EW has already been used in combination with a number of chemicals, such as organic acids, essential oils, calcium oxides, benzalkonium chloride, ozone, carbon monoxide, etc., for the decontamination of food products and food contact surfaces (Vázquez-Sánchez et al. 2014). Some chemicals, such as benzalkonium chloride, in combination with EW have been used for disinfecting food contact surfaces. The selection of chemicals to be used along with EW depends on various factors, such as the time of exposure to food products with EW, the food matrix, treatment conditions (ACC, ORP, pH, and temperature), the interval between contamination and decontamination, etc. It has been reported that longer time intervals between contamination and decontamination and higher temperatures allow firm attachment of bacterial cells to the fresh produce, thus making it harder to decontaminate (Lianou et al. 2012). In the following sections, we will describe the use of different chemicals in combination with EW for the use of food decontamination.

10.3.1 EW in Combination with Organic Acids

Organic acids are weak acids and have a wide pKa range; the lowest pKa value (3) has been reported for carboxylic and the highest (9) has been recorded for phenolic acid (Deruiter 2005; Liptak et al. 2002). The use of organic acids such as acetic acid, lactic acid, citric acid, and fumaric acid is extensive in food decontamination both alone and in combination with other intervention technologies, including EW. Although, the bactericidal mechanism is not fully understood (Ricke 2003), depending on the physiological condition of the microorganisms and the physiochemical characteristics of the external medium, organic acids are capable of demonstrating both bacteriostatic and bactericidal properties. Most of these compounds are weakly acidic in nature; therefore, pH is considered as a primary determinant of activity, because it affects the concentration of undissociated acid. The undissociated forms of organic acids can easily enter through the lipid membrane of the microbial cell, and once internalized in the neutral pH of the cell cytoplasm, dissociate into anions and protons (Davidson 2001; Ngnitcho et al. 2018; Ricke 2003).

Lactic acid is a naturally occurring organic acid and is generally used as a food preservative and decontaminating and flavoring agent because of its GRAS property (Oh and Marshall 1993). The use of lactic acid on food products such as meat and poultry has already been investigated (Woolthuis and Smulders 1985; Zeitoun and Debevere 1990). Lactic acid in combination with EW has also been investigated with food produce. Wang (2011) reported the combination of lactic acid and EW at various concentrations and dipping times to decontaminate grape tomatoes and baby carrots. The results showed a significant difference between the combined treatment of lactic acid and EW and that of lactic acid or EW alone. By increasing the dipping time and/or concentration, the microbial reduction with the hurdle treatment was high compared to that of lactic acid or EW alone. This can be explained by the fact that combinations of chemical disinfectants may perform better in both the sensory and microbial qualities of products, due to possible synergistic effects during treatment (Rahman et al. 2010).

Among the organic acids used as antimicrobial agents on meat and meat products, fumaric acid (FA) has demonstrated a strong bactericidal activity when compared to that of lactic and acetic acid (Mansur et al. 2015; Podolak et al. 1995; Podolak et al. 1996). We reported the sanitizing effectiveness of EW and FA at different dipping temperatures (25–60 °C), times (1–5 min), and concentrations (5–30 ppm for EW and 0.125–0.5% for FA) on pure cultures of *S. aureus*, *L. monocytogenes*, *E. coli* O157:H7 and *S.* Typhimurium (Tango et al. 2015). For instance, FA at 0.25% exhibited the highest sanitizing effect, yielding complete inactivation of *E. coli*, *S.* Typhimurium, *L. monocytogenes*, and *S. aureus*, and reduced viability by 3.95–5.76 log CFU/mL at 25–60 °C after 1 min of treatment. However, EW complete inactivation was obtained with the highest ACC reading (20 ppm) at 40 °C for 3 and 5 min. These results indicated suitable treatment conditions for the FA and EW treatment against foodborne pathogens. The effectiveness of the EW and FA combination against foodborne pathogens in pork meat was investigated. The shelf life of pork meat was extended by 6 and 4 days at 4 and 10 °C, respectively (Mansur et al. 2015). Recently, Tango et al. (2017) investigated the hurdle enhancement of SAEW with FA, calcium oxide, ultraviolet irradiation, ultrasonication, and microbubbles with or without mild heat treatment. The results showed that the microbial decontamination efficacy of SAEW increased by adding a second and/or third treatment. The addition of ultrasonication to SAEW+FA was more effective in terms of microbial reduction compared to that of the microbubble treatment. The microbial reductions were significantly more pronounced when SAEW+FA was used with CaO and the ultrasonication treatment. However, at the same time, the use of ultrasonication adversely affected the quality of tomatoes. It was suggested that the use of CaO followed by SAEW+FA could significantly reduce the microbial count while maintaining the quality of apples and tomatoes. Ngnitcho et al. (2018) also reported the effectiveness of hurdle enhancement of EW by FA in soybean sprouts, and it was found that the combined SAEW+FA and SAEW+FA+ultrasonication treatments were more effective compared to that of single treatments, maintaining the quality of the food at the end of storage (7 days).

The effect of citric acid (an organic acid) in combination with EW has already been studied in the food industry as a sanitizing solution. Rahman et al. (2011) reported the hurdle treatment of AlEW and citric acid with mild heat to ensure the microbial safety, shelf life, and sensory quality of shredded carrots. The citric acid (1% concentration) and AlEW treatment at 50 °C exhibited an approximately 3.97 log CFU/g reduction in the total bacterial count and yeast and molds. This treatment also reduced *L. monocytogenes* by 3.97 log CFU/g and *E. coli* O157:H7 by 4 log CFU/g. It was suggested that combinations of AlEW and citric acid could be used as a sanitizer, as it maintained the sensory and microbial quality of the fresh-cut carrots and enhanced the overall shelf life of the fresh produce. Due to the green applications of EW, various combinations have been tried to address the consumer demand for fresh-like food without chemical preservatives. *B. cereus* is a Gram-positive, spore-forming bacterium, causing emetic and diarrheal food poisoning, and could contaminate cereal grains in the processing units from farm to fork (Park et al. 2009). An outbreak of *B. cereus* infection has been directly linked to the consumption of contaminated cereal grains. In this regard, the combined effect of EW and 1% citric acid against *B. cereus* vegetative cells and spores has been evaluated. The results indicated that the combined action of EW and 1% citric acid was considerable in inhibiting *B. cereus* on cereal grains and could be used as a sanitizer in various food products (Park et al. 2009).

10.3.2 EW in Combination with Other Chemicals

Hence, it is an established fact that the effectiveness of EW improves when combined with other decontamination technologies. Recently, the hurdle enhancement of EW with sodium chloride (2%) was able to maintain the quality of fresh-cut 'Nanglae' pineapples and could be used to extend the shelf life of pineapples (Lasomboon et al. 2016). In another study, EW in combination with carbon monoxide gas treatment improved the quality and freshness of tuna steaks, extending the shelf life under refrigeration storage (Huang et al. 2006; Rasco and Ovissipour 2015). Recently, Zhang and Yang (2017) studied the effect of EW with H_2O_2 on fresh-cut lettuce and reported that the combination of EW with 1% H_2O_2 achieved 1.69 and 0.96 log CFU/g reductions in aerobic mesophilic counts and yeasts and molds, respectively. It was suggested that 1% H_2O_2 combined with 4 mg/LEW could be a promising approach for treating fresh-cut lettuce.

10.4 Electrolyzed Water in Combination with Nonthermal Techniques

Hurdle technology is the intelligent combination of different technologies that achieves the maximum reduction or elimination of microorganisms on food products with reduced quality deterioration. The efficient application of hurdle technology requires the use of antimicrobial agents at an optimum level to ensure the microbial safety and stability of the final product. The application of hurdle technology involving EW and nonthermal technologies has become more prevalent now because in food preservation, the quality of the final product is as important as microbial reduction. Moreover, the understanding of the mechanisms behind nonthermal technologies and their influence on the physiology and behavior of microorganisms has increased (Singh and Shalini 2016).

Nonthermal methods include high pressure (high hydrostatic pressure and high-pressure processing), gases (cold plasma, ozone, and carbon dioxide), ionizing radiation (gamma irradiation and electron beam), and light (ultraviolet and pulsed light) and are the physical forces that have the potential to inactivate microorganisms at ambient or sublethal temperature while maintaining the sensory and nutritional qualities of fresh produce. Some of these technologies such as ultrasonication, high pressure, and pulsed electric field (PEF) need mild heat for generating internal energy (Pereira and Vicente 2010). The other methods including cold plasma and irradiation are considered to be purely nonthermal (Li and Farid 2016). The advantages of nonthermal technologies are that, in general, they use less energy than conventional thermal processes and are nontoxic and environmentally friendly. However, it is known that the nonthermal treatments cause sublethal injury to cells. The injured cells might be able to recover from the damage and begin to survive when the storage conditions are suitable. Thus, the application of hurdle technologies based on the use of nonthermal technology in combination with a sanitizing agent capable of interfering with the sustainability of injured cells is possible (Condón et al. 2011). Recently, tremendous attempts and progress have been made in the combination of EW and nonthermal technologies. In recent years, the satisfactory results of this combination have led to a global trend of increased scientific research in the fresh produce industry (Ma et al. 2017; Oliveira et al. 2015). This information becomes important since the physical-based preservation can disrupt the cell structure and possibly facilitate the penetration of hypochlorous acid into the cytoplasmic membrane (Li et al. 2017; Tango et al. 2014a, b). Each nonthermal technology has its advantages and disadvantages, but in many cases, the combination with EW is essential to prolonging shelf life by microbial inactivation and to improving the quality of food products. The combination of EW and nonthermal technologies may appear to be the best method to reach the results that nonthermal treatment individually has not been able to accomplish (Morris et al. 2007).

10.4.1 EW in Combination with Ultrasonication

Ultrasonication is the most widely applied nonthermal technology in combination with EW. The application of ultrasonication in the food industry depends on the process and power and frequency used. Low-power (<1 W/cm^2) and high-frequency (10.01 MHz) ultrasonication can easily pass through a media without damaging the materials (Barbosa-Cánovas et al. 2005; Patil et al. 2013). Such powers and frequencies are often used to improve the physicochemical characteristics of dairy, meat, and poultry products (Carrillo-Lopez et al. 2017). The combination of EW and high-power (10–1000 W/cm^2) and low-frequency (20–100 kHz) ultrasound is successfully used in food processing to improve microbial inactivation (Barbosa-Cánovas et al. 2005; Carrillo-Lopez et al. 2017), maintain the quality of produce (Ding et al. 2015), enhance the germination of seeds and γ-aminobutyric acid accumulation on cereal (Kim et al. 2006; Liu et al. 2013), and assist postharvest disease control (Khayankarn et al. 2013). The range of published combined EW and ultrasonication applications is summarized in Table 10.2, and the descriptions of these applications in more detail are given in the following sections.

The effectiveness of combinations of EW and ultrasonication was intensively studied under in vitro and in vivo conditions. The in vitro investigation of SAEW combined with ultrasonication on the inactivation of foodborne pathogens has been studied to obtain a better understanding of the killing mode and the synergetic effect of both technologies. This investigation was based on the dynamic changes in the

Table 10.2 Experimental procedures applied for investigation of the effectiveness of electrolyzed water combined with ultrasonication treatment

Objective	Treatment condition	Type of combination	Microorganisms	Fresh produce	References
Mechanism of combined SAEW–US treatment.	SAEW (2 mg/L, pH: 6.1) + US (400 W) for 1–10 min at 20 °C	Simultaneous	*S. aureus*	*In vitro* study	Li et al. (2017)
SAEW-US on decontamination, germination, γ-aminobutyric acid accumulation of brown rice.	AlEW (pH 8.9) soaking 8h and AlEW–US (100 W) treatment.	Simultaneous	Total aerobic count	Brown rice	Liu et al. (2013)
Synergetic effect of EW–US to control post-harvest disease.	AEW (100 mg/L) + US (1 MHz) for 10 min. Packaged and stored at 13 °C for 20 day.	Simultaneous	Fusarium sp.	Pineapple	Khayankarn et al. (2013)
Synergetic effect of EW–US for microbial remove and impacting quality factors.	1kg fruits treated with 10 L SAEW (33.67 mg/L) + US (240 W) for 10 min.	Simultaneous	Total aerobic bacteria Yeast and mold	Tomato and strawberry	Ding et al. (2015)

(continued)

Table 10.2 (continued)

Objective	Treatment condition	Type of combination	Microorganisms	Fresh produce	References
Synergetic effect of SAEW, US, mild heat and developing survival model.	1. SAEW (30 mg/L) – US (400 W) + DW at 40 °C for 6 min. 2. SAEW (30 mg/L) + US (400 W) at 40 °C for 3 min.	Subsequent	*E. coli* O157:H7	Fresh kale	Mansur et al. (2014)
Synergetic effect of SAEW, US, mild heat and developing survival model.	SAEW (30 mg/L) + US (400 W/L) for 3 min.	Simultaneous	*B. cereus*	Brown rice	Tango et al. (2014a, b)
Synergetic effect of SAEW and US at different time and temperature and developing inactivation model.	SAEW (30 mg/L) + US (40 kHz) at 40–60 °C for 1, 3, 5 min.	Simultaneous	*L. monocytogenes* *S.* Typhimurium	Bell pepper	Luo and Oh (2016)
Synergetic effect of SAEW, US, furamic acid (FA) at different temperature and developing inactivation model.	SAEW (30 mg/L) + FA (0.5%) + US (400 W) at 23, 30, and 40 °C for 3 min.	Simultaneous	*L. monocytogenes* *E. coli* O157:H7	Soybean sprouts	Ngnitcho et al. (2018)

(continued)

Table 10.2 (continued)

Objective	Treatment condition	Type of combination	Microorganisms	Fresh produce	References
Synergetic effect of SAEW and US on microbial reduction.	SAEW (30 mg/L) + US (400 W) at 23±0.2 °C for 3 min	Simultaneous	Total bacteria Count Yeast and mold *E. coli* O157:H7 *L. monocytogenes*	Lettuce Sesame leaf Cabbage Spinach	Forghani and Oh (2013)
Synergetic effect of SAEW and US at different pHs, free chlorines, mechanical agitations.	Neutral EW (155–200 mg/L) + US () for 15 or 30 min with agitations (40, 60, 100 rpm).	Simultaneous	*S.* Typhimurium *E. coli* O157:H7	Tomato Lettuce	Afari et al. (2016)
Synergetic effect of SAEW and US at different orders of treatments (LcEW and US), times, temperatures, and US powers.	LcEW (5–10 mg/Ln pH 6.8)–US (100–400 W/L) or US (100–400 W/L)–LcEW ((5–10 mg/Ln pH 6.8) for 1–5 min at 24±2 °C for each step.	Subsequent	Total bacteria count *E. coli* O157:H7	Lettuce	Forghani et al. (2013)
Comparison of effectiveness of SAEW–US and sodium chlorite and sodium hypochlorite.	SAEW+US, sodium chlorite (100 mg/L), and sodium hypochlorite (100 mg/L) at 40 °C for 3 min.	Simultaneous	Aerobic mesophilic bacteria *Enterobacteriaciae* *Pseudomonas* spp. Yeast and mold *L. monocytogenes*	Fresh cut-kale	Mansur and Oh (2015)
Synergetic effect of SAEW and US at 40 °C for 3 and 5 min.	SAEW (pH 5.3–5.5, 20–22 mg/L) – US (400 W/L)	Simultaneous	*E. coli* O157:H7*S. aureus* *B. cereus* *A. funmigatues*	Kashk	Forghani et al. (2015)

(continued)

Table 10.2 (continued)

Objective	Treatment condition	Type of combination	Microorganisms	Fresh produce	References
Synergetic effect of SAEW and US for microbial decontamination	NEW (pH 7.0, 4 mg/L) 10 min –US (80 W) for 10 min.	Subsequent	*E. coli* *Pichia pastoris* *A. pullulans*	Stainless steel coupons	Zhao et al. (2017)
Comparison of effect of UV–AEW–US to UV–AEW and UV–US	AEW (65 mg/L, pH 2.2) 1 min – UV (254 nm) 1 min – US (200 W/L) 5 min	Subsequent and Simultaneous	Total viable count Yeast and mold Coliforms *L. monocytogenes*	Raw salmon	Mikš-Krajnik et al. (2017)
Comparison of additive effect of US, UV, Microbubbles on CaO–SAEW+FA.	SAEW (30 mg/L) + FA (0.5) 3 min followed by US (400 W/L, 3 min), UV (254 nm, 10 min), or Microbubbles 3 min.	Subsequent and Simultaneous	Total aerobic bacteria *L. monocytogenes* *E. coli* O157:H7	Apple Tomato	Tango et al. (2017)

AEW: Acidic electrolyzed water; AlEW: Alkaline electrolyze d water; SAEW: Slightly acidic electrolyzed water; NEW: neutral electrolyzed water; LcEW: Low concentration electrolyzed water; FA: Fumaric acidic; CaO: calcium oxide; US: ultrasonication; UV: ultraviolet. (+): simultaneous and (–): subsequent treatment physiological characteristics of bacterium cells after a treatment using flow cytometric analysis, scanning electron microscopy, and transmission electron microscopy (Li et al. 2017; Liao et al. 2017). It has been reported that SAEW and ultrasonication alone induced damage to *S. aureus* cells, and this damage depended on the treatment time. The single-treatment time should be long enough to induce cell death, while the damage to cells caused by their combination was permanent and irreversible after a short time. As shown by the electron microscopy analysis, morphological changes due to the combined treatment reveal that greater damage to cells and the important release of cell contents were observed compared that from the single treatment (Li et al. 2017). The simultaneous treatment of a pure culture of *S. aureus* with SAEW (2 mg/L) and ultrasonication achieved a reduction of 3.06 log CFU/mL after a 10-min treatment (Li et al. 2017). However, a significant difference appeared when the experiment was performed under in vivo conditions. The most frequently studied combination of ultrasonication and EW improved the safety profile of food produce and achieved the target 3–4 log reduction in spoilage and foodborne bacteria with SAEW containing an ACC of approximately 30–40 mg/L (Forghani and Oh 2013; Luo and Oh 2016).

EW combined with ultrasonication treatment has been shown to improve the germination process and γ-aminobutyric acid accumulation of sprouts. An ultrasonic pretreatment with AlEW followed by AEW treatment has been found to promote a reduction in total microbial count on the surface of brown rice, the length of sprouts, and γ-aminobutyric acid accumulation compared to the single treatment on the 4th day of germination (Liu et al. 2013). The acidic environment induced by AEW might enhance the activity of glutamic acid decarboxylase, enzymes catalyzing alpha-decarboxylation of glutamate (Watchararparpaiboon et al. 2010). It has to be noted that the γ-aminobutyric acid content on germinated brown rice depends more on the acidity than the chlorine concentration of EW (Liu et al. 2013). Another beneficial effect of the EW combined with ultrasonication treatment was investigated for the incidence of postharvest disease in pineapple caused by fungus. The combination treatment of EW (100 ppm and pH 2.3) and ultrasonication significantly inhibited the decay incidence and prolonged the shelf life of the fruit for 20 days without an impact on the quality of pineapple fruit (Khayankarn et al. 2013).

10.4.2 EW in Combination with Ultraviolet Irradiation

UV-C wavelengths of 200–280 nm have demonstrated an antimicrobial effect, since these wavelengths are near the peak efficiency of UV absorption by DNA (Bintsis et al. 2000; Kim et al. 2013). The inactivation mechanism of UV illumination is due to the cross-linking of pyrimidine dimers in DNA, which inhibits microbial reproduction. The application of UV has been limited because of the weak penetrability of UV wavelengths and shadowing effects (Lim and Harrison 2016). UV radiation might be applied in combination with EW to improve the safety and quality of fresh produce.

The combination treatment of EW and ultraviolet radiation has been investigated for microbial reduction and survival, changes in bioactive components, and the physicochemical characteristics and sensorial quality of raw salmon fillets and date palm (Jemni et al. 2014; Mikš-Krajnik et al. 2017). Mikš-Krajnik et al. (2017) evaluated the decontamination of natural microbiota and the inactivation of *L. monocytogenes* on raw salmon fillets using acidic electrolyzed water and ultraviolet light. Combining 5 min of UV light (254 nm from a distance of approximate 8 cm) followed by AEW at 65 mg/L of available chlorine and pH 2.6 for 1 min resulted in reductions of 0.43 and 0.58 log CFU/g for the total viable count and *L. monocytogenes*, respectively. The combined treatments were effective compared to the corresponding individual treatments (Mikš-Krajnik et al. 2017). The combined NEW and 6.22 kJ/m^2 UV treatments were effective against total mesophilic bacteria, total coliform bacteria, and yeasts and molds on date palm with a reduction of approximately 1.05, 0.82, and 1.33 log CFU/g, respectively (Jemni et al. 2014). The phenolic concentration increased slightly during storage at 20 °C, especially for untreated dates compared with those treated by EW combined with UV–C light. These results demonstrated that the UV–C treatment combined with EW washing stabilized the phenolic concentration during storage of dates at 20 °C. Moreover, the date fruits treated with UV–C light combined with AEW maintained a better color compared to that of the fruits observed as a control and individually treated (Jemni et al. 2014).

10.4.3 EW in Combination with High-Pressure Processing

The lethal effect of high-pressure processing (HPP) on spore-forming bacteria has been investigated, and the results demonstrated that HPP is an effective technology for the inactivation of vegetative bacterial cells; however, its antibacterial effect on spores is limited (Akasaka and Matsuki 2015; Wang et al. 2017a, b). Pressure levels of approximately 100–200 MPa induce receptors depending germination, while high-pressure levels of approximately 500–600 MPa can directly cause the germination of spores independent of germinant receptors (Akasaka and Matsuki 2015; Luu-Thi et al. 2014; Setlow 2003). Low-pressure levels during HPP can be assisted by EW to enhance the sanitization effect and improve the physicochemical properties of fresh produce (Wang et al. 2018a, b).

Wang et al. (2017a, b) investigated the effect of combining high HHP and SAEW treatment on the inactivation of a pure culture of *B. cereus* spores. The authors treated *B. cereus* in an ethanesulfonic acid buffer (pH 6.1) or SAEW (24, 35, 44 and 55 mg/L) using a multicycle HHP treatment. The effective conditions for inactivation of *B. cereus* spores were found at 300 MPa HPP + SAEW with 44 mg/L ACC or 200 MPa HPP + SAEW with 44 mg/L ACC + 500 MPa HPP, resulting in reductions of 3.27 and 3.99 log CFU/mL, respectively. The results indicated that the 300 MPa HPP + SAEW treatment might involve non-nutrient germination processes. A pressure of 200 MPa was sufficient to induce germination and allowed the EW to inactivate the vegetative cells. Moreover, EW might cause oxidative damage to the spore's protein during the first cycle, which facilitates germination at 200 MPa (Cortezzo et al. 2004), and then the germinated spores were easily destroyed by the second cycle of HPP (Reineke et al. 2013; Wang et al. 2017b). These optimal conditions of multipulsed high-pressure-assisted SAEW on the microbial and physicochemical properties of mud snails were also evaluated (Wang et al. 2018a). The results showed that the multicycle treatment (200 MPa HPP + SAEW with 44 mg/LACC + 500 MPa HPP) was effective, with a reduction of approximately 5 logs of psychotropic bacteria and total microbial counts. The HPP-SAEW treatment did not induce significant changes in essential free amino acids, but a decrease of 1.679 mg/g dry weight in the nonessential free amino acids. In addition, there was no significant change in the chewiness and resilience when mud snail samples were subjected to the HPP-SAEW treatment (Wang et al. 2018a).

The combined HPP and SAEW treatment was studied to understand the germination and inactivation pathways of *B. cereus* spores. The authors used label-free quantitative proteomic, gene ontology, and KEGG pathway analyses to comprehensively investigate the impact of the HPP combined with SAEW treatment on protein expression changes (Wang et al. 2018b). The change in the spore's structure after HPP-SAEW treatment was also investigated using imaging techniques, including TEM, SEM, SMCM combined with flow cytometric (FLM) analysis, and DPA determination (Wang et al. 2017a, b). The proteomic analysis of *B. cereus* spores subjected to HPP–SAEW demonstrated that the protein degradation, signaling, metabolic, and biosynthesis pathways were involved in HPP–SAEW-mediated spore inactivation. Quorum sensing and amino acid-biosynthetic routes were the main KEGG pathways mediated by SAEW under HPP conditions. Specifically, the phosphorylation of the Spo0A transcription factor was mediated by downregulation of the ABC transporter ATP-binding protein, which was associated with inactivation of *B. cereus* spores. Therefore, the reduced phosphorylation level was the key factor that influenced *B. cereus* spore inactivation of the HHP–SAEW treatment. (Wang et al. 2018b). The morphology of *B. cereus* spores exposed to HPP–SAEW treatment showed a large fraction of damage and an incomplete exosporium. Additionally, spores subjected to the HPP–SAEW treatment had irregular surface bumps and a critical distortion. Approximately 90% of DPA was released in the groups treated by HPP-SAEW. The FCM and SLCM analyses showed that the HPP-SAEW treatment was not mainly dependent on germination, but rather, directly killed. Therefore, the inactivation mechanism of the HPP–SAEW treatment was possibly that one of the proteins was particularly susceptible to the combined HPP-SAEW treatment (Wang et al. 2017a, b).

10.4.4 EW in Combination with Modified Atmosphere Packaging Technology

The evaluation of the posttreatment effect of EW treatment often demonstrated the dynamics of food microbiota associated with a decrease in the organoleptic characteristic of fresh produce during storage (Botta et al. 2018). To avoid posttreatment microbial survival, the combination of EW treatment with modified atmosphere packaging technology has been applied to limit the complex dynamics of spoilage and pathogenic bacteria and the associated decrease in quality of meat, shrimp muscle, and sweet cherries (Eda and Seckin 2015; Posada-Izquierdo et al. 2014; Zhang et al. 2015). Modified atmosphere packaging containing a concentration of 40–80% CO_2 is effective in the preservation of fresh produce safety and quality, primarily due to CO_2 retardation of growth by extension of the survival lag phase of microorganisms during storage in MAP containing high CO_2 levels (Qian et al. 2013; Zhang et al. 2015). The CO_2 concentration inside food packages plays a particularly important role in decreasing the respiration rate, reducing the production of ethylene, and retarding senescence (Eda and Seckin 2015; Izumi and Inoue 2018).

Water ice-glazing and freezing storage inhibit microbial growth and slow down enzymatic activity and thus are common methods for the long-term preservation of shrimp products. However, water ice-glazing or freezing processes cannot completely inactivate the bacteria on fresh shrimp (Wang et al. 2014). A negative effect on the quality of products at the end of the freezing storage of shrimp has also been observed (Gonçalves and Junior 2009; Soares et al. 2013). Zhang et al. (2015) studied the potential of weakly acidic electrolyzed water (WAEW) ice-glazing (pH 6.4–6.6 and an ACC of 6.4–6.5 mg/L) combined with MAP on the quality of peeled shrimp during frozen storage. The chemical analysis results demonstrated that combining WAEW ice-glazing and MAP generated a great synergistic effect of antibacterial and antioxidant properties capable of maintaining lower TMA, TBARS, and TVBN values in frozen samples. The combined treatment successfully retarded the degradation of the physical structure of shrimp muscle and maintained the stability of color during frozen storage (Zhang et al. 2015). The results of another study indicated that an EW concentration between 25 and 100 mg/L combined with passive atmosphere packaging was effective at extending the shelf life of sweet cherries (Eda and Seckin 2015). Another study investigated the effect of CO_2 in controlled atmospheres on sublethally injured coliform bacteria during the refrigerated storage (5 and 10 °C) of shredded cabbage (Izumi and Inoue 2018). Washing with EW (pH 2.7 and 25 ppm available chlorine) resulted in a reduction of 0.4–1.1 log and caused 42–77% sublethal injury to cells of spoilage microflora on shredded cabbage. A significant reduction of coliform growth during storage at 5 °C was observed on shredded cabbage treated with EW and stored under MAP with CO_2 of 10 and 15% compared to that of shredded cabbage unwashed with EW. The injury slightly increased to 49–84% of coliform cells present on samples treated with EW–MAP at the end of storage. The degree of injured cells of coliform bacteria on shredded cabbage during storage was significantly dependent on the CO_2 atmosphere and temperature tested (Izumi and Inoue 2018).

The survivors of a pure culture of *E. coli* O157:H7 cells after NEW (80 mg/L free chlorine) treatment were exposed to different atmospheres (0, 30, and 60% CO_2 balanced with N_2, and air) on tryptone soy agar yeast extract and stored at 10–15 °C for 24 h. The survival data revealed that the fraction of sublethally injured *E. coli* O157:H7 continuously underwent an additional inhibitory effect during the storage period under modified atmosphere conditions. The observed extension of the lag growth phase was a direct consequence of the prior sublethal injury. The effects of the NEW–MAP treatment were highly pronounced in comparison to that of chlorine dioxide (50 mg/L free chlorine) combined with MAP (Singh and Shalini 2016). The effect of storage under modified atmosphere packaged lettuce on the survival of *E. coli* O157:H7 subjected to NEW (50 mg/L free chlorine and a pH of 6.5) treatment was also investigated at different temperatures. Subsequently, a survival of *E. coli* O157:H7 model was developed to describe the combined treatment effect (Posada-Izquierdo et al. 2014). A wash processing for 30 s performed with NEW resulted in a reduction of 1.0 log CFU/g of *E. coli* O157:H7 on lettuce. The surviving *E. coli* O157:H7 population was able to grow at 8–16 °C, and the viability decreased below 1.1 log CFU/g in the first 15 days at 4 °C. At 8 °C, *E. coli* O157:H7 exhibited a long lag phase of 19 days. A comparison with existing data in the literature indicated that the growth predicted by the model for *E. coli* O157:H7 treated with EW-MAP was lower than those observed with other sanitizing treatments or packaging conditions (untreated product, heat treatment, chlorinated water, etc.). These results indicated that the NEW–MAP treatment affected the physiological state of *E. coli* O157:H7 cells and extended the lag time beyond the typical shelf-life time of this type of product, especially during storage at 8 °C (Posada-Izquierdo et al. 2014).

10.5 EW Combination with Treatments

Biological sanitizing agents, such as chitosan and nisin, have been applied in combination with EW because of their antibacterial and antioxidative activities, film-forming characteristics, biodegradability and biocompatibility, nontoxic nature, and solubility under acidic conditions (Chang et al. 2015; García-Almendárez et al. 2008; Zhou et al. 2008). Several studies have been performed to investigate the potential of combined treatments of EW and chitosan or nisin to improve the quality and safety of fresh produce (Arevalos-Sánchez et al. 2012; Brychcy et al. 2015; Chen et al. 2016; Luan et al. 2017; Xu et al. 2014; Zhou et al. 2011). The changes in the myofibrils during refrigerated storage were studied to determine the mechanism of the changes in the texture of puffer fish flesh subjected to the combined treatment of EW and chitosan (Zhou et al. 2011). Wash processing with EW (pH 6.1 and ACC 21 mg/L) followed by immersion in a chitosan coating solution showed that the combined treatment was more efficient than that of single treatments at inhibiting bacterial growth and maintaining the firmness, trimethylamine, total volatile basic nitrogen, and thiobarbituric acid. The myofibril degradation was studied by the analysis of myofibril length, and it was observed that the lengths on samples treated with EW–chitosan were longer those on samples subjected to single treatments. This result indicated that the combined treatment was effective at inactivating certain enzymes including trypsin, calpain, lipase, protease, and DNase, which are often associated with the shortening of myofibrils in fish muscle (Taylor et al. 1997). Moreover, the microscopic analysis of the EW–chitosan-treated flesh demonstrated better myofibril integrity, which might contribute to maintaining the physicochemical properties of the obscure puffer fish (Zhou et al. 2011). Other studies have also shown that the EW–Chitosan treatment was effective at inhibiting the protein decomposition and lipid oxidation on fish flesh during refrigerated storage, therefore extending the shelf life by 7–10 days compared to single-treatment results (Luan et al. 2017; Xu et al. 2014). E. Brychcy et al. (2015) proposed a new type of edible biopolymeric film containing acidic EW for improving the microbial and sensory quality of fresh produce. They prepared and investigated the physicochemical characteristics of chitosan–hydroxypropyl methylcellulose (HPMC) films incorporated with lysozyme and AEW. The results showed that the presence of EW in the film decreased the solubility, increased the concentration of the sodium chloride solution and extended the electrolysis time, which increases the elasticity of the biopolymeric films (Wijffels et al. 1996). The fusion of AEW and chitosan–HPMC may provide better incorporation with coated fresh produce and a multidirectional protection effect, including microbial, physicochemical, and sensory qualities (Brychcy et al. 2015). A study investigated the potential effect of NEW and nisin on the inactivation of *L. monocytogenes* biofilms on stainless steel and glass surfaces and listeriolysin O activity from planktonic and biofilm cells (Arevalos-Sánchez et al. 2012). The authors reported that the combination of neutral EW (60 mg/L)–Nisin (6976 IU/coupon) resulted in higher planktonic and biofilm cell reductions than did the single treatments at 37 °C for 15 min on both surfaces. However, biofilm cells subjected to sublethal NEW concentrations induced the highest pore-forming toxin lysteriolyin O activity (11 HU/mg), which indicated a more potentially virulent phenotype than nontreated biofilm or planktonic cells. They concluded that changes in the biofilm development environment may be responsible for the activation of the transcriptional regulator PrfA, which in turn influences the transcription of virulence factors such as listeriolysin O (Leistner and Gorris 1995).

10.6 Combination of Alkaline and Acidic EW During Decontamination Processes

Several studies have reported the reduction of EW antimicrobial efficacy in contact with organic materials, especially proteins and lipids, during decontamination processing (Gómez-López et al. 2017; Jo et al. 2018; Park et al. 2008). To overcome the detrimental effect of chlorine in contact with organic materials, AlEW has been used as a prewashing treatment to enhance AEW or NEW effectiveness (Hao et al.

2015; Jadeja and Hung 2014; Koseki et al. 2004). The prewashing of food products using AlEW results in the oxidative reduction of organic materials and subsequently increases the vulnerability of microorganisms on food surfaces due to the great exposure to chlorine in the washing solution. AlEW has been applied as a pretreatment to replace trisodium phosphates, since it has a great antioxidative effect on highly unsaturated fats and oils (Miyashita et al. 1999). Moreover, it has been shown to have a valuable effect on the prevention and removal of feces on chicken carcass surfaces (Kim et al. 2005). To reduce human exposure and the cross-contamination of beef carcasses, Jadeja and Hung (2014) have studied the possibility of inactivating the aforementioned pathogens from cattle hides. They used different combinations of AlEW, NEW, and mild heat treatment to reduce the natural contaminated bacteria and artificially inoculated *E. coli* O157:H7 and *S.* Typhimurium from 20-cm pieces of beef hide. The effectiveness and suitability of the combination as a microbial decontamination process in fresh cattle hide wash processing were demonstrated using 32 mL of hot AlEW (pH 11.60) at 43 °C followed by 28 mL of NEW spray (150 mg/L available chlorine at room temperature). This combination was more effective in the reduction of *E. coli* O157:H7 and *S.* Typhimurium than were the individual treatments, 0.02% peroxyacetic acid, and tap water.

Another study (Hao et al. 2015) used the combined effect of AlEW and AEW in the microbial reduction of fresh-cut cilantro. The objective here was to evaluate the disinfection potency of the different combinations of AlEW and AEW in comparison with that of single SAEW. The results demonstrated that the combination of AlEW (5 min) and AEW (5 min) had a greater disinfection potency than did individual AEW and SAEW treatments. The scanning electron microscopy analysis showed that little bacterial population was observed on the surfaces of *E. coli* O78-inoculated fresh-cut cilantro treated by all the combinations of AlEW and AEW treatments. The pretreatment of AlEW enhanced the sanitization potency of AEW by removing debris and other organic materials on the sample surface. In addition, the alternating change in alkaline and acidic conditions exceeded the bacterial resistance (Hao et al. 2015). The effectiveness of this combination has also been proven during the decontamination process of *E. coli* O157:H7 and *Salmonella* inoculated on lettuce using mildly heated AlEW for 5 min and subsequent washing with chilled AEW for 5 min (Koseki et al. 2004).

10.7 Research and Industrial Updates and Challenges of EW Hurdle Technology

Even though sodium hypochlorite remains the most used sanitizing technology in fresh produce decontamination processes, there is a constantly increasing interest in hurdle technology using EW with physical, biological, and chemical treatments due to the fact that the food industry can be provided with an effective technology that has a role in environmental sustainability. Previously, hurdle enhancement treatments were focused on the combination of EW with one other treatment, for instance, EW in combination with ultrasonication, organic acid, or ultraviolet treatment. Currently, there is a tendency toward increasing the number of technologies to apply in combination to enhance the sanitization effect of EW (Mikš-Krajnik et al. 2017; Ngnitcho et al. 2018; Pang and Hung 2016; Tango et al. 2017; Tirawat et al. 2016). Recent research has demonstrated that the hurdle enhancement of EW treatments can play a great role in fresh produce decontamination processing. The application of more than two treatments in one hurdle process should be carefully studied to avoid fresh produce quality deterioration. The challenge of hurdle-enhanced EW treatments remains the optimization of the parameters of treatments associated with EW to determine the effectiveness and suitability of the process. The adequate types of combinations and dose, time, and temperature of physical, biological, and chemical treatments are a challenge for the establishment of effective hurdle technologies associated with EW. More studies are required to use EW hurdle technology for decontamination purposes in the fresh produce industry.

10.8 Conclusions

EW hurdle treatment has potential applications in the fresh produce industry to control microbial contamination without changing their sensory and nutritional quality. Some research have demonstrated that EW hurdle treatment could inactivate spores and certain enzymes. The application of EW hurdle treatment can be effective when parameters (type of combination and dose, time, temperature, etc.) are optimized to allow the efficiency of sanitization without changing the nutritional and sensory qualities of the food. The biological and physical treatments may add to EW missing or additional effects that induce severe damage to the cell wall and breakage of DNA, which result in complete inactivation of microbial cells. Thus, EW hurdle treatment can be an effective technique for quality assurance and food safety. Studies evaluating the use of EW hurdle treatment on fresh produce should be encouraged to provide a better understanding of the decontamination process and promote an efficient and sustainable alternative technology for the food industry.

References

Afari GK, Hung Y-C, King CH et al (2016) Reduction of *Escherichia coli* O157:H7 and *Salmonella* Typhimurium DT 104 on fresh produce using an automated washer with near neutral electrolyzed (NEO) water and ultrasound. Food Control 63:246–254

Akasaka K, Matsuki H (2015) High pressure bioscience. Springer

Arevalos-Sánchez M, Regalado C, Martin S et al (2012) Effect of neutral electrolyzed water and nisin on *Listeria monocytogenes* biofilms, and on listeriolysin O activity. Food Control 24:116–122

Barbosa-Cánovas G, Góngora-Nieto M, Rodriguez J et al (2005) Nonthermal processing of foods and emerging technologies. Food engineering. Encyclopaedia of life support sciences. EOLSS Publishers/UNESCO, Paris, pp 575–593

Bilek SE, Turantaş F (2013) Decontamination efficiency of high power ultrasound in the fruit and vegetable industry, a review. Int J Food Microbiol 166:155–162

Bintsis T, Litopoulou-Tzanetaki E, Robinson RK (2000) Existing and potential applications of ultraviolet light in the food industry—a critical review. J Sci Food Agric 80:637–645

Block SS (2001) Disinfection, sterilization, and preservation. Lippincott Williams and Wilkins

Botta C, Ferrocino I, Cavallero MC et al (2018) Potentially active spoilage bacteria community during the storage of vacuum packaged beefsteaks treated with aqueous ozone and electrolyzed water. Int J Food Microbiol 266:337–345

Brychcy E, Kulig D, Zimoch-Korzycka A et al (2015) Physicochemical properties of edible chitosan/hydroxypropyl methylcellulose/lysozyme films incorporated with acidic electrolyzed water. Int J Polym Sci

Carrillo-Lopez L, Alarcon-Rojo A, Luna-Rodriguez L et al (2017) Modification of food systems by ultrasound. J Food Qual

Chang S-H, Lin H-TV, Wu G-J et al (2015) pH Effects on solubility, zeta potential, and correlation between antibacterial activity and molecular weight of chitosan. Carbohyd Polym 134:74–81

Chen J, Xu B, Deng S et al (2016) Effect of combined pretreatment with slightly acidic electrolyzed water and botanic biopreservative on quality and shelf life of bombay duck (*Harpadon nehereus*). J Food Quality 39:116–125

Condón S, Mañas P, Cebrián G (2011) Novel technologies in combined processes. In: Nonthermal processing technologies for food. Blackwell Publishing and IFT, pp 379–405

Cortezzo D, Koziol-Dube K, Setlow B et al (2004) Treatment with oxidizing agents damages the inner membrane of spores of *Bacillus subtilis* and sensitizes spores to subsequent stress. J Appl Microbiol 97:838–852

Davidson PM (2001) Chapter 29. Chemical preservatives and natural antimicrobial compounds, pp 593–627. In: Doyle MP, Beuchat LR, Montville TJ (eds) Food microbiology—fundamentals and frontiers, 2nd edn. American Society for Microbiology, Washington, DC, pp 765–801

Deruiter J (2005) Carboxylic acid structure and chemistry: part 2. Principles Drug Action 1:1–10

Ding T, Ge Z, Shi J et al (2015) Impact of slightly acidic electrolyzed water (SAEW) and ultrasound on microbial loads and quality of fresh fruits. LWT-Food Sci Technol 60:1195–1199

Ding T, Xuan X-T, Li J et al (2016) Disinfection efficacy and mechanism of slightly acidic electrolyzed water on *Staphylococcus aureus* in pure culture. Food Control 60:505–510

Eda H, Seckin AM (2015) The effect of different electrolyzed water treatments on the quality and sensory attributes of sweet cherry during passive atmosphere packaging storage. J Appl Polym Sci 102:32–41

Farkas J, Mohácsi-Farkas C (2011) History and future of food irradiation. Trends Food Sci Technol 22:121–126

Forghani F, Eskandari M, Oh D-H (2015) Application of slightly acidic electrolyzed water and ultrasound for microbial decontamination of kashk. Food Sci Biotechnol 24:1011–1016

Forghani F, Oh D-H (2013) Hurdle enhancement of slightly acidic electrolyzed water antimicrobial efficacy on Chinese cabbage, lettuce, sesame leaf and spinach using ultrasonication and water wash. Food Microbiol 36:40–45

Forghani F, Rahman S, Park M-S et al (2013) Ultrasonication enhanced low concentration electrolyzed water efficacy on bacteria inactivation and shelf life extension on lettuce. Food Sci Biotechnol 22:131–136

García-Almendárez BE, Cann IK, Martin SE et al (2008) Effect of *Lactococcus lactis* UQ2 and its bacteriocin on *Listeria monocytogenes* biofilms. Food Control 19:670–680

Giannattasio S, Guaragnella N, Ždralević M et al (2013) Molecular mechanisms of *Saccharomyces cerevisiae* stress adaptation and programmed cell death in response to acetic acid. Front Microbiol 4

Gómez-López VM, Gil MI, Allende A (2017) A novel electrochemical device as a disinfection system to maintain water quality during washing of ready to eat fresh produce. Food Control 71:242–247

Gonçalves AA, Junior CSGG (2009) The effect of glaze uptake on storage quality of frozen shrimp. J Food Eng 90:285–290

Guerrero SN, Ferrario M, Schenk M et al (2017) Hurdle technology using ultrasound for food preservation. In: Ultrasound: advances for food processing and preservation. Elsevier, pp 39–99

Hao J, Li H, Wan Y et al (2015) Combined effect of acidic electrolyzed water and alkaline electrolyzed water (AlEW) on the microbial reduction of fresh-cut cilantro. Food Control 50:699–704

Huang YR, Shiau CY, Hung YC et al (2006) Change of hygienic quality and freshness in tuna treated with electrolyzed water and carbon monoxide gas during refrigerated and frozen storage. J Food Sci 71

Izumi H, Inoue A (2018) Viability of sublethally injured coliform bacteria on fresh-cut cabbage stored in high CO_2 atmospheres following rinsing with electrolyzed water. Int J Food Microbiol 266:207–212

Jadeja R, Hung Y-C (2014) Efficacy of near neutral and alkaline pH electrolyzed oxidizing waters to control *Escherichia coli* O157:H7 and *Salmonella* Typhimurium DT 104 from beef hides. Food Control 41:17–20

Jemni M, Gómez PA, Souza M et al (2014) Combined effect of UV-C, ozone and electrolyzed water for keeping overall quality of date palm. LWT-Food Sci Technol 59:649–655

Jo H-Y, Tango CN, Oh D-H (2018) Influence of different organic materials on chlorine concentration and sanitization of slightly acidic electrolyzed water. LWT-Food Sci Technol 92:187–194

Khan I, Tango CN, Miskeen S et al (2017) Hurdle technology: a novel approach for enhanced food quality and safety—a review. Food Control 73:1426–1444

Khayankarn S, Uthaibutra J, Setha S et al (2013) Using electrolyzed oxidizing water combined with an ultrasonic wave on the postharvest diseases control of pineapple fruit cv. 'Phu Lae'. Crop Protect 54:43–47

Kim C, Hung Y-C, Russell S (2005) Efficacy of electrolyzed water in the prevention and removal of fecal material attachment and its microbicidal effectiveness during simulated industrial poultry processing. Poultry Sci 84:1778–1784

Kim HJ, Feng H, Kushad MM et al (2006) Effects of ultrasound, irradiation, and acidic electrolyzed water on germination of alfalfa and broccoli seeds and *Escherichia coli* O157:H7. J Food Sci 71:M168–M173

Kim Y-H, Jeong S-G, Back K-H et al (2013) Effect of various conditions on inactivation of *Escherichia coli* O157:H7, *Salmonella Typhimurium*, and *Listeria monocytogenes* in fresh-cut lettuce using ultraviolet radiation. Int J Food Microbiol 166:349–355

Koide S, Shitanda D, Note M et al (2011) Effects of mildly heated, slightly acidic electrolyzed water on the disinfection and physicochemical properties of sliced carrot. Food Control 22:452–456

Koseki S, Yoshida K, Kamitani Y et al (2004) Effect of mild heat pre-treatment with alkaline electrolyzed water on the efficacy of acidic electrolyzed water against *Escherichia coli* O157:H7 and *Salmonella* on lettuce. Food Microbiol 21:559–566

Lasomboon S, Setha S, Suthiluk P (2016) Effect of acidic electrolyzed water combined with packaging on quality changes of fresh-cut 'Nanglae' pineapple. J Food Sci Agric Technol (JFAT) 2

Lee D, Heinz V, Knorr D (2003) Effects of combination treatments of nisin and high-intensity ultrasound with high pressure on the microbial inactivation in liquid whole egg. Innov Food Sci Emerg 4:387–393

Leistner L (2000) Basic aspects of food preservation by hurdle technology. Int J Food Microbiol 55:181–186

Leistner L (1992) Food preservation by combined methods. Food Res Int 25:151–158

Leistner L, Gorris LG (1995) Food preservation by hurdle technology. Trends Food Sci Technol 6:41–46

Li X, Farid M (2016) A review on recent development in non-conventional food sterilization technologies. J Food Eng 182:33–45
Li J, Ding T, Liao X et al (2017) Synergetic effects of ultrasound and slightly acidic electrolyzed water against *Staphylococcus aureus* evaluated by flow cytometry and electron microscopy. Ultrason Sonochem 38:711–719
Lianou A, Koutsoumanis K, Sofos J (2012) Organic acids and other chemical treatments for microbial decontamination of food. In: Microbial decontamination in the food industry. Elsevier, pp 592–664
Liao LB, Chen WM, Xiao XM (2007) The generation and inactivation mechanism of oxidation–reduction potential of electrolyzed oxidizing water. J Food Eng 78:1326–1332
Liao X, Xuan X, Li J et al (2017) Bactericidal action of slightly acidic electrolyzed water against *Escherichia coli* and *Staphylococcus aureus* via multiple cell targets. Food Control 79:380–385
Lim W, Harrison MA (2016) Effectiveness of UV light as a means to reduce *Salmonella* contamination on tomatoes and food contact surfaces. Food Control 66:166–173
Liptak MD, Gross KC, Seybold PG et al (2002) Absolute p K a determinations for substituted phenols. J Am Chem Soc 124:6421–6427
Liu R, He X, Shi J et al (2013) The effect of electrolyzed water on decontamination, germination and γ-aminobutyric acid accumulation of brown rice. Food Control 33:1–5
Liu Q, Tan CSC, Yang H et al (2017) Treatment with low-concentration acidic electrolysed water combined with mild heat to sanitise fresh organic broccoli (*Brassica oleracea*). LWT-Food Sci Technol 79:594–600
Luan L, Wu C, Wang L et al (2017) Protein denaturation and oxidation in chilled hairtail (*Trichiutus haumela*) as affected by electrolyzed oxidizing water and chitosan treatment. Int J Food Prop 20:S2696–S2707
Luo K, Oh DH (2016) Inactivation kinetics of *Listeria monocytogenes* and *Salmonella enterica* serovar Typhimurium on fresh-cut bell pepper treated with slightly acidic electrolyzed water combined with ultrasound and mild heat. Food Microbiol 53:165–171
Luu-Thi H, Grauwet T, Vervoort L et al (2014) Kinetic study of *Bacillus cereus* spore inactivation by high pressure high temperature treatment. Innov Food Sci Emerg Technol 26:12–17
Ma L, Zhang M, Bhandari B et al (2017) Recent developments in novel shelf life extension technologies of fresh-cut fruits and vegetables. Trends Food Sci Technol 64:23–38
Manas P, Pagán R (2005) Microbial inactivation by new technologies of food preservation. J Appl Microbiol 98:1387–1399
Mansur AR, Oh DH (2015) Combined effects of thermosonication and slightly acidic electrolyzed water on the microbial quality and shelf life extension of fresh-cut kale during refrigeration storage. Food Microbiol 51:154–162
Mansur AR, Wang J, Park MS et al (2014) Growth model of *Escherichia coli* O157:H7 at various storage temperatures on kale treated by thermosonication combined with slightly acidic electrolyzed water. J Food Protect 77:23–31
Mansur AR, Tango CN, Kim GH et al (2015) Combined effects of slightly acidic electrolyzed water and fumaric acid on the reduction of foodborne pathogens and shelf life extension of fresh pork. Food Control 47:277–284
Mikš-Krajnik M, Feng LXJ, Bang WS et al (2017) Inactivation of *Listeria monocytogenes* and natural microbiota on raw salmon fillets using acidic electrolyzed water, ultraviolet light or/and ultrasounds. Food Control 74:54–60
Miyashita K, Yasuda M, Ota T et al (1999) Antioxidative activity of a cathodic solution produced by the electrolysis of a dilute NaCl solution. Biosci Biotech Bioch 63:421–423
Morris C, Brody AL, Wicker L (2007) Non-thermal food processing/preservation technologies: a review with packaging implications. Packag Technol Sci 20:275–286
Mukhopadhyay S, Gorris LGM (2014) Hurdle Technology∗ A2—Batt, Carl A. In: Tortorello ML (ed) Encyclopedia of food microbiology, 2nd edn. Academic Press, Oxford, pp 221–227
Ngnitcho P-FK, Khan I, Tango CN et al (2017) Inactivation of bacterial pathogens on lettuce, sprouts, and spinach using hurdle technology. Innov Food Sci Emerg Technologi 43:68–76

Ngnitcho P-FK, Tango CN, Khan I et al (2018) The applicability of Weibull model for the kinetics inactivation of *Listeria monocytogenes* and *Escherichia coli* O157:H7 on soybean sprouts submitted to chemical sanitizers in combination with ultrasound at mild temperatures. LWT-Food Sci Technol 91:573–579

Oh D-H, Marshall DL (1993) Antimicrobial activity of ethanol, glycerol monolaurate or lactic acid against *Listeria monocytogenes*. Int J Food Microbiol 20:239–246

Oliveira M, Abadias M, Usall J et al (2015) Application of modified atmosphere packaging as a safety approach to fresh-cut fruits and vegetables—a review. Trends Food Sci Technol 46:13–26

Pang YH, Hung YC (2016) Efficacy of slightly acidic electrolyzed water and UV-ozonated water combination for inactivating *Escherichia Coli* O157:H7 on Romaine and Iceberg lettuce during spray washing process. J Food Sci 81

Park EJ, Alexander E, Taylor G et al (2008) Effects of organic matter on acidic electrolysed water for reduction of foodborne pathogens on lettuce and spinach. J Appl Microbiol 105:1802–1809

Park YB, Guo JY, Rahman S et al (2009) Synergistic effect of electrolyzed water and citric acid against *Bacillus cereus* cells and spores on cereal grains. J Food Sci 74:M185–M189

Patil R, Bhoir P, Deshpande P et al (2013) Relevance of sonochemistry or ultrasound (US) as a proficient means for the synthesis of fused heterocycles. Ultrason Sonochem 20:1327–1336

Pereira R, Vicente A (2010) Environmental impact of novel thermal and non-thermal technologies in food processing. Food Res Int 43:1936–1943

Podolak R, Zayas J, Kastner C et al (1995) Reduction of *Listeria monocytogenes*, *Escherichia coli* O157:H7 and *Salmonella* typhimurium during storage on beef sanitized with fumaric, acetic, and lactic acids. J Food Saf 15:283–290

Podolak R, Zayas J, Kastner C et al (1996) Reduction of bacterial populations on vacuum-packaged ground beef patties with fumaric and lactic acids. J Food Protect 59:1037–1040

Posada-Izquierdo GD, Pérez-Rodríguez F, López-Gálvez F et al (2014) Modeling growth of *Escherichia coli* O157:H7 in fresh-cut lettuce treated with neutral electrolyzed water and under modified atmosphere packaging. Int J Food Microbiol 177:1–8

Qian Y-F, Xie J, Yang SP et al (2013) Study of the quality changes and myofibrillar proteins of white shrimp (*Litopenaeus vannamei*) under modified atmosphere packaging with varying CO_2 levels. Eur Food Res Technol 236:629–635

Rahman S, Ding T, Oh DH (2010) Inactivation effect of newly developed low concentration electrolyzed water and other sanitizers against microorganisms on spinach. Food Control 21:1383–1387

Rahman SME, Jin Y-G, Oh D-H (2011) Combination treatment of alkaline electrolyzed water and citric acid with mild heat to ensure microbial safety, shelf-life and sensory quality of shredded carrots. Food Microbiol 28:484–491

Rahman SME, Wang J, Oh D-H (2013) Synergistic effect of low concentration electrolyzed water and calcium lactate to ensure microbial safety, shelf life and sensory quality of fresh pork. Food Control 30:176–183

Rasco B, Ovissipour M (2015) Electrolyzed water applications in aquaculture and the seafood industry. J Aqua Res Develop 6:1

Reineke K, Schlumbach K, Baier D et al (2013) The release of dipicolinic acid—the rate-limiting step of Bacillus endospore inactivation during the high pressure thermal sterilization process. Int J Food Microbiol 162:55–63

Ricke S (2003) Perspectives on the use of organic acids and short chain fatty acids as antimicrobials. Poult Sci 82:632–639

Rostami Z, Ahmad MA, Khan MU et al (2016) Food preservation by hurdle technology: a review of different hurdle and interaction with focus on foodstuffs. J Pure Appl Microbiol 10:2633–2639

Setlow P (2003) Spore germination. Curr Opin Microbiol 6:550–556

Singh S, Shalini R (2016) Effect of hurdle technology in food preservation: a review. Crit Rev Food Sci 56:641–649

Soares NM, Mendes TS, Vicente AA (2013) Effect of chitosan-based solutions applied as edible coatings and water glazing on frozen salmon preservation—a pilot-scale study. J Food Eng 119:316–323

Tango CN, Mansur AR, Kim GH et al (2014a) Synergetic effect of combined fumaric acid and slightly acidic electrolysed water on the inactivation of food-borne pathogens and extending the shelf life of fresh beef. J Appl Microbiol 117:1709–1720

Tango CN, Wang J, Oh DH (2014b) Modeling of *Bacillus cereus* growth in brown rice submitted to a combination of ultrasonication and slightly acidic electrolyzed water treatment. J Food Protect 77:2043–2053

Tango CN, Mansur AR, Oh DH (2015) Fumaric acid and slightly acidic electrolyzed water inactivate Gram positive and Gram negative foodborne pathogens. Microorganisms 3:34–46

Tango CN, Khan I, Kounkeu PFN et al (2017) Slightly acidic electrolyzed water combined with chemical and physical treatments to decontaminate bacteria on fresh fruits. Food Microbiol 67:97–105

Taylor RG, Papa I, Astier C et al (1997) Fish muscle cytoskeleton integrity is not dependent on intact thin filaments. J Muscle Res Cell M 18:285–294

Tirawat D, Phongpaichit S, Benjakul S et al (2016) Microbial load reduction of sweet basil using acidic electrolyzed water and lactic acid in combination with mild heat. Food Control 64:29–36

Torlak E (2014) Inactivation of *Alicyclobacillus acidoterrestris* spores in aqueous suspension and on apples by neutral electrolyzed water. Int J Food Microbiol 185:69–72

Vázquez-Sánchez D, Cabo M, Rodríguez-Herrera J (2014) Single and sequential application of electrolyzed water with benzalkonium chloride or peracetic acid for removal of *Staphylococcus aureus* biofilms. J Food Saf 34:199–210

Wang J (2011) Evaluation of hypochlorous acid (electrolyzed water), lactic acid, and peroxyacetic acid as sanitizers for fresh vegetables. Oklahoma State University

Wang JJ, Lin T, Li JB et al (2014) Effect of acidic electrolyzed water ice on quality of shrimp in dark condition. Food Control 35:207–212

Wang L, Xia Q, Li Y (2017a) The effects of high pressure processing and slightly acidic electrolysed water on the structure of *Bacillus cereus* spores. Food Control 79:94–100

Wang L, Xia Q, Li Y (2017b) Synergistic effects of high pressure processing and slightly acidic electrolysed water on the inactivation of *Bacillus cereus* spores. Int J Food Sci Technol 52:2429–2435

Wang L, Tao H, Li Y (2018a) Multi-pulsed high pressure assisted slightly acidic electrolyzed water processing on microbe, physical quality, and free amino acids of mud snail (*Bullacta exarata*). J Food Process Pres 42:e13509

Wang L, Xia Q, Li Y (2018b) Label free-based proteomic analysis of proteins in *Bacillus cereus* spores regulated by high pressure processing and slightly acidic electrolyzed water treatment. Food Control 90:392–400

Watchararparpaiboon W, Laohakunjit N, Kerdchoechuen O (2010) An improved process for high quality and nutrition of brown rice production. Food Sci Technol Int 16:147–158

Waters BW, Hung YC (2014) The effect of pH and chloride concentration on the stability and antimicrobial activity of chlorine-based sanitizers. J Food Sci 79:M622–M627

Wijffels R, Buitelaar R, Bucke C (1996) Stable support materials for the immobilization of viable cells. In: Immobilized cells: basics and applications, vol 11, pp 55

Woolthuis CH, Smulders FJ (1985) Microbial decontamination of calf carcasses by lactic acid sprays. J Food Protect 48:832–837

Xie J, Sun X, Pan Y et al (2012a) Combining basic electrolyzed water pretreatment and mild heat greatly enhanced the efficacy of acidic electrolyzed water against *Vibrio parahaemolyticus* on shrimp. Food Control 23:320–324

Xie J, Sun XH, Pan YJ et al (2012b) Physicochemical properties and bactericidal activities of acidic electrolyzed water used or stored at different temperatures on shrimp. Food Res Int 47:331–336

Xu G, Tang X, Tang S et al (2014) Combined effect of electrolyzed oxidizing water and chitosan on the microbiological, physicochemical, and sensory attributes of American shad (*Alosa sapidissima*) during refrigerated storage. Food Control 46:397–402

Yamaner C, Ayvaz M, Konak R et al (2016) Efficacy of neutralised electrolysed water and mild heat against foodborne pathogens isolated from Ficus carica. Ital J Food Sci 28:208–220

Zeitoun A, Debevere J (1990) The effect of treatment with buffered lactic acid on microbial decontamination and on shelf life of poultry. Int J Food Microbiol 11:305–311

Zhang J, Yang H (2017) Effects of potential organic compatible sanitisers on organic and conventional fresh-cut lettuce (*Lactuca sativa* Var. *Crispa L*). Food Control 72:20–26

Zhang B, Ma L-K, Deng S-G et al (2015) Shelf-life of pacific white shrimp (*Litopenaeus vannamei*) as affected by weakly acidic electrolyzed water ice-glazing and modified atmosphere packaging. Food Control 51:114–121

Zhao L, Zhang Y, Yang H (2017) Efficacy of low concentration neutralised electrolysed water and ultrasound combination for inactivating *Escherichia coli* ATCC 25922, *Pichia pastoris* GS115 and *Aureobasidium pullulans* 2012 on stainless steel coupons. Food Control 73:889–899

Zhou R, Liu Y, Xie J et al (2011) Effects of combined treatment of electrolysed water and chitosan on the quality attributes and myofibril degradation in farmed obscure puffer fish (*Takifugu obscurus*) during refrigerated storage. Food Chem 129:1660–1666

Zhou R, Mo Y, Li Y et al (2008) Quality and internal characteristics of Huanghua pears (*Pyrus pyrifolia* Nakai, cv. Huanghua) treated with different kinds of coatings during storage. Postharvest Biol Technol 49:171–179

Chapter 11
Safety Evaluation of Electrolyzed Water

Donghong Liu and Ruiling Lv

11.1 Introduction

Electrolyzed water (EW) is a sanitizer with strong bactericidal activity due to hypochlorous acid. The application of EW is a sustainable and green concept and has several advantages over traditional cleaning systems including cost-effectiveness, ease of application, effective disinfection, on-the-spot production and safety for human beings and environment (Huang et al. 2008). These features make it an appropriate sanitizing and cleaning system for use in high-risk settings such as in hospitals and other healthcare facilities as well as in food processing and environments (Rahman et al. 2016). Interestingly, EW has been applied in Japan for several years as an antimicrobial agent. EW exhibits antimicrobial activity against a variety of microorganisms and eliminates most common types of viruses, bacteria, fungi, and spores in a relatively short amount of time (usually within 5–20 s) in food products, food processing surfaces, and nonfood surfaces (Ding et al. 2015; Hao et al. 2015). But there are still some disadvantages and limitations about the EW, such as phytotoxicity, irritation to hands, and corrosion caused by the free chlorine content or high Oxidation-Reduction Potential (ORP), reduction in the concentration of chlorine over time, which reduces the bactericidal activity.

D. Liu (✉) · R. Lv
Department of Food Science and Nutrition, National Engineering Laboratory of Intelligent Food Technology and Equipment, Zhejiang University, Hangzhou 310058, Zhejiang, China
e-mail: dhliu@zju.edu.cn

D. Liu · R. Lv
Key Laboratory for Agro-Products Postharvest Handling of Ministry of Agriculture, Zhejiang Key Laboratory for Agro-Food Processing, Hangzhou 310058, Zhejiang, China

D. Liu
Fuli Institute of Food Science, Zhejiang University, Hangzhou 310058, Zhejiang, China

T. Ding et al. (eds.), *Electrolyzed Water in Food: Fundamentals and Applications*,
https://doi.org/10.1007/978-981-13-3807-6_11

In Japan, EW is designated as a food additive in June 2002, and EW includes AEW and SAEW in the current ingredient standard. In addition, as the same sterilization agent of halogen type, sodium hypochlorite is designated in the year 1955, and advanced sarashmeal powder is designated as food additive in 1931 (Japan Food Safety Commission 2006). This chapter is about the safety of EW, including biological safety and chlorine residue.

11.2 Residual Chlorine in Foods During Food Processing

11.2.1 Chlorine Residue

Spinach (leaves) was dipped for 10 min in SAEW (pH 6.5, ACC of 70.2 mg/kg), and spinach was thoroughly rinsed with tap water to be used as a sample. As a result, ACC in the sample was measured, and no effective chlorine was detected (Japan Food Research Laboratories 2002) (detection limit 0.5 mg/kg).

Vegetables (cucumbers and cabbage) were cut, rinsed, and drained for about 2 min with running water. Then they were immersed in AEW (pH 3.0, 3.1, 4.5; ACC of 10, 20, 20 mg/kg; each 200 mL), and then rinsing for 10 min. Residual chlorine was measured by gas chromatography–mass spectrometry. However, the analysis sample was not subjected to water treatment after washing and rinsing with the tap water. As a result, no residual chlorine was detected in the samples treated with any SAEW immediately after and after 5 min (Unicorn 2006).

11.2.2 Formation of Trihalomethane

In order to verify the generation and survival of the trihalomethane by the sterilization treatment using EW, the following experiment was carried out. The same experiment was carried out using tap water instead of EW as the control experiment (Fig. 11.1).

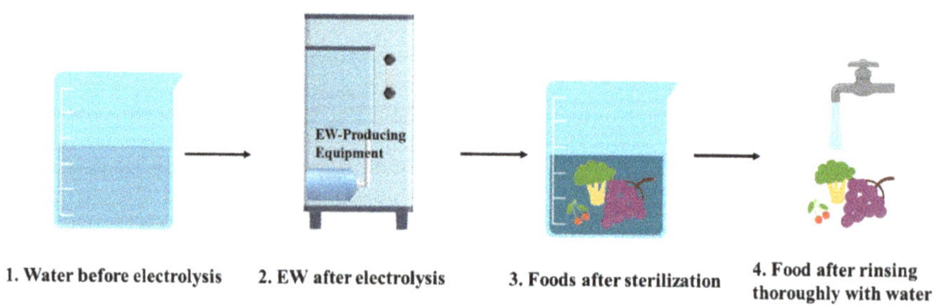

Fig. 11.1 The steps for verifying the generation and survival of the trihalomethane

(1) Using a SAEW (pH 5.9, ACC of 78 mg/kg), spinach (1 bundle) was soaked for 10 min and rinsed for 1 min with tap water to measure residual chlorine and trihalomethane. Sodium hypochlorite solution and tap water were used as the control experiment. The measurement points are as follows:

  i. Tap water
  ii. SAEW, sodium hypochlorite solution, tap water
  iii. Spinach and spinach after sterilization
  iv. Spinach after rinsing

In order to confirm the amount of trihalomethane generated during the formation of SAEW, the removal of residual chlorine and trihalomethane by the activated carbon treatment was carried out and the SAEW was produced, and the effect of the generated SAEW was effective. Chlorine concentration and trihalomethane were measured.

The amount of trihalomethane in the foods treated with the SAEW was about one-fourth of that of the tap water, and it was low compared with the sodium hypochlorite treatment, and it was thought that the residual amount of the trihalomethane was low in the food. The amount of trihalomethane (0.0037 mg/L) produced by tap water after the removal of trihalomethane was less than the amount of trihalomethane (0.046 mg/L) produced in the formation of EW. It seems to be influenced by the trihalomethane contained in tap water. No effective chlorine was detected from any of the foods which had been sterilized by any aqueous solution (Japan Food Research Laboratories 2006).

(2) SAEW (raw chlorine concentration of 50 mg/kg) using raw water as raw water and SAEW (ACC of 55 mg/kg) with raw water as a pure water was dipped into 2 mm cabbage for 30 s, followed by water for 15 s the residual chlorine, total trihalomethane, chloroform, dibromochloromethane, bromoform, and bromochloromethane were measured after washing and cutting. The measurement points are as follows:

  i. Tap water
  ii. Weak acidic hypochlorous acid water with raw water as raw water
  iii. Cabbage after sterilization and rinsing

As a result, no trihalomethane was detected in cabbage treated with any SAEW. In addition, trihalomethanes of 0.017 mg/kg were detected from the tap water and of 0.016 mg/L from the SAEW which used tap water as the raw water. But there was no detection from the SAEW which was pure water, and the formation of the trihalomethane was the result of the SAEW which used raw water as the tap water was thought to be the existence of the trihalomethane derived from tap water (Unicorn 2006).

11.2.3 Radical Formation

The possibility of radical formation is considered to be caused by the decomposition of EW by the contact with the food. It is considered that the decomposition of hypochlorous acid in foods is related to vitamin and lipid.

The amount of ascorbic acid in the mesophyll was measured after sterilization of 100 g of spinach (100 grams) for 10 min using sterilized water (3 EW, 3 mixed diluted water) and 2 L of tap water. As the result, the cleaning of EW (ACC of about 70 mg/L) for spinach single leaves was equivalent to the ascorbic acid (all reduction type) equivalent to the cleaning of tap water or mixed sodium hypochlorite (ACC of about 50–100 mg/L) Faculty of biology oriented science and technology, Kinki University (2002). These results suggest that EW does not affect the ascorbic acid content in the cleaning of spinach. In addition, the amount of the radical generation in the food is considered to be small, because the trihalomethane in the food by the SAEW treatment does not increase significantly.

Using 600 mL of AEW (pH 2.5 ± 0.1, ACC 42.3 ± 1.4 mg/L), sodium hypochlorite (pH 9.3 ± 0.2, ACC of 153.6 ± 3.4 mg/L) and tap water (pH 7.0 ± 0.2, ACC 0.3 ± 0.1 mg/kg) 600 mL of cabbage and carrots. The amount of ascorbic acid content was measured by dipping the cucumber of 3–5 cm^2 lettuce and 2–3 mm thick cucumber. As the result, there was no difference in the ASCO Lubin acid content compared with the sodium hypochlorite solution treatment and the tap water treatment in the case of the AEW treatment of both vegetables. From this fact, it is not considered that the radicals are generated at harmful concentrations (Koseki and Itoh 2001).

11.3 Biological Safety

11.3.1 Acute Toxicity

In ICR mice (five males and five females each) the use of SAEW (pH 5.0–5.5, ACC 50–80 mg/kg, 50 mL/kg) resulted in no confirmed death in either sex, nor was it considered an animal with toxic symptoms (Food and drug safety evaluation center 1995a).

11.3.2 Mutagenic and Genetic Toxicity

NEW (ACC: 0.90–1.14 mg/L; ORP: 80 mV), which was prepared with a water-electrolysis apparatus equipped with a non-diaphragm cell and a highly compressed activated carbon block, was evaluated for the mutagenic and genotoxic potentials, at concentrations up to 100% dose/plate (Saitoh et al. 2010). The results showed that EW did not induce reverse mutations in *Salmonella typhimurium* strains and *Escherichia coli* strain, in either the absence or presence of rat liver S9 for exogenous metabolic activation. Also, EW did not cause chromosome aberrations in Chinese hamster lung fibroblast cells. Specific pathogen-free rats were administered with EW at a dose of 20 mL/kg/day for 28 days via intragastric infusion (Storkey et al. 2014). There was no significant EW related toxic change in terms of any items such as clinical symptoms, body weight, food consumption, urinalysis, hematology, blood chemistry, necropsy, each organ weight, and histopathology. Thus, the no-observable-adverse-effect level (NOAEL) for EW was estimated to be greater than 20 mL/kg/day under the conditions examined, demonstrating the consistency with the expected safety for a human with a body weight of 60 kg to drink the EW up to at least 1.2 L/day (Forghani et al. 2013).

11.3.3 Cytotoxicity

The inhibitory effect of SAEW (pH 5.0–5.5, ACC of 50–80 mg/kg) on Chinese hamster cell cultures (V79 cells) revealed a clear cytotoxic effect. In the case of 50.0% or more, the colony was not observed, and the IC50 value calculated from the test was 20.0% or less (Food and drug safety evaluation center 1995b).

11.3.4 Stimuli and Sensitization

Abnormalities in both animals were detected in skin primary stimuli test using SAEW, skin accumulation stimuli test and eye irritation test, and sensitization test using Hartley guinea pigs in female New Zealand white rabbits. No abnormalities were detected in any one (Food and drug safety evaluation center 1995c).

11.3.5 Sub-chronic Oral Toxicity

The safety of EW from the perspective of its use in mouthwash was demonstrated (Morita et al. 2011). With regard to its safety, in addition to the localized effect in the oral cavity, its effect on the gastrointestinal tract, including the oesophagus, and its systemic effect must be demonstrated. The water with pH of 7.5 and ACC of 0.5 mg/L was used. We provided mice with access to EW as drinking water for 8 weeks and investigated its effect on periodontal soft tissues, tooth morphology, the digestive tract, and body weight. The study found that drinking EW had no effect on mouse growth in terms of changes in body weight, and no abnormal findings were observed on visual or histological inspections of the gastrointestinal tract. This suggests that EW has no systemic effects and was safe when used in mouthwash. But as its causative relation with attrition is unclear, more detailed studies are required (Kubota et al. 2015).

11.4 Conclusions

As for the safety of EW, there were many reports on the AEW (pH 2.5, ACC of 50–60 mg/kg) and the SAEW (pH 5.5, ACC of 70 mg/kg), about the acute oral toxicity test, the skin irritation test, the acute eye stimulation test. No changes were observed in skin sensitization test, oral mucosal stimulation test, return mutation test, and chromosomal abnormality test. In addition, cytotoxicity was found to suppress the proliferation of cells at a high concentration, but it was found to be less toxic than other commercial disinfectants. Rats were administered with sodium hypochlorite (500–2000 mg/kg) for 104 weeks, and sodium hypochlorite (500, 1000 mg/kg) was administered to rats for 103 weeks and carcinogenic. Together with these results and other toxicity tests and literature, the safety of EW under practical use conditions is considered to be satisfactory (Morita et al. 2011).

References

Annual report of Japan Food Analysis Center (2002) Residual chlorine residues in spinach treated with sodium hypochlorite (207 mg/kg) and hypochlorous acid (pH 6.5 effective chlorine concentration: 70.2 mg/kg) (in Japanese)

Ding T, Ge Z, Shi J et al (2015) Impact of slightly acidic electrolyzed water (SAEW) and ultrasound on microbial loads and quality of fresh fruits. LWT-Food Sci Technol 60(2):1195–1199

Faculty of biology oriented science and technology, Kinki University (2002) Effect of sterilized water on nutrient composition in foods (in Japanese)

Food and drug safety evaluation center (1995a) Toxicity test of soft acidic water in mice (in Japanese)

Food and drug safety evaluation center (1995b) Return mutation test using bacteria of soft acid water (in Japanese)

Food and drug safety evaluation center (1995c) Skin primary stimulation test of rabbit using soft acidic water (in Japanese)

Forghani F, Rahman SME, Park MS et al (2013) Ultrasonication enhanced low concentration electrolyzed water efficacy on bacteria inactivation and shelf life extension on lettuce. Food Sci Biotechnol 22(1):131–136

Hao JX, Li HY, Wan YF et al (2015) Combined effect of acidic electrolyzed water (AEW) and alkaline electrolyzed water (AlEW) on the microbial reduction of fresh-cut cilantro. Food Control 50:699–704

Huang YR, Hung YC, Hsu SY et al (2008) Application of electrolyzed water in the food industry. Food Control 19(4):329–345

Japan Food Research Institute (2006) Research on residual amount of trihalomethane (in Japanese)

Japan Food Safety Commission Additives research group (2006) Additives evaluation: hypochlorous acid water (in Japanese)

Koseki S, Itoh K (2001) Effect of electrolyzed acidic electrolyzed water on the quality of cut vegetables. J Food Sci Technol 48:365–369 (in Japanese)
Kubota A, Goda T, Tsuru T et al (2015) Efficacy and safety of strong acid electrolyzed water for peritoneal lavage to prevent surgical site infection in patients with perforated appendicitis. Surg Today 45(7):876–879
Morita C, Nishida T, Ito K (2011) Biological toxicity of acid electrolyzed functional water: effect of oral administration on mouse digestive tract and changes in body weight. Arch Oral Biol 56(4):359–366
Rahman SME, Khan I, Oh DH (2016) Electrolyzed water as a novel sanitizer in the food industry: current trends and future perspectives. Compr Rev Food Sci Food Saf 15(3):471–490
Saitoh Y, Harata Y, Mizuhashi F et al (2010) Biological safety of neutral-pH hydrogen-enriched electrolyzed water upon mutagenicity, genotoxicity and subchronic oral toxicity. Toxicol Ind Health 26(4):203–216
Storkey C, Davies MJ, Pattison DI (2014) Reevaluation of the rate constants for the reaction of hypochlorous acid (HOCl) with cysteine, methionine, and peptide derivatives using a new competition kinetic approach. Free Radical Biol Med 73:60–66
Unicorn (2006) Trihalomethane analysis in hypochlorous acid treated cabbage (in Japanese)

Chapter 12
Future Trends of Electrolyzed Water

Tian Ding

12.1 Introduction

EW, as a new sanitizer, is getting popularity in food industry among many countries in recent years. The main advantage of using EW is the ability for on-site production, thus circumventing problems associated with chlorination including the transport, storage and handling of dangerous chlorine (Jeong et al. 2007). The acceptance of EW as a sanitizer is evident from its use in a number of applications in various fields including agriculture, medical sterilization, food sanitation, livestock management, and other fields that employ antimicrobial techniques. At present, it is important to identify the elements that are limiting the advancement of the EW technology to a full-fledge commercial scale. The former chapters in this book have discussed the fundamentals and applications of EW technology. In this closing chapter, the challenges and future trends of EW technology for wider applications will be provided.

T. Ding (✉)
Department of Food Science and Nutrition, National Engineering Laboratory of Intelligent Food Technology and Equipment, Zhejiang University, Hangzhou 310058, Zhejiang, China
e-mail: tding@zju.edu.cn

T. Ding
Key Laboratory for Agro-Products Postharvest Handling of Ministry of Agriculture, Zhejiang Key Laboratory for Agro-Food Processing, Hangzhou 310058, Zhejiang, China

T. Ding
Fuli Institute of Food Science, Zhejiang University, Hangzhou 310058, Zhejiang, China

T. Ding et al. (eds.), *Electrolyzed Water in Food: Fundamentals and Applications*,
https://doi.org/10.1007/978-981-13-3807-6_12

12.2 Regulatory Approval

There are diverse opinions and regulations across different countries on the application of EW as a sanitizing agent. As early as 2002, EW has been designated as a food additive by the Ministry of Health, Labour and Welfare of Japan (Official Gazette No. 3378 MHLW Notification No. 212; MHLW 2002). In the USA, the hypochlorous acid generated by electrolyzed water has been allowed as a sanitizer in organic crop production, organic livestock production and organic handling by the United States Department of Agriculture National Organic Program (USDA-NOP 2017). Organic Federation of Canada has also recognized EW (which may contain hypochlorous acid as a byproduct of either electrolysis or from the dissolution of chlorine compounds in water) as approved for organic use in CAN/CGSB-32.311-2015 of the Permitted Substance List Table 7.3 (OFC 2015). In the European Union (EU), EW can only be applied to drinking water, whereas its use on food products such as meat and fish has not acquired the permission because of its properties of protein inactivation (Maricica and Galanakis 2018). In other countries and areas, the biggest challenge for the adoption and extension of EW technology is the regulatory approval.

12.3 Commercial Development

Recently, a continuously growing trend for the commercialization of EW around the world is being observed. Keeping in view the importance of EW, many companies have been established and are producing EW. Examples of such companies include Envirolyte, KEWS, RVD Corporation, EcoLogic Solutions Inc., Viking pureTM, EcoloxTech and eWater Advantage, which are based in Estonia, South Korea, Latvia, and the USA. These companies claim to produce 100% pure, eco-friendly, nontoxic, and less expensive EW for domestic and industrial use (Rahman et al. 2016). Another US-based company, EAU Technologies Inc., completed a trial by using EW with a leading international beverage bottling company in September 2009, which aided the approval of the use of EW technology as a sanitation process for CIP (Clean-in-Place) applications. EW produced by EAU Technologies is marketed under the brand name Empowered WaterTM and is an environmentally friendly and highly effective solution for CIP applications for industrial use. In particular, in food industry, CIP applications are used to clean stationary equipment during changing over the products and system start-up (Rahman et al. 2016).

12.4 Consumer Acceptance

Novel food technologies enable the development of food industry, but not all the emerging technologies can be accepted by consumers. The consumer confidence is a key point and should be addressed. The attitudes of consumers to EW technology is still unclear. The data about the consumers' opinion is difficult to obtain. More organizations and government should participate in this assessment of consumers' opinion about the EW technology. The advertisement can assist in the improvement of public awareness of the advantages of EW technology. The beneficial effect of EW technology on food and the economic and environmental advantages should be clearly demonstrated.

12.5 Concluding Remarks

Many aspects of EW that need to be considered include: (1) relatively high initial cost of equipment with brief shelf life; (2) the tendency for EW to lose its antimicrobial potential quickly, if it is not continuously supplied with Cl_2, H^+, HOCl by electrolysis; (3) the corrosion caused by the free chlorine content or high ORP during the use of acidic EW; (4) discomfort to the operator caused by the pungent chlorine gas generated by some EW generators when operated at pH<5; (5) phytotoxicity, irritation of hands, and corrosion caused by the free chlorine content or high ORP during the use of AEW; (6) reduction in antimicrobial activity by inappropriate storage and the presence of organic matter in EW; (7) the standards for safety assessment of EW technology. These are the relevant topics in EW technology deserve future research.

In the near future, most of the industry is likely to start using EW, owing to its benefits. The EW cleaning and sanitizing at ambient temperature reduces the energy consumed for the process. In addition, the EW solution is relatively simple in composition and easy to handle. Humans are also unaffected by the exposure to EW, as long as the eyes are not exposed and it is not orally ingested. Many chemical industries do not possess sufficient knowledge on EW, and more advertisement is required to introduce it to the industry. There is a good future for EW, and over the next 10 years, most of the food plants will start using EW, owing to its benefits (Rahman et al. 2016).

EW exhibits strong bactericidal, fungicidal and virucidal effects in various sectors such as medicine and dentistry as well as on fruits, vegetables, seafood, eggs, poultry, utensils, agriculture, raw meat and carcasses. However, EW with very low pH is corrosive in nature and affects the organoleptic properties of some foods, which limit its use. With the development of newer types of EW such as SAEW and SAlEW, some of these issues have been solved. The application of combination of multiple techniques has shown advantages over individual treatment in terms of synergistic microbial reduction, enhanced shelf life, and food quality maintenance. Various factors in addition to the physio-chemical properties of EW are found to govern the efficacy of EW such as water temperature, ACC, ORP, pH, types of electrolyte, flow rates of water and electrolyte, storage conditions, and concentration of salts, which need to be monitored during its production and applications. Further, the bactericidal activity of EW is also influenced by the presence of organic matter, water hardness, and pollutants in the product. Therefore, an advanced and dynamic EW production system that is capable of overcoming all the current limitations can be developed through further research in the field. These may include procedures for expanding the usage of EW in food processing settings and the application in HACCP and Sanitation SOP systems. The future looks very promising for the application of EW technology in the food industry.

References

Jeong J, Kim JY, Cho M et al (2007) Inactivation of *Escherichia coli* in the electrochemical disinfection process using a Pt anode. Chemosphere 67(4):652–659

Ministry of Health, Labour and Welfare of Japan (2002). https://jewa.org/en/what-is-electrolyzed-water/. Accessed 21 Sept 2018

Maricica S, Galanakis CM (2018). http://scitechconnect.elsevier.com/sustainable-sanitation-in-the-food-industry/

Organic Federation of Canada (2015). http://organicfederation.ca/canadian-organic-standards#electro Accessed 21 Sept 2018

Rahman S, Khan I, Oh D (2016) Electrolyzed water as a novel sanitizer in the food industry: current trends and future perspectives Comprehensive. R Food Sci Food Safety 15(3):471–490

United States Department of Agriculture National Organic Program (2017). https://www.gpo.gov/fdsys/pkg/FR-2018-01-17/pdf/2017-28172.pdf. Accessed 21 Sept 2018

Index

T. Ding et al. (eds.), *Electrolyzed Water in Food*,
https://doi.org/10.1007/978-981-13-3807-6

Lightning Source UK Ltd.
Milton Keynes UK
UKHW021940050219
336778UK00001B/9/P

9 789811 338069